To my family, especially my husband Styvens. Thank you for your love, unwavering support and your sacrifices as I pursue my purpose.

Monica Gray

Table of Contents

Foreword *by* **Noel Blackburn** xvii

Foreword *by* **Stephen R. Cox** xix

Preface xxii

Acknowledgment xxix

Section 1
High Impact Higher Education Pedagogical Frameworks and Practices

Chapter 1
Improving Minority Student Persistence: An Institutional Factors Approach....... 1
Yvonne R. Hilton, Lincoln University, USA
Monica Gray, Lincoln University, USA

Chapter 2
Minority Recruitment and Retention Among Gifted Students 26
Chandra A. Stallworth, Independent Researcher, USA
Ken D. Thomas, Auburn University, USA

Chapter 3
Engineering Study Abroad: High Impact Strategy for Increasing Access........... 42
Monica Gray, Lincoln University, USA
Connie Lundy, Lincoln University, USA

Section 2
Micro-Community or Target Approaches to Diversifying Engineering

Chapter 4
Empowering Women in STEM: Embedding STEM in K-12 Education 61
Gretchen Dietz, University of Mount Union, USA
Julie Hessedence, University of Mount Union, USA
Terry Long, University of Mount Union, USA
Helen E. Muga, University of Mount Union, USA

Chapter 5
Developing a Learning Community of Engineers Through an Honors First-Year Seminar 88
Melissa L. Johnson, University of Florida, USA
Kristy Spear, University of Florida, USA

Chapter 6
Engineering Teams: Supporting Diversity in Engineering Education............... 106
Jennifer Loy, Griffith University, Australia
Simon Howell, Griffith University, Australia
Rae Cooper, Griffith University, Australia

Chapter 7
Addressing Cultural and Gender Project Bias: Engaged Learning for Diverse Student Cohorts.. 130
Jennifer Loy, Griffith University, Australia
Rae Cooper, Griffith University, Australia

Section 3
Critiques and Administrative Suggestions to Restructure Engineering Education

Chapter 8
Revisioning the Engineering Profession: How to Make It Happen!.................. 156
Judith Gill, University of South Australia, Australia
Mary Ayre, University of South Australia, Australia
Julie Mills, University of South Australia, Australia

Chapter 9
A Globally Focused, Experiential Educational System for STEM Fields: Measures for Intentionally Promoting Diversity 176
Aaron Sakulich, Worcester Polytechnic Institute, USA
Amy Peterson, Worcester Polytechnic Institute, USA

Chapter 10
Diversifying Engineering Education: A Transdisciplinary Approach From RWTH Aachen University 201
Linda Steuer, RWTH Aachen University, Germany
Anna Bouffier, RWTH Aachen University, Germany
Sonja Gaedicke, RWTH Aachen University, Germany
Carmen Leicht-Scholten, RWTH Aachen University, Germany

Chapter 11
Engineering Pathways in a U.S. Public Institution of Higher Education: A Strategy for Fostering Student Diversity 236
Fabiola Ehlers-Zavala, Colorado State University, USA
Anthony Maciejewski, Colorado State University, USA

Chapter 12
Getting Off the Engineering Enrollment Rollercoaster: Interaction Between Academia and Upstream Petroleum Industry 260
Tatyana Plaksina, University of Calgary, Canada

Compilation of References 287

About the Contributors 314

Index 323

Detailed Table of Contents

Foreword *by* **Noel Blackburn** xvii

Foreword *by* **Stephen R. Cox** xix

Preface xxii

Acknowledgment xxix

Section 1
High Impact Higher Education Pedagogical Frameworks and Practices

Chapter 1
Improving Minority Student Persistence: An Institutional Factors Approach....... 1
Yvonne R. Hilton, Lincoln University, USA
Monica Gray, Lincoln University, USA

Student persistence in college is a major concern for every institution of higher education. Statistics show that the greatest percentage of attrition occurs after the freshman year. Many studies surrounding this phenomenon tend to focus on pre-college predictors to gain knowledge into mass premature departure during the freshman to sophomore transition. However, very few have looked at institutional factors and how they may explain the problem. Further, most research studies have been done at relatively large and predominantly white institution. This chapter investigates student satisfaction with institutional factors at a small Historically Black College and University (HBCU). Specifically, it looks at how the campus environments may impact freshman students' decisions to return for their sophomore year. While freshman students were dissatisfied with certain aspects of the university they consider to be important, the results precluded good prediction in a practical sense if they influence their decision not to return.

Chapter 2
Minority Recruitment and Retention Among Gifted Students 26
Chandra A. Stallworth, Independent Researcher, USA
Ken D. Thomas, Auburn University, USA

Consistent with the national goal implemented by our current government, Auburn University is also working to recruit and retain underrepresented minorities in higher education. The rationale for this is simple, that is to allow a greater advantage when competing against others. One of the ways to foster this competition is by nurturing our gifted underrepresented minority students. In the 2010-2011 school year, the Honors College, which serves as a gateway for underrepresented minority students, developed a distinct focus on helping our students reach and their educational/academic goals. Within this paper we will go over some of the steps we have begun to take to reach our goal, in addition to future plans we have to continue these efforts.

Chapter 3
Engineering Study Abroad: High Impact Strategy for Increasing Access........... 42
Monica Gray, Lincoln University, USA
Connie Lundy, Lincoln University, USA

Successful engineers must be technically savvy, self-confident as well as culturally competent. Cultural competence is the ability to tolerate ambiguity and empathize with the socio-cultural nuances of different people groups. This calls for a diverse engineering workforce especially in today's increasingly global economy. In the United States, Minorities and Females constitute only 4% and 15% of the engineering workforce respectively. Research shows that women and students of color, dropout due to feelings of not belonging coupled with low self-efficacy. To change the profession's diversity portfolio requires a plethora of high impact approaches. Common among successful retention strategies is the provision of structured opportunities for all students to develop self-directing competencies in both the cognitive and affective learning domains. This chapter demonstrates that the study abroad experience engenders, facilitates and fosters these very aptitudes as well as cultural literacy, and advocates for its inclusion in discussions on increasing under-represented participations in engineering.

Section 2
Micro-Community or Target Approaches to Diversifying Engineering

Chapter 4
Empowering Women in STEM: Embedding STEM in K-12 Education 61
Gretchen Dietz, University of Mount Union, USA
Julie Hessedence, University of Mount Union, USA
Terry Long, University of Mount Union, USA
Helen E. Muga, University of Mount Union, USA

This chapter covers a project that was part of an I3 (Ideas, Innovation, Invention) challenge within a global engineering, junior course at University of Mount Union. The course exposes engineering students to global societal challenges and their solutions. Of importance in this project is the need to boost numbers in the STEM (Science, Technology, Engineering, and Mathematics) and the need to increase diversity. With this in mind, the team developed and tested five different activities that involved different areas of STEM that were both fun and didactic in nature. Each activity had an associated module to allow for assessment of understanding. An orange buoyancy module, a strength of paper module, a soil erosion module, a simple circuit module, and a simple electric motor module were developed. They were tested on K-5 students at Washington Elementary School, Alliance, OH and on K-5 students at Ollas Arriba Elementary School, Panama City, Panama.

Chapter 5
Developing a Learning Community of Engineers Through an Honors First-Year Seminar... 88
Melissa L. Johnson, University of Florida, USA
Kristy Spear, University of Florida, USA

First-year engineering majors face a myriad of obstacles as they begin college. Taking challenging foundational coursework, navigating new expectations for performance and experience, and understanding the broader impact of their academic interests are just a few of those obstacles. In addition, female students sometimes face additional barriers to success, particularly as some question their own competence in the field. This chapter focuses on a first-year seminar for honors students that highlights the high impact practices students should participate in throughout their undergraduate career. These practices include global engagement, undergraduate research, and internships that are essential for early exposure to future career interests. By developing both formal and informal learning communities within the seminar, first-year students are exposed to opportunities, mentoring, and support that help them make informed decisions about their major and career.

Chapter 6
Engineering Teams: Supporting Diversity in Engineering Education............... 106
Jennifer Loy, Griffith University, Australia
Simon Howell, Griffith University, Australia
Rae Cooper, Griffith University, Australia

Engineering education increasingly involves working in groups. This is partly because of a growing value placed on graduate attributes relating to effective team working, and partly a response to the practicalities of working with large groups in an educational environment and the emphasis on peer learning. This chapter argues that a superficial approach to understanding the drivers for establishing and managing groups during first year activities can have negative outcomes, including re-enforcing majority dominance. This will potentially contribute to attrition amongst minority students and undermine the outcomes for the engineering cohort as a whole. This chapter provides strategies for building groups in the first year focussing on team building, valuing diversity and cultural awareness. It emphasises the importance of transferable skills for students and of understanding themselves, their heritage, attitudes and values and their contribution to a team, building an approach to support diversity in teams throughout the engineering degree program.

Chapter 7
Addressing Cultural and Gender Project Bias: Engaged Learning for Diverse Student Cohorts.. 130
Jennifer Loy, Griffith University, Australia
Rae Cooper, Griffith University, Australia

Engaged student learning is based on creating significant learning experiences for every student. Attracting a more diverse student body into Engineering requires a re-evaluation of the conventional project topics that dominate the discipline. Recognising and addressing cultural and gender bias in the development of project work allows for the education of Engineering faculty on the development of a range of project work opportunities that support the learning for a more diverse cohort. The selection of set project work has the potential to negatively impact the learning experience of minority students. This chapter considers the elements influencing set project work and provides strategies for understanding cultural and gender bias, and for redesigning project work that provides for a more diverse cohort.

Section 3
Critiques and Administrative Suggestions to Restructure Engineering Education

Chapter 8
Revisioning the Engineering Profession: How to Make It Happen! 156
Judith Gill, University of South Australia, Australia
Mary Ayre, University of South Australia, Australia
Julie Mills, University of South Australia, Australia

Beginning with a brief account of the value of diversity and inclusivity in a globalizing world, this chapter presents an overview of the current situation of the engineering profession in some English-speaking countries. The starting point addresses the enduring difficulty encountered by attempts to increase and diversify professional engineering. Drawing on a series of studies of engineering education, engineering workplaces and people, both in Australia and beyond, this chapter outlines barriers to entering engineering for anybody other than white mainstream males. Access and retention have long been recognized as serious impediments to increasing numbers of women in engineering. The particular breakthrough in this chapter describes the ways in which some Australian women engineers are working to sustain and enrich their professional status within the workplace by developing strategies that enable them to continue as professionals without diminishing other important features of their life worlds. The implications for all sectors of education, and employers, to emerge from this study offer a basis for redesigning engineering as a more diverse and inclusive profession.

Chapter 9
A Globally Focused, Experiential Educational System for STEM Fields: Measures for Intentionally Promoting Diversity.. 176
Aaron Sakulich, Worcester Polytechnic Institute, USA
Amy Peterson, Worcester Polytechnic Institute, USA

The diversification of the STEM labor force is essential to the health of the American economy, both because diverse teams are more adept at achieving project goals and because of shifting national demographics. Although the professional STEM workforce is currently more diverse than ever, sustained increases in diversity over the last 20 years has been relatively limited and may be decelerating. The focus of this chapter is specific programmatic components that can be implemented to intentionally supporting STEM students of diverse backgrounds. The hub of such an educational system would be Individual Development Plans (IDPs), which would empower students to follow an intentional path during their studies, combined with a strong focus on reflection and connection through ePortfolios and experiential learning both in academic projects and extracurricular professional activities such as internships.

Chapter 10
Diversifying Engineering Education: A Transdisciplinary Approach From RWTH Aachen University 201
Linda Steuer, RWTH Aachen University, Germany
Anna Bouffier, RWTH Aachen University, Germany
Sonja Gaedicke, RWTH Aachen University, Germany
Carmen Leicht-Scholten, RWTH Aachen University, Germany

Engineers and therefore engineering education are challenged by the increasing complexity of questions to be answered globally. The education of future engineers therefore has to answer with curriculums that build up relevant skills. This chapter will give an example how to bring engineering and social responsibility successful together to build engineers of tomorrow. Through the integration of gender and diversity perspectives, engineering research and teaching is expanded with new perspectives and contents providing an important potential for innovation. Aiming on the enhancement of engineering education with distinctive competencies beyond technical expertise, the teaching approach introduced in the chapter represents key factors to ensure that coming generations of engineers will be able to meet the requirements and challenges a changing globalized world holds for them. The chapter will describe how this approach successfully has been implemented in the curriculum in engineering of a leading technical university in Germany.

Chapter 11
Engineering Pathways in a U.S. Public Institution of Higher Education: A Strategy for Fostering Student Diversity........ 236
Fabiola Ehlers-Zavala, Colorado State University, USA
Anthony Maciejewski, Colorado State University, USA

This chapter relates a strategy that emerged from a larger effort of a land-grant institution in the U.S. to more rapidly increase the number of international students on campus and diversify its student body through the development and implementation of pathway programs. Pathway students are international students that do not meet the criteria for direct entry into a university due to lower levels of English language proficiency and/or GPA. The authors discuss strategies for ensuring success in these endeavors.

Chapter 12
Getting Off the Engineering Enrollment Rollercoaster: Interaction Between Academia and Upstream Petroleum Industry 260
Tatyana Plaksina, University of Calgary, Canada

Traditionally, petroleum engineering programs trained young professionals for the oil and gas industry. After years, this strong tie made the universities completely dependent on petroleum companies for their student placement and research funding. Similar to the operators and service companies, the universities became sensitive to changes in demand. In the last decade, in response to a rapid increase in crude oil prices and unconventional production, the industry experienced a hiring surge which caused a subsequent agitation among engineering freshmen to claim petroleum engineering as their major. As a result, this short period of high demand for petroleum engineering graduates created tremendous pressure on the departments making them expand beyond their educational capacities. In some cases, that led to doubling of the number of undergraduates. Simultaneously, this rapid expansion of the student enrollment created a growing demand for faculty who are now experiencing serious demographic and research gaps.

Compilation of References 287

About the Contributors 314

Index 323

Foreword

"Glass Ceiling," a term used to identify an environment in which advancement of women or minorities is restricted, strongly suggests that the members of these groups are already part of an organization. Perhaps another term, possibly "Glass Door", should describe the obvious disparity of female and minority participation the engineering profession. All conditions being equal, the type of preparation, alma mater and social network, when the employment door actually opens are essential to actually landing a position in an establishment. If we analyze the processes involved in preparing under-represented groups, a number of concerns will surface - ones that have been adequately addressed and others, not so much. Minority students have low participation rates in international programs and internships. Experiential educational opportunities that increase self-confidence and other soft skills such as respect and appreciation for other cultures. Thus, under-represented cohorts may not possess interview skills to clearly articulate their abilities and impress employers. Therefore, while affirmative action might facilitate getting an interview, there is no policy in place guaranteeing a call back for the job.

The issue of under-representation in Science, Technology, Engineering and Mathematics (STEM) fields seem to be deeply entrenched across various societal contexts. In countries like the United States, this deep-seated issue continues to baffle stakeholders. While women comprise more than half the U.S. population, they represent less than 13% of the engineering workforce (*NSF, 2015 Women, Minorities, and Persons with Disabilities in Science and Engineering*). Contemporary challenges in educating minorities and other under-represented cohorts in engineering and other STEM fields include low enrollment and graduation rates, as well as psychological, social, and environmental barriers.

Over the years, although several strategies including first-year seminars, internships, learning communities, and capstone projects have been promoted to curb the minority brain-drain from engineering programs, these measures, typically implemented piecemeal, target under-preparedness without consideration for other competing and causative factors such as the lack of self-directing competencies. To sustainably increase the number of women, minorities, and other under-represented groups completing engineering degrees and successfully navigating professional career paths, there needs to be 1) research into international perspectives on student preparation and success; 2) elucidation of the sets of common socio-cultural, psychological and environmental factors that facilitate engineering students' success in the major education systems globally; and 3) the proposal of high impact, non-traditional strategies such as study and interning abroad that harness multiple factors and address several student deficiencies simultaneously.

This book provides a global perspective on the importance of diversity and inclusivity for the sustainable growth of the engineering profession, which can also be extended to other STEM fields. It is timely and adds new dimensions to this ongoing and globally germane conversation.

Noel Blackburn
Brookhaven National Laboratory, USA

Noel D. Blackburn *is the Manager of University Relations and DOE Internship Programs in the Office of Educational Programs (OEP) at Brookhaven National Laboratory (BNL), a multi-disciplinary national laboratory owned by the Department of Energy (DOE) and home to seven Nobel Prize Laureates in numerous fields of research. Mr. Blackburn is responsible for designing, implementing and managing workforce development research programs for undergraduates, graduates, and faculty at BNL, while creating access opportunities for under-represented groups into BNL and the DOE enterprise with such programs as the Interdisciplinary Consortium for Research and Educational Access in Science and Engineering (InCREASE). His passion is the removal of the locked "Glass Door" which hinders access for the well-qualified, less connected pool of intellectuals who are unable to stay ahead of the innovation curve while educating the next generation. His philosophy is based on a simple mantra – 'rising tides raise all ships'. Mr. Blackburn has served as an Adjunct Lecturer for the Physical, Environmental and Computer Science Department at Medgar Evers College. He is a member of the Trinidad and Tobago Professional Engineers Association, the American Society of Civil Engineers and American Society for Engineering Education. Blackburn earned his Master of Engineering in Environmental Engineering, where he received a university citation for outstanding performance, and a Bachelors of Engineering in Civil Engineering from the City College of New York, CUNY. He lives in New York with his lovely wife Kayla and daughter Adafih.*

Foreword

According to the National Center for Science and Engineering Statistics, the number of engineering bachelor degrees awarded to females increased from 0.4% to 19% from 1966 to 2012 respectively. On the surface this seems to be a giant leap forward, however, this victory is eclipsed by the seemingly entrenched demographic disparity. Women constitute more than half the general population but is grossly under-represented in the engineering profession. African Americans represent only 4% of the engineering workforce while representing 13% of the population. There is still a need to further increase female and minority representation in the STEM enterprise and institutions of higher education must seek ways to address the present disparity by implementing ways to remove barriers to access and retention. In addition, the national shift towards knowledge-intensive economic activity, and the anticipated retirement of STEM professionals in the government sector in the near future, continue to increase the need for well-trained individuals in these fields. When certain higher education researchers suggest that the issue of female and minority under-representation in STEM resides primarily on the poor academic mathematics/ science preparation of minority students, and their lack of interest and motivation toward STEM, without, in most cases, changing the quality of their learning environment, the outcomes have not produced positive results.

The Bureau of Labor Statistics had identified the shortage of engineers and scientists at the turn of the century and recommended increased production take place if the U.S. is to be competitive in global markets. Agencies such as the National Science Foundation, the National Institutes of Health, the American Association for the Advancement of Science and many others, have increased their focus on programs and initiatives to address the current shortage and the potential damage it will create for industry and the nation at large. The largest untapped segment of the population that exists to educate in the STEM disciplines are minorities, (African American, Hispanic American and Native Americans) and women.

However, the notion still prevails that the mission of higher education is one that focuses on "fixing" the deficits in its students. Some college professors still view minority students as academically at risk, lacking abilities, unmotivated and culturally deprived, instead of seeing them as capable, motivated and resilient learners able to build upon their cultural strengths. Much of the statistical information published regarding minority students' performance looks more at the mechanism of how the population performs using the norms established by their non-minority counterparts. In more recent years, engineering and science educators have begun to peel away the previous misconceptions of minority student's performance and actually capture those skills that are natural and can be enhanced to allow them to perform at the very highest levels.

With this type of emphasis from the government, industry and higher educational institutions, there are a growing number of initiatives that are now strategizing methods to stop the minority STEM brain drain to other disciplines and investigate newer methodologies to increase the quality and quantity of under-represented students to become the scientist for the next millennium. Some of the new strategies include: first year seminars, internships, learning communities and capstone projects. These are typically implemented utilizing the classic fixing of the "broken student" methodology. There are currently educators across the country that are researching and testing methods that show the true understanding of the causative factors that impact both cognitive learning and self-directing competencies will facilitate scientific learning for all.

During the last 20 years coordinated efforts across the country have convened and established metrics and environments that have proven successful including: 1. the introduction of more esoteric topics in STEM to create interest in future scientific advances particularly in middle schools, 2. Introduction of undergraduate students to research as soon as possible, 3. Understanding the common phobias and discontinuities for underrepresented students, (e.g., do they belong in this environment), 4. Pedagogical approaches that utilize the common knowledge of the students and translates abilities to more sophisticated topics, 5. Introduction to graduate students of similar background, 6. International research opportunities for a more global perspective of their discipline.

This book is a compilation of the above-mentioned efforts, detailing successes, pitfalls and future direction to continue the current momentum. It will provide strategies for educators and policy makers who are addressing these issues on a daily basis. Congratulations.

Stephen R. Cox
Drexel University, USA

Stephen R. Cox, *Ph.D., is the Project Director of the Greater Philadelphia Region Louis Stokes Alliance for Minority Participation (LS AMP) headquartered at Drexel University, Pennsylvania since 1990. The mission of LS AMP is to substantially increase the quantity and quality of African American, Hispanic and Native American students receiving baccalaureate degrees in science, technology, engineering and mathematics (STEM), and subsequently, entering graduate school to attain doctoral degrees. An engineer and physicist by training, Dr. Cox has spent the last three decades advocating for increase participation of women and minorities in STEM. His many honors and awards have included; Benjamin Benneker Legacy Award for African American Inclusion in Science, Men Making a Difference in the African American Community, the Equal Access to S.E.M Education through creative and Innovative Program Development and Foundation Award by the National Organization for the Advancement of Black Chemists and Chemical Engineers for the advancement of STEM in the Delaware Valley.*

Preface

In today's global economy, there is an urgent need for a diverse engineering workforce representing self-confident and culturally literate individuals who are able to tolerate ambiguity as well as empathize with the socio-cultural nuances of different people groups. However, the engineering workforce does not typically mirror the demographic profile of their host society. For example, in the US, the high attrition rates from engineering programs for women, minorities and under-represented groups range from 49% – 62% and result in an engineering workforce that consist of 0.3% – 13% representation from this cohort. In stark contrast, women and minorities comprise more than half the U.S. population.

This issue of underrepresentation seems to be deeply entrenched in certain societal contexts; however, there are some global instances where women are in the engineering workforce. In places like the US where this deep-seated issue continues to baffle stakeholders, several strategies have been promoted. These include first-year seminars, internships, learning communities, and capstone projects. These are typically implemented piecemeal, targeting underpreparedness, without consideration for the other competing causative factors such as the lack of self-directing competencies.

The American Association of Engineering Education (ASEE) Board authorized 2014-2015 as the Year of ACTION on Diversity. ASEE is the world's leading entity in engineering education and thus this designation of 2014-2015 was a mandate for the engineering world to include diversity as one the National Academy of Engineering's Grand Challenges of Engineering. Diversity is inclusivity and so very much more, an idea that was highlighted by William A. Wulf in his annual report to the National Academy of Engineering in the early 2000's. Diversity is strength in creativity, broadness of new ideas, and embracing new perspectives to arrive at the most truly innovative, resource-smart solutions possible. Diversity refers to "gender, race, ethnic background, disability, sexual orientation, gender expression, age, socio-economic status, nationality and other non-visible differences" such as personality, aspiration, learning style and motivation that Dr. Wulf refers to as individual differences (quotes from the ASEE Definition of Diversity).

With the engineering academy currently focused on minorities and underrepresented groups within the engineering profession, this publication is very timely. The key objective of this publication is to serve as a repository for global findings with respect to engineering education promotion across all education levels. This publication has the sub-objectives of investigating:

1. The impact of industry-engineering interactions to improve inclusion and diversity through all stages of the engineering profession
2. The use of specialized programs (such as study abroad) to retain diverse populations in engineering
3. The impact of student affairs on retention of minorities in engineering (e.g., student organizations)
4. The faculty-staff models that allow for successful recruitment-retention-graduation of diverse populations in engineering

The lack of minority and female representation in engineering majors and subsequent careers continues to be a seemingly entrenched challenge in the face of numerous well intentioned attempts. Changing the current trajectory and landscape requires the application of coordinated high impact strategies from global and diverse perspectives. This publication seeks success stories from around the world that involve combinations of K-12 preparation, re-engineering the current curriculum, the incorporation of experiential educational experiences such as study abroad, providing more underrepresented-friendly work environments and addressing barriers to women and minorities. This book seeks to integrate causation factors, barriers and challenges together with sets of holistic solutions that have shown successful application of a systems approach to addressing this challenge. Across twelve (12) chapters, the views and expertise of twenty six (26) authors are presented in this book. Their contributions span studies in Australia, North America, Europe, and Central America.

ORGANIZATION OF THE BOOK

Scholarly Value and Impact of Book

This book expands on the field of the engineering education. The main themes that the chapters in this book attempt to expand upon include:

- Diversity
- Inclusion

- Women in engineering
- K-12 STEM outreach
- Gender and cultural bias
- Cultural awareness
- Teamwork and team construction
- Support system development
- High impact approaches
- Education-industry interaction

Target Audience

This book is a great resource for academic advisors and counselors in high school and colleges of engineering; faculty advisors in colleges of engineering; directors of multicultural and engineering student centers at the college level; students undertaking research in the area of administration of higher education; recruitment officers and student affairs/involvement professionals at colleges. As a secondary audience this publication would make for an excellent textbook resource for students that are studying for masters and doctorates in the administration of higher education.

The Content

There are twelve (12) original chapters in this book. These chapters were divided among three (3) sections. These sections are:

- **Section 1: High Impact Higher Education Pedagogical Frameworks and Practices** (three chapters)
- **Section 2: Micro-Community or Target Approaches to Diversifying Engineering** (five chapters)
- **Section 3: Critiques and Administrative Suggestions to Restructure Engineering Education** (four chapters)

The forewords are written by Mr. Noel Blackburn and Dr. Stephen R. Cox. Both of these gentlemen are, and have been, avid champions for diversity in engineering for many years at different institutions. In his foreword, Mr. Blackburn highlights the well understood concept of the "glass ceiling" and makes a case for making widespread the term "glass door" to expose, more fully, the disparity in participation of women and minorities in the engineering profession. Dr. Cox, in his foreword, confronts the ideology and thinking of "the broken" student and provides insights into proven metrics for diversifying STEM fields.

Section 1: High Impact Higher Education Pedagogical Frameworks and Practices

In Chapter 1, Yvonne R. Hilton presents her study entitled "Improving Student Persistence". Throughout higher education, student persistence is of major concern. Statistics show that the greatest percentage of attrition occurs after the freshmen year. Most studies that have been done took place at large predominately white institutions (PWIs). This study observed student satisfaction with institutional environment at Lincoln University. Lincoln is a small, historically black university (HBU) with less than 3000 matriculated students. Specifically, this study it looked at how the institutional environment may impact freshmen students' decisions to return for their sophomore year. The results indicate that although freshmen students are dissatisfied with certain aspects of the environment they consider to be important, this generally had little if any influence on their decision to return.

Chandra A. Stallworth and Ken D. Thomas tackle the issue of recruitment and retention of underrepresented minorities at a large predominantly white institution (PWI) in Chapter 2 which is titled "Minority Recruitment and Retention Among Gifted Students". In this chapter Stallworth and Thomas focus on the efforts of the Auburn University Honors College to recruit and retain underrepresented minorities at the highest academic echelons. The chapter delves into the creation of a minority support system that has now grow into a chartered student organization.

In Chapter 3, Monica Gray and Connie Lundy explore their study titled "Engineering Study Abroad: High Impact Strategy for Increasing Access". Changing the current engineering workforce's diversity portfolio is of grave national importance and requires a plethora of high impact approaches. One overlooked approach is study aboard. In this chapter the authors challenge and make their striking case for the inclusion of study abroad on the list of high impact approaches. They do this by discussing the benefits, opportunities and challenges in sending engineers abroad, exploring strategies to internationalizing engineering education, and examining how these can be achieved to ensure affordability, broad access, retention and program efficiency.

Section 2: Micro-Community or Target Approaches to Diversifying Engineering

"Empowering Women in STEM: Embedding STEM in K-12 Education" is the title of Chapter 4 which was co-written by Gretchen Dietz, Julie Hessedence, Terry Long, and Helen E. Muga. This group of civil and mechanical engineering students along with an engineering faculty member from University of Mount Union seeks to tackle the number of women involved in engineering by getting girls excited about science, technology, engineering and math (STEM) in the K-12 school system. This chapter details the approaches used by this team in both the Ohio education system and the Panamanian education system. The impact of their work is also highlighted in this chapter.

Judith Gill, Mary Ayre, and Julie Mills of University of South Australia, Australia authored Chapter 5: "Revisioning the Engineering Profession: And How to Make It Happen!" In this chapter the authors argue that by there is a great need to develop equity and inclusivity in engineering with particular emphasis on women in the workplace. This chapter highlights the barriers to women and minorities in the engineering profession and suggests a path to best get there based on the experiences of some Australian women engineers.

In Chapter 6, Melissa L. Johnson and Kristy Spear elaborated on a learning community as a high impact approach in their work titled "Developing a Learning Community of Engineers Through an Honors First-Year Seminar". This chapter focuses on a first-year seminar for honors students that highlights the high impact practices students should participate in throughout their undergraduate career. By developing both formal and informal learning communities within the seminar, first-year students are exposed to opportunities, mentoring, and support that help them make informed decisions about their major and career. The authors found that providing this level of support and clarifying the opportunities that exist within the field of engineering through a first-year seminar course was a successful way to address this challenge and retain and graduate more engineering students.

"Engineering Teams: Supporting Diversity in Engineering Education Through Attention to Detail in Engineering Group Work Experiences in First Year" by Jennifer Loy and Simon Howell is presented in Chapter 7. Working in groups is an inevitable part of the engineering education experience. This chapter argues that an artificial method to establishing these groups during first year activities can reinforce majority dominance which only further goes against the inclusion and diversity that is being hoped for. This chapter provides alternative strategies for building groups in the first year that focus on team building, valuing diversity and cultural awareness. It also provides a framework for building on the initial activities that will support diversity throughout the degree program.

Jennifer Loy of Griffith University, Australia, presents her study titled "Addressing Cultural and Gender Bias in Set Project Work: Creating Project Diversity and Encouraging" as Chapter 8. This chapter considers the type of projects chosen for marketing engineering in schools and the set project work typically forming the basis of first and second year engineering education. This chapter provides a review of possible cultural and gender bias project work in Engineering education and provides strategies for redesigning project work that provides for a more diverse cohort.

Section 3: Critiques and Administrative Suggestions to Restructure Engineering Education

In Chapter 9, Aaron Sakulich and Amy Peterson explored "A Globally Focused, Experiential Education System for the STEM Fields". In this study, the authors propose an outline for an educational system that would intentionally support STEM students regardless of gender, race, ethnic background, disability, sexual orientation, gender expression, age, socioeconomic status, nationality, or individual differences such as learning style and career aspirations. The authors purport that such a system would use Individual Development Plans (IDPs) in order to empower students to follow an intentional path during their studies as they explore career options, identify a mentoring network, and set overall career goals.

Linda Steuer, Anna Bouffier, and Sonja Gaedicke's study titled "Diversifying Engineering Education: A Transdisciplinary Approach From RWTH Aachen University", as presented in Chapter 10, critically analyzes an innovative approach of the Gender and Diversity in Engineering Institute (GDI) at RWTH Aachen University. RWTH Aachen University wants to open the way towards a more holistic and transdisciplinary approach of engineering. Through the integration of gender and diversity into engineering research and teaching there has been expansion with new perspectives and content providing an important potential for innovation generation. This allows the authors to conclude that engineers bear social responsibility and should take this into account by considering social diversity. However, this reasoning suggests that, in addition to the core engineering subjects, engineering curricula have to include disciplines which enable engineering students to see the related questions within societies they are developing their products for.

Chapter 11 makes the case for the emergence pathway programs for internationalizing US higher education campuses in the study titled "Engineering Pathways in a US Public Institution of Higher Education: A Strategy for Fostering Student Diversity". Authors Fabiola Ehlers-Zavala and Anthony Maciejewski remark that "A pathway program is a program that has been designed to attend to the needs of a specific group of international students in their first year of education in the US. Pathway students are students that do not meet the criteria for direct entry into a university due to lower levels of English language proficiency and / or GPA. Therefore, these students need greater academic and language support to be positioned for success upon matriculation into the degree programs of their choice." The author concluded that there is most certainly a need for these types of programs in US Colleges of Engineering to better prepare US students to enter the real world with greater cultural awareness and experience with diversity and inclusion.

In the book's last chapter, Chapter 12, "Getting Off the Engineering Enrollment Rollercoaster: Example of Interaction Between Academia and Upstream Petroleum

Industry", by Tatyana Plaksina, there was a main aim to examine the petroleum industry in the US and to determine how petroleum engineering programs can better interact with industry partners. This interaction is in hopes to allow for better diversity in the petroleum sector. In this chapter the author elaborates on the diversity of the students, inclusive of on a gender basis, as well as the faculty that are involved in teaching and research in petroleum engineering programs. In conclusion, the author suggests strategies for enhancing diversity of students and faculty based on integration approaches.

REFERENCES

American Society of Engineering Education (ASEE). (2015). *ASEE: Action on Diversity*. Retrieved May 12, 2015 from http://diversity.asee.org/

Acknowledgment

First and foremost, the editors would like to thank each one of the twenty-six (26) authors for their contributions. Their sincere gratitude goes to the authors who contributed their time and expertise to this book.

The editors also wish to acknowledge the valuable contributions of the reviewers regarding the improvement of quality, coherence, and content presentation of chapters. Without their support, this edited volume would not have become a reality.

To Mr. Noel Blackburn, Office of Educational Program at Brookhaven National Laboratory and Dr. Stephen Cox, Project Director of the Greater Philadelphia Region Louis Stokes Alliance for Minority Participation at Drexel University, the editors wish to express a multitude of thanks for writing the foreword to this volume.

The editors would like to give a heartfelt thank you to their IGI Global Development Editor, Ms. Elena Wehr and Editorial Assistant, Ms. Maria Rhode, for their unwavering patience and support through an arduous process.

Finally, the editors would like to express gratitude for the support of their affiliate institutions: Chemistry & Physics Department, Lincoln University, Pennsylvania, USA; and The Honors College & Biosystems Engineering Department, Auburn University, Alabama, USA.

Monica Gray
Lincoln University, USA

Ken D. Thomas
Auburn University, USA

Section 1
High Impact Higher Education Pedagogical Frameworks and Practices

Chapter 1
Improving Minority Student Persistence:
An Institutional Factors Approach

Yvonne R. Hilton
Lincoln University, USA

Monica Gray
Lincoln University, USA

ABSTRACT

Student persistence in college is a major concern for every institution of higher education. Statistics show that the greatest percentage of attrition occurs after the freshman year. Many studies surrounding this phenomenon tend to focus on pre-college predictors to gain knowledge into mass premature departure during the freshman to sophomore transition. However, very few have looked at institutional factors and how they may explain the problem. Further, most research studies have been done at relatively large and predominantly white institution. This chapter investigates student satisfaction with institutional factors at a small Historically Black College and University (HBCU). Specifically, it looks at how the campus environments may impact freshman students' decisions to return for their sophomore year. While freshman students were dissatisfied with certain aspects of the university they consider to be important, the results precluded good prediction in a practical sense if they influence their decision not to return.

DOI: 10.4018/978-1-5225-2212-6.ch001

INTRODUCTION

Retention and attrition rates in higher education are at the forefront of discussions among stakeholders such as students, parents, educators, administration and policy makers. Attrition is a matter with detrimental results for the individual, colleges and universities, and the country as a whole. For example, in today's knowledge-driven economy, most jobs require some postsecondary education and/or training. Regardless of the type and size of the tertiary institution, retention and attrition issues are common to all. Therefore, these discussions will continue as strategies are developed to encourage students to persist through to graduation.

Over the past four decades, many studies have been conducted in an attempt to identify possible predictors of degree completion. For the most part, studies have been carried out at mid-western and southwestern institutions in the United States (U.S.) where African-American students are in the minority. In addition, the vast majority of studies reported in literature have focused on the use of pre-college predictors such as high school rank, grade point average (GPA), standardized test scores (e.g., SAT/ACT) and socio-economic profile to determine the probability of student success. While these are important factors in the selection of students, they may not reveal much about what an institution can do to retain students once they have been admitted. The objectives of this chapter are to explore the factors of retention and attrition at a small Historically Black College and University (HBCU) and to elucidate how institutional environment impact freshman to sophomore college departure decisions.

Background

Only about fifty percent of students entering U.S. colleges and universities graduate (Athreya & Eberly, 2015). Degree completion affects not only the individual but the society as a whole. In terms of earnings differentials, on average, persons who complete at least an undergraduate degree make about one million dollars more than high school graduates over their lifetime (Tinto, 2012). On a larger scale, college completion translates to human capital accumulation at the national level. Societal benefits range from informed voting, low unemployment and incarceration rates and improved community health to how well-prepared the next generation will be to enter formal schooling (Baker, 2014; Hout, 2012; Oreopoulos & Petronijevic, 2013). Institutions with knowledge of the catalysts for premature college departure can begin to make the changes necessary to help increase student retention and simultaneously decrease attrition rates. In the long run, these changes will serve to protect and strengthen the academic and financial integrity of the institution, improve the future viability of the students and their respective communities and increase the economic efficiency of the nation (Lotkowski et al., 2004). Given all the individual and national advantages of an educated society, why then is the attrition rate so high and what are the causative and/or predictive factors?

More than four decades ago, Nicholson (1973) proposed the following academic (e.g., high school rank and standardized test scores) and nonacademic (e.g., motivation and academic promise, and socio-economic profile) variables as predictors of postsecondary academic success. Over the next twenty years, similar studies were described in the literature: Tinto (1975) modeled individual background (e.g., social status and high school experiences), personal characteristics (e.g., sex and ethnicity) and individual motivation (including commitment to the institution); Lunneborg (1975) found the best predictor of graduation and credits earn was high school GPA; Lenning, Beal, Sauer (1980) and Bean & Metzner (1985) investigated the usual student characteristics, but also included institutional climate; Jacobi (1991) reviewed the literature on the association between mentoring and student success; and Cabrera, Nora and Castaneda (1993) attempted to reconcile the two major frameworks that dominated this research area, Tinto's "Student Integration Model" and Bean's "Student Attrition Model".

At the intersection of these two major models, an interesting commonality emerges, namely, organizational factors such as a successful match between institution and student is a major determinant of persistence. That is, proximal factors, characteristic of the institutional setting, facilitate student satisfaction and predicts students' success. Barefoot (2004) agrees that the impact of institutional environment on retention and attrition requires further examination. In addition, while many studies have been conducted on the issues of retention and attrition among college students, overall, relatively few of them centered on the HBCUs and/or the type of student they attract; under-represented, under-served minorities (UUMs).

From an institutional support perspective, the research indicates intervention programs, such as intrusive advising and first year experience programs, have had positive impact on student academic success and retention (Gandara & Maxwell-Jolly, 1999; Higgins & Glanville, 1983; Pew Charitable Trusts, 2007; Tobolowsky, 2005; Turnbo, 2002; Wilcox, 2005). However, research examining the role of the institutional environment on freshmen students remains virtually untapped (Barefoot, 2004). Further, the paucity of data is especially realized at HBCUs, as most studies are done at majority institutions. As discussed above, a college educated workforce builds a stronger nation and provide a better quality of life for all citizens. HBCU's draw a significant number of first generation UUM students and face a number of challenges (Nettle, Wagener, Millett & Killenback, 1999). Lotkowski, Robbins, and Noeth (2004) reported that African Americans are less like to attend and complete college compared to their White peers, it is especially important for HBCUs to retain those who enter their doors.

Most HBCUs, like Lincoln University attract and enroll at-risk, first generational, African American young adults. Only 41.6% of the institution's freshmen class continue to graduation. Retention is at the forefront of its strategic plan, as an average of 30.5% of its freshmen do not persist beyond their first year (Lincoln University, 2016). This is about seven percentage points below the 77.1% national freshmen-to-sophomore year retention rate (NCHEMS, 2016). On the surface this statistic gives the impression of attrition parity, but closer inspection reveals that of the 17.3 million undergraduate students, Blacks constitute only 13.9% compared to 55.5% Whites. Outside of the usual pre-college contributors to retention and attrition, it appears that higher education institutions are uncertain about how other factors may impact student persistence. Institutional attributes play an important role in student integration and satisfaction with the college they are attending (Berger & Braxton, 1998). When students feel comfortable with the campus environment and begin to experience a sense of belonging, they become motivated and engaged with the institution on social as well as academic levels. This subsequent engagement tends to lead to higher levels of student achievement, commitment and persistence (Thomas & Galambos, 2004; Strauss & Volkwein, 2004; Elliott, 2002; DeWitz & Walsh, 2002; Berger & Braxton, 1998; Browne, Kaldenberg, Brown & Brown, 1998).

METHODOLOGY

Research Setting and Participants

This study was conducted at Lincoln University. Located in Southeastern Chester County in Pennsylvania, USA, Lincoln University is the nation's first degree-granting HBCU. The average enrollment is about 2,000 students from the surrounding communities, within or around the large metropolitan areas of the Northeastern United States, and other parts of the world namely Africa and the Caribbean. Lincoln is a liberal arts institution and is accredited by the Middle States Association of Colleges and Schools. It offers undergraduate programs in the arts and sciences; and graduate programs in human services, education, and business administration.

HBCUs have for the last 160^{+} years provided access to higher education with missions to "*prepare Blacks for participation in American society*" (Nichols, 2004). HBCUs really began to materialize following the civil war, in large part as the result of racism. In the late 1800's locations earmarked for the "*education of Black students evolved from the occasional use of churches and abandoned train cars to a handful of modest state-supported colleges and universities*" (National Science Foundation, 2002). They were established wherever large populations of Black people lived.

In 1989, African American students attending HBCUs persisted in college at a rate almost 10% higher than that of their African American peers attending predominately white institutions (Nettles et al., 1999). This is remarkable in light of the fact that "HBCUs were not designed to succeed. On the contrary, they were merely established to appease Blacks and to serve as "holding institutions" to prevent Black students from matriculating into majority institutions (Evans, Evans, & Evans, 2002). Nevertheless, over the years, they have produced the vast majority of Black professionals and leaders, including Martin Luther King, Jr., Mary McLeod Bethune, Thurgood Marshall, Toni Morrison and Langston Hughes (Nichols, 2004). They have done this despite the fact that most students attending HBCU's receive less academic preparation for college than their peers attending other four-year colleges (Nettles et al., 1999).

HBCUs are first choice institutions for many Black students. One reason is, they make Black students feel comfortable - creating an environment in which they can stay true to themselves and their culture. In addition, they have a reputation for producing graduates that are very competitive in their fields of study (Evans et al., 2002). Also, since they help their students realize that everyone can have the opportunity to go to college and be successful, *"private HBCU's are the primary source of postsecondary education for economically disadvantaged African American students...with over 98% of the private HBCU students receiving need-based federal financial aid"* (Nettles et al., 1999).

Research Design

A longitudinal (panel) study of the freshman class was conducted during the 2007-2008 academic year using a survey research method. Data collection employed a Noel-Levitz Student Satisfaction Inventory (SSI) questionnaire with a standardized set of items developed by the Noel-Levitz Center for Institutional Effectiveness, an authority in student recruitment and retention. The Noel-Levitz SSI is a 70+ statement-based survey instrument designed to obtain student rated importance and satisfaction levels with the institution where they are enrolled. The SSI has been in use since 1994, and about two thousand institutions have used it to survey over two-million students since 2005. Students rated how important their expectations as well as their satisfaction with the institutional performance in those areas. The questionnaire's statements were categorized into eleven groups called scales. Information on the students' personal expectations and perceptions of, and experiences with various components of the university was collected at the start and conclusion of the research period. This tool was used to examine what Lincoln's freshmen students deemed as important factors on a college campus as well as their satisfaction with Lincoln's performance in meeting those factors. The initial and year-end surveys were administered at the beginning of fall 2007 and end of spring 2008 semesters respectively. The sample size were representative and reasonable conclusions about the entire freshman student body based on participant responses can be drawn.

- **The Initial Survey:** Was comprised of two sections. The first section consisted of eleven questions requesting standard demographic information such as race/ethnicity, gender and age as well as parental education, reasons for choosing Lincoln University, college choice ranking and intended major. The second section consisted of the "importance" portion of the Noel-Levitz Student Satisfaction Inventory. 386 Freshmen representing 61% of the total first year class participated.
- **The Year-End Survey:** Also had two sections. Section one was the complete Noel-Levitz Student Satisfaction Inventory. Section two contained four questions regarding age, major, plans to return, and college choice (generated from ACT Survey of Student Opinions). A fifth question was an open-ended question that asked students to share any thoughts and feelings about their Lincoln University experience. 216 or 34% of the first year class participated.

Importance of Institutional Factors to Incoming Students

The factors or scales observed in this study were Academic Advising Effectiveness, Campus Climate, Campus Life, Campus Support Services, Concern for the Individual, Instructional Effectiveness, Recruitment and Financial Aid Effectiveness, Registration Effectiveness, Safety and Security, Service Excellence and Student Centeredness. These eleven factors were referred to as scales. Each scale was made up of several items. A brief description of each scale is given in Table 1 . The importance of each scale item was ranked by participants using a Likert-like scale ranging from 1 (not important at all) to 7 (very important).

Data Analysis

Participant responses were analyzed using various descriptive statistics, such as measures of central tendency and dispersion (mean, median, mode, range, variance and standard deviation), as well as frequencies and percentages. From the collected data, the researcher made comparisons with the level of importance (importance rating) students indicated that various institutional factors held for them comparable to their actual experiences. Data on student demographics and academic performance was also considered in determining the influence of satisfaction with various aspects of the institution on student retention between the freshman and sophomore years.

Table 1. Institutional factors or scales and their descriptions

Factor or Scales	Description
Academic Advising Effectiveness	Assessed the comprehensiveness of the institution's academic advising program
Campus Climate	Assessed the institution's promotion of campus pride and sense of belonging
Campus Life	Assessed the institution's comprehensiveness in providing comfortable living conditions, services and activities
Campus Support Services	Assessed the quality of support programs and services
Concern for the Individual	Assessed the institution's commitment to treating each student as an individual
Instructional Effectiveness	Assessed institution's commitment to academic excellence
Recruitment and Financial Aid Effectiveness	Assessed the comprehensiveness of the institution's admissions and financial aid programs
Registration Effectiveness	Assessed the quality of registration processes
Safety and Security	Assessed the quality of institution's safety/security services
Service Excellence	Assessed the quality of service and personal concern for students in various areas of campus
Student Centeredness	Assessed the extent to which the institution made the student feel valued

Three statistical tests were used to analyze the collected data; the chi-square test, the t-test for independent samples and one-way analysis of variance (ANOVA). Chi-square (crosstabs) was used to look at differences when the responses were discrete and when comparing percentages or proportions. The chi-square test allowed for the comparison of the distribution of rankings (1-7) across groups. For example, differences in female-male responses were determined comparing the proportion of responses in each ranking (1-7). The t-test and ANOVA were used when the responses were continuous and when comparing means of two groups (t-test) and three or more groups (ANOVA). The t-test for independent samples was utilized to obtain the mean scale scores for gender and residency. A one-way ANOVA was utilized to obtain the mean scale scores of school of intended major, college choice ranking and grade point average. The information generated from these processes was used to compare the differences within and between the specific groups in their importance and satisfaction ratings of Lincoln University.

RESULTS

Comparison of students' perceptions of actual experiences provided information about the degree to which their expectations were met. Furthermore, the data indicated whether students' experiences with various factors of the university related to their intentions to return to Lincoln University for the subsequent fall semester.

Freshmen Profile

The size of the freshmen class in fall 2007 consisted of a total of 633 students; 271 males and 362 females. The racial makeup of this group was 96.5% African-American; 0.9% Hispanic; 0.2% Caucasian; and 2.4% non-resident of the United States. 251 of these students were Pennsylvania (in-state) residents. The other 382 were out-of-state residents including international students. Of the 386 freshmen students who completed the initial survey study there were 173 males and 213 females. 44% percent of the sample were in-state residents and 56% were out-of-state residents.

78% of the initial survey respondents reported at least one parent did not pursue post-secondary education. Of both parents, mothers tended to be better educated on average. About 25% participants came from families where one or more parents completed college. On the other end of the continuum, 11% of students reported coming from homes where at least one parent did not complete high school.

Lincoln University was the third choice institution for about 60% of the students. About the same proportion stated that the main purpose for coming to the university was to earn an undergraduate degree, followed by, no definite purpose, other reasons such as sport, to take courses to transfer to another institution and finally family pressure. The Pearson Chi-Square analysis showed no significant difference between the responses for female and male or in- and out-of-state residents.

Attributes of the institutions that influenced students' decisions to attend included program of study, courses, entrance requirements, social climate, ethnic makeup, male/female ratio, facilities and friends attending. Two-sample t-test showed no significant difference for in- and out-of-state residents but a significant difference between males and females on program of study, male/female ratio and friends. While a high percentage of both male and females deemed program of study important, females tended to rank it higher. Conversely, for the social items (ratio of male to female students on campus and friends attending Lincoln), males ranked these as important to a much greater percentage than their female counterparts. They also selected the not important choices less frequently. While they do come for an education, it would appear that the men are more strongly motivated to attend Lincoln because of social opportunity, whereas the women have a more academically oriented purpose for attendance.

Importance of Institutional Factors at the Beginning of Freshman Year

The mean scores for all eleven scales of the Noel-Levitz SSI, ranged from a low of 5.58 to a high of 6.11. This indicates that all scale factors were of some importance to students. However, the top five were academic advising, instructional effectiveness, campus support services, student centeredness, and campus climate.

As seen in Table 2 below, the data from the eleven scales were analyzed to capture any differences as they relate to four student characteristics; gender, residency, college choice ranking and intended major. Student responses varied by gender much more frequently than any of the others. As a matter of fact, there were no significant differences related to residency and intended major on any of the eleven institutional factors scales.

- **Gender Difference:** Across all eleven scales, female students tended to rate the institutional factors higher than their male peers. While the rank order for each of the scales was not very different between the genders, a t-test for independent samples revealed that there were significant differences at a significance level of 0.01 between males and females for all scales except safety/ security. Registration effectiveness had the largest mean difference (0.49)

Table 2. Mean importance of institutional factors for incoming LU students

Overall Rank	Scale Items	Mean	Female Rank	Female Mean	Male Rank	Male Mean	p-value
1	Academic Advising Effectiveness	6.11	1	6.23	1	5.95	.021**
2	Instructional Effectiveness	6.07	2	6.19	2	5.90	.005**
3	Campus Support Services	6.04	3	6.17	4	5.86	.006**
4	Student Centeredness	6.03	4	6.14	3	5.88	.009**
5	Campus Climate	6.00	5	6.14	6	5.82	.001**
6	Recruitment/ Financial Aid Effect.	5.97	7	6.07	5	5.83	.045**
7	Concern for the Individual	5.93	8	6.06	9	5.75	.007**
8	Registration Effectiveness	5.93	6	6.07	10	5.65	.000**
9	Service Excellence	5.93	9	6.04	7	5.78	.015**
10	Campus Life	5.90	10	5.99	8	5.75	.028**
11	Safety and Security	5.58	11	5.66	11	5.47	.182**

*Indicates a significant difference at the .05 alpha level via t-test for independent samples.
** Indicates a significant difference at the .01 alpha level via t-test for independent samples.

between genders and ranked 6th and 10th by female and male respondents respectively.

- **Choice Rank Difference:** One-way ANOVA analysis of student responses showed that the student centeredness and concern for the individual scales had significant differences as they relate to college choice ranking. These two scales were more important to those students that had Lincoln University as their first or second choice. A subsequent Tukey analysis revealed differences between those who indicated institution was in their top two choices and third choice participants on the student centeredness scale. In the concern for the individual scale there was only a significant difference between those who indicated Lincoln as first and third choice. This may suggest that for these categories, student expectations are lower when the institution is not their top choices.

Importance of Institutional Factors at the End of Freshman Year

Similar to fall semester, the spring data was analyzed to capture differences as they relate to specific student characteristics. However, the intended major characteristic was replaced with grade point average (GPA). Despite the significant difference in the initial and final survey sample sizes, the importance rankings of the eleven institutional scales were, in general, not very different between fall and spring. Academic advising remained the top institutional concern for student while service excellence remained number nine. All the scales, moved up or down one to three places with the exception of registration effectiveness as shown in Table 4 below. Student responses varied by gender and residency more frequently than the others as show in Table 5 below.

Table 3. Choice rank means for fall significant importance scales

Scales	Choice Rank Means 1st 2nd 3rd			p-value
Student Centeredness	6.11	---	5.78	.028
	---	6.11	5.78	.025
Concern for the Individual	6.03	---	5.67	.035
	---	6.00	5.67	.053

Table 4. Change in perceived importance of institutional factors from fall to spring

Scale Items	Spring Rank	Fall Rank	Rank Diff.	Spring Mean	Fall Mean	Mean Diff.
Academic Advising Effectiveness	1	1	0	6.27	6.11	-0.16
Registration Effectiveness	2	8	+6	6.13	5.93	-0.20
Recruitment and Financial Aid Effectiveness	3	6	+3	6.11	6.07	-0.04
Instructional Effectiveness	4	2	-2	6.11	5.97	-0.14
Campus Support Services	5	3	-2	6.09	6.04	-0.05
Campus Climate	6	5	-1	6.03	6.00	-0.03
Student Centeredness	7	4	-3	6.03	6.03	0.00
Concern for the Individual	8	7	-1	6.02	5.90	-0.12
Service Excellence	9	9	0	6.02	5.93	-0.09
Safety and Security	10	11	+1	6.02	5.58	-0.44
Campus Life	11	10	-1	6.02	5.93	-0.09

Table 5. Gender and residency mean importance scores for spring

Scale	Gender			Residency		
	Female Mean	Male Mean	Diff	IS Mean	OS Mean	Mean Diff.
Academic Advising Effectiveness	6.38	6.03	0.35*	6.08	6.36	0.28*
Registration Effectiveness	6.19	5.86	0.32*	5.90	6.16	0.27*
Recruitment/Financial Aid Effect.	6.20	5.83	0.37*	5.79	6.21	0.42*
Instructional Effectiveness	6.15	5.80	0.34*	5.78	6.15	0.36*
Campus Support Services	6.17	5.80	0.37*	5.84	6.15	0.31*
Campus Climate	6.06	5.83	0.23	5.74	6.09	0.35*
Student Centeredness	6.06	5.90	0.16	5.80	6.11	0.31*
Concern for the Individual	6.11	5.77	0.33*	5.78	6.09	0.31*
Service Excellence	6.04	5.79	0.25	5.72	6.07	0.35*
Safety and Security	5.90	5.63	0.27	5.71	5.85	0.14*
Campus Life	6.05	5.74	0.30	5.71	6.06	0.35*
*Mean difference significant at the 0.05 level.						

- **Registration Effectiveness:** Increase from 8th to 2nd from fall to spring. This is a huge change and could be due to the fact that between the two opportunities to respond to these statements, students had occasion to go through the registration process at least once (pre-registration for the spring semester in October) and possibly twice (pre-registration for the next fall semester in March). Campus life, concern for the individual, safety and security, and service excellence remained lower in importance than the other scale items.
- **Gender Differences:** There were five scales that show a marked difference between male and female responses; instructional effectiveness, recruitment and financial aid, campus support services, academic advising and concern for the individual. Like the fall, female participants ranked each of these scales higher than males, but both indicated that the scale was important to them.
- **Residency Differences:** Table 5 displays that residency has a difference in importance. Seven of the eleven scales show significant differences between in-state (n=62) and out-of-state or international students (n=128) and what is important to them on campus. Although all items are important to students, in each scale, out-of-state students indicated that the particular scale was higher in importance than in-state students. The largest difference occurs with recruitment and financial aid.
- **GPA Differences:** GPA was divided into three categories (0 – 1.99; 2.0 – 2.99; 3.0 and above). One-way ANOVA was performed to observe any differences as it pertains to GPA. A significant difference existed in two scales – student centeredness and instructional effectiveness. A Tukey analysis was performed to see in which specific GPA categories these differences occurred. The Tukey analysis showed that there is significant difference between those students who said that their GPA was below a 2.0 and those who said their GPA was 3.0 and higher. This analysis also showed a significant difference between the same individuals and the recruitment and financial aid effectiveness scale. These analyses showed that the higher the GPA, the more important the scales are to students.

Satisfaction of Freshmen Students with Lincoln University's Institutional Factors

In addition to asking students to rank the importance of various institutional factors, the year-end survey asked students to indicate how satisfied they were with the eleven factors. Student satisfaction ratings of the institutional factors scales provided answers to the question, how satisfied were freshmen students with the institution? Table 6 displays the mean score for each scale based on participant's responses to the SSI. Students responded to each item on a 1 to 7 Likert-like scale, with 7 being high. Mean satisfaction scores are typically in a range of 4 to 5" (Noel & Levitz, n.d.). Lincoln University scored within the average range for all of the eleven observed scales.

- **The Importance to Satisfaction Gap – Scale Level:** Although Lincoln scored within the typical satisfaction range, analysis of Importance to Satisfaction convergence seemed to indicate that the institution was not quite meeting student needs or expectations across the board. Performance gap scores were calculated by subtracting mean importance scores from mean satisfaction scores. The larger the gap, the greater the discrepancy between what students expect and their level of satisfaction with the current situation. The smaller the gap, the better the institution is doing at meeting student expectations (Noel & Levitz, 2006). All scales had a gap score greater than 1.00 (see Table 7). Two aspects of the institution showed gap scores of 1.5 or greater. They are safety and security (1.64) and campus life (1.50). This warranted a closer look at individual items within each scale.
- **The Importance to Satisfaction Gap – Within Scale:** Table 8 and Table 9 show the smaller and larger performance gaps within each scale respectively. Under safety and security, the largest areas of dissatisfaction for students were security response times to emergencies (1.79) and adequate parking (1.67). For campus life the largest areas of dissatisfaction were with general

Table 6. Student satisfaction scores

Scale	Mean Score
Academic Advising	5.07
Registration Effectiveness	4.70
Recruitment/Financial Aid	4.62
Instructional Effectiveness	4.83
Campus Support Services	4.83
Campus Climate	4.70
Student Centeredness	4.76
Concern for the Individual	4.72
Service Excellence	4.71
Safety/Security	4.38
Campus Life	4.52

Table 7. Performance gap scores

Scale	Importance	Satisfaction	Gap
Academic Advising	6.27	5.07	-1.20
Registration Effectiveness	6.13	4.70	-1.43
Recruitment/Financial Aid	6.11	4.62	-1.49
Instructional Effectiveness	6.11	4.83	-1.28
Campus Support Services	6.09	4.83	-1.26
Campus Climate	6.03	4.70	-1.33
Student Centeredness	6.03	4.76	-1.27
Concern for the Individual	6.02	4.72	-1.30
Service Excellence	6.02	4.71	-1.31
Safety/Security	6.02	4.38	-1.64
Campus Life	6.02	4.52	-1.50

Table 8. Smaller performance gap scores of items within scales

Scale / Item	Gap Score
Campus Life • A variety of intramural activities are offered. • Males and females have equal opportunities to participate in intercollegiate athletics. • The student handbook provides helpful information about campus life. • New student orientation services help students adjust to college. • The intercollegiate athletic programs contribute to a strong sense of school spirit.	-0.76 -0.80 -0.84 -0.91 -0.95
Campus Climate • There is a strong commitment to racial harmony on this campus.	-0.92
Campus Support Services • Tutoring services are readily available. • Bookstore staff are helpful.	-0.91 -0.99
Registration Effectiveness • Class change (drop/add) policies are reasonable.	-0.93
Instructional Effectiveness • Nearly all of the faculty are knowledgeable in their field.	-0.99

living conditions in residence halls (2.85) and food selections in the cafeteria (2.75). These data on residence halls and food selections were reinforced by student comments on the second part of the year-end survey. Although most student comments were generalized and non-specific, dissatisfaction with housing conditions was noted ten times and thirteen times for food choices. However, by looking at the individual items within the scales it is possible to

Table 9. Large performance gap scores of items within scales

Scale / Item	Score
Safety and Security • Security staff respond quickly to emergencies. • The amount of student parking space on campus is adequate. • The campus is safe and secure for all students.	-1.79 -1.67 -1.61
Campus Life • Living conditions in the residence halls are comfortable (adequate space, light, heat, air, etc.) • There is an adequate selection of food in the cafeteria. • Residence hall regulations are reasonable. • There are a sufficient number of weekend activities for students. • The student center is a comfortable place for students to spend their leisure time. • Student activity fees are put to good use.	-2.85 -2.75 -2.04 -1.95 -1.77 -1.77

see areas, where Lincoln does appear to be meeting students' needs/expectations (Table 8). Some of these expectations include athletic programs, freshman orientation programs, and tutoring services.

- **Importance to Satisfaction: Student Characteristics**. Satisfaction ratings were analyzed with respect to student characteristics; gender, residency, choice rank and GPA. There was no significant difference in the eleven institutional scale scores related to gender or residency. A relationship was noted between GPA and the academic advising effectiveness scale using one-way ANOVA. The mean gap for students with a GPA lower than a 2.0 was 1.01. The gap for those with a GPA of 2.0 to 2.99 and 3.0 and above was 0.01 and 0.15 respectively. Therefore, students who had a GPA that was below 2.0 were not satisfied with the academic advising process. For those with a GPA of 2.0 and above, the academic advising process was meeting with their expectations. Generally, students' satisfaction increased as their GPA increased. One-way ANOVA indicated a significant difference for choice ranking and, student centeredness, instructional effectiveness, registration effectiveness, campus climate and campus support. The mean satisfaction scores for all factors decreased as choice rank decreases. After running a Tukey analysis, there was no notable difference in responses between those students that indicated Lincoln was their first and second choice but for those whom this was their third choice had a significantly lower satisfaction score. Further, the difference in importance-satisfaction gap exists between those who indicated that the institution was their first choice and those who said it was their third choice. A correlations analysis supports the observation of the individual items within these scales - the frequency of neutral and not satisfied responses increases as choice rank decreases. These observations are shown on Table 10.

Table 10. Choice correlations

		Choice	Student Center.	Instruct. Effect.	Registr. Effect.	Campus Climate	Camp. Supp. Services
Choice	Pearson	1.00	-.248**	-.178**	-.190**	-.238**	-.185**
	P-value		.001 *	.014**	.008**	.001**	.010**
	N	192	191**	191**	191**	191**	191**

*Correlation is significant at the .05 level (2-tailed).
**Correlation is significant at the .01 level (2-tailed).

Freshman Satisfaction at Lincoln University Compared to Other HBCUs

Overall, Lincoln University is very comparable to other HBCUs in the United States. According to comparison information obtained from Noel-Levitz, in nine of the eleven institutional scales there was no statistical difference in satisfaction between Lincoln and other HBCUs. However, Lincoln freshmen were significantly more satisfied than freshmen at other HBCUs with safety and security and significantly less satisfied with campus support services. See Table 11 for details.

Despite the fact that Lincoln freshmen felt the campus was not as safe as it could or should be, Lincoln had an importance-satisfaction gap score that is 0.28 lower than the nationwide average for other HBCUs. This means that, on the whole, Lincoln students perceive to experience better security than their peers at other HBCUs. This may be due largely to the location of the university – a very rural, agricultural community in Southern Chester County, Pennsylvania.

On the other hand, according to its freshmen, the institution lacked campus support services. Their ratings suggest that campus technology is not adequate to meet the needs of the students. Student open-ended comments affirm the source of the negative ratings about campus support services. Some of those comments were:

- "Lincoln is a good school…it just needs better technology."
- "There needs to be…improvements regarding…student academic enrichment."
- "This school is not advanced… it lacks technology."
- "The lack of working computers and printers is a hassle."

The final survey asked three general summary questions designed to gauge student satisfaction with their entire university experience. Table 12 indicates that the freshmen experience at Lincoln University was comparable to other HBCUs.

Table 11. Lincoln vs. four-year HBCU freshmen importance and satisfaction scores

Scale	Lincoln University Freshmen				Four-Year HBCU Freshmen				Mean Diff.
	IMP	SAT	SD	Gap	IMP	SAT	SD	Gap	
Academic Adv.	6.27	5.07	1.36	1.20	6.17	5.03	1.35	1.14	0.04**
Registration Eff.	6.13	4.70	1.15	1.43	6.07	4.65	1.23	1.42	0.05**
Recruit/Fin. Aid	6.11	4.62	1.14	1.49	6.09	4.47	1.31	1.62	0.15**
Instructional Eff.	6.11	4.83	1.10	1.28	6.08	4.84	1.13	1.24	-0.01**
Campus Support	6.09	4.83	1.09	1.26	6.04	5.02	1.13	1.02	-0.19**
Campus Climate	6.03	4.70	1.05	1.33	5.99	4.67	1.15	1.32	0.03**
Student Centered.	6.03	4.76	1.13	1.27	5.96	4.73	1.21	1.23	0.03**
Concern for the Ind.	6.02	4.72	1.12	1.30	6.00	4.68	1.20	1.32	0.04**
Service Excellence	6.02	4.71	0.97	1.31	5.97	4.63	1.14	1.34	0.08**
Safety/Security	6.02	4.38	1.25	1.64	6.07	4.10	1.36	1.97	0.28**
Campus Life	6.02	4.52	1.06	1.50	5.90	4.61	1.15	1.29	-0.09**

IMP – Importance; SAT – Satisfaction; SD – Standard Deviation
*Difference statistically significant at the .05 level
**Difference statistically significant at the .01 level
Lincoln University Freshmen vs. HBCU Four-Year Freshmen 2008.

Table 12. Freshmen overall satisfaction at Lincoln University vs. Other HBCUs

Summary	LU	HBCU
So far, how has your college experience met your expectations?	Mean =4.16	Mean = 4.10
• Worse than I expected	24%	16%
• About what I expected	39%	40%
• Better than I expected	32%	31%
Rate your overall satisfaction with your experience here thus far.	Mean =4.61	Mean = 4.52
• Not Satisfied	13%	10%
• Somewhat Dissatisfied	8%	12%
• Neutral	19%	22%
• Somewhat Satisfied	24%	23%
• Satisfied	31%	29%
All in all, if you had to do it over, would you enroll here again?	Mean =4.49	Mean = 4.47
• Definitely No	10%	8%
• Probably/Maybe Not	15%	20%
• I don't know	16%	19%
• Probably/Maybe Yes	38%	35%
• Definitely Yes	16%	15%

Table modified from Lincoln University Freshmen vs. HBCU Four-Year Freshmen 2008. Results of the Noel-Levitz Student Satisfaction Inventory. November, 2008.

Table 13 shows the institution's strengths and challenges according to the analysis of SSI instrument. Strengths were items with high importance and high satisfaction while challenges were items with high importance and low satisfaction. They were identified as items above the mid-point in importance and in the upper and lower 25% of satisfaction scores respectively, or the top 25% of performance gap scores (challenges only). Both strengths and challenges are listed in descending order of importance.

Table 13. Lincoln University's strengths and challenges

Strengths
The content of the courses within my major is valuable.
My academic advisor is approachable.
My academic advisor is knowledgeable about requirements in my major.
My academic advisor is concerned about my success as an individual.
Tutoring services are readily available.
The instruction in my major field is excellent.
Major requirements are clear and reasonable.
Faculty are usually available after class and during office hours.
The staff in the health services area are competent.
Nearly all of the faculty are knowledgeable in their field.
I can easily get involved in campus organizations.
Challenges
Living conditions in the residence halls are comfortable (adequate space, light, heat, air, etc.)
Security staff respond quickly in emergencies.
There is an adequate selection of food available in the cafeteria.
Computer labs are adequate and accessible.
The campus is safe and secure for all students.
Financial aid awards are announced to students in time to be helpful in college planning.
Adequate financial aid is available for most students.
Residence hall regulations are reasonable.
The student center is a comfortable place for students to spend their leisure time.
Tuition paid is a worthwhile investment.
Lincoln University Freshmen vs. HBCU Four-Year Freshmen 2008. Results of the Noel-Levitz Student Satisfaction Inventory. November, 2008.

Relationship Between Freshmen Satisfaction and Persistence to Sophomore Year

Analysis to determine the relationship between freshmen institutional satisfaction and their persistence to their sophomore year precluded good prediction in a practical sense. The majority of participants did not disclose their student identification number, which at that time was their Social Security Number (SSN), on survey instruments making it impossible to connect student survey responses with data from school records on their persistence to the sophomore year.

Nevertheless, three questions on the second part of the year-end survey provide some indication of students' intentions about returning for the following school year and how their freshman experience influenced their decision to persist or not at Lincoln University. These questions were:

1. Do you plan to return for the next fall semester?
2. Have your experiences with the university influenced your decision to return to or withdraw from Lincoln?
3. If you could start college over again, would you choose to attend this college?

This second part of the survey also asked participants to indicate the three specific areas within the institution with which they have had the best and the worst experiences. One hundred, ninety-six freshmen students completed part one of this survey, but 216 freshmen students (consisting of 146 females and 70 males) completed the second part of the survey. Most of these 216 students completed the first part as well. However, because of the lack of student identification numbers, it is not possible to know exactly how many completed both parts. 88% of participants responded in the affirmative that they plan to return for their sophomore year. Of those, 28.4% said that their experiences with the various aspects of the institution influenced this decision and 71% said these experiences did not influence this decision. The actual freshman to sophomore retention rate for the 2007 incoming class was 69.5%.

FUTURE RESEARCH DIRECTIONS

This study was a foundational construct for further research into student satisfaction and retention on Lincoln University's campus. Continued research is necessary to collect and analyze longitudinal data that will be useful in guiding the institution in its planning processes toward the enhancement of all its students and the entire academy. More research needs to be conducted to capture the best possible representation of the students who actually do not return.

Retention of all students is of importance to the continued success of individual majors and to the institution as a whole. This particular study was specific to freshman retention only, and freshmen are just one cohort of the entire student body. Therefore, a more comprehensive research study should be conducted to see if institutional factors impact student retention on every level (sophomores, juniors and seniors).

Only 33.7% of participants' indicated that Lincoln University was their first choice. That is well under half of the fall 2007 cohort. Therefore, a research study should be conducted to identify what institution or types of institutions were and are students first choice and why. This would yield valuable information that could be integrated into market, recruitment and strategic planning. That data would also furnish administration with information of services and programs that will attract and sustain the interest of prospective and current students.

Implications for University Policy and Practice

- **Student Identification Numbers:** At the time of this study, students used their SSNs as their student identification (ID) numbers. As of the fall 2008, the university's new system for student registration utilizes an eight-digit student ID numbers independent of SSNs. Now, students can feel a bit more assured of their safety from identity theft on campus and will feel less apprehensive about disclosing this information during future research studies.
- **Student Expectations and Experiences:** This study has brought to light some of the areas of dissatisfaction students have with Lincoln University. In this document, they have been referred to as challenges or worst experiences. The vast majority of these challenges and experiences came from four areas within the university – security, housing, food services and financial aid/bursar. This dissatisfaction is not unique to Lincoln University alone (Noel-Levitz, 2006). Beginning in the fall of 2009, the meal plan changed from a "flex" plan to a "block" plan. Students had more food choices without having to pay more money. It would be interesting to see if this change increased satisfaction.
- **Student Satisfaction Inventory:** Perhaps Lincoln University should consider making the Noel-Levitz Student Satisfaction Inventory (or a similar instrument) a regular part of its assessment and strategic planning process. The Noel-Levitz firm has been a partner with higher-education institutions for over 30 years. It works with over 2000 campuses and specializes in strategic planning for enrollment and student success (www.noellevitz.com). Now that Lincoln has an idea of where its strengths and challenges lie, it can begin to make the necessary adjustments to maintain momentum and/or improve services. Making the SSI a part of the university's long term evaluation pro-

cesses could help the institution measure and track the impact of specific changes. The data and subsequent information can also become a part of the process improvement program that will be provide valuable artifacts for institutional reaccreditation.

CONCLUSION AND RECOMMENDATION

Lincoln University scored within the average satisfaction ratings on the Noel-Levitz SSI. However, there were divergences between importance and satisfaction at and within scales. These gaps indicate students' perception on how well the institution was meeting their needs - the smaller the gap, the greater the satisfaction; the larger the gap the greater the dissatisfaction. With that being said, freshmen students seem to be most satisfied with the academic advising, instructional effectiveness, campus support and student centered factors. Conversely, they were not as satisfied with safety and security, campus life and, recruitment and financial aid factors.

The results of this study also indicated that the institutional factors of Lincoln University may not be a major factor in explaining the attrition of freshman students, although follow-up research with students who did not return to the University may be necessary to confirm this conclusion. It appears that the experiences freshman students have had at the university had little to do with their decision to persist or not beyond their first year. Therefore, other factors might be at work which motivates students to stay or leave the institution. Nevertheless, more research with a more representative sample of freshmen students at the end of their freshman year needs to be done before definitive information about the impact of institutional factors on freshmen to sophomore retention can be obtained.

Males and female tended to find different characteristics of the institution more important. These findings are a great resource for the marketing and recruitment. The Importance of factors vary over the academic year. This may also be true from one academic level to the next. These should be noted and incorporated into a customer satisfaction plan or program.

Based on the results of the SSI, the following are recommended actions to be considered by the university:

1. Items under the safety and security, campus life, campus climate, campus support services and recruitment and financial aid scales had high importance but low satisfaction ratings. This indicates that these particular items should be given immediate attention. It is recommended that institution examine these items, employ actions necessary to remedy the current status as well as monitor them over time.

2. Items under the instructional effectiveness, academic advising, service excellence and campus support services scales received high importance and high satisfaction ratings indicating institutional strength. This information should be recognized and shared among the stakeholders of the academy. Momentum in these areas should be sustained in order to maintain or improve current status.
3. The items under the scales for campus life and recruitment and financial aid had low importance and high satisfaction ratings. This indicates that the institution is doing well with these items, but they are not as important to the students. Therefore, it is recommended that it may benefit from redirecting resources from these areas to other areas of higher importance to students.

REFERENCES

Athreya, K., & Eberly, J. C. (2015). *The College Premium, College Noncompletion, and Human Capital Investment.* Federal Reserve Bank of Richmond Working Paper (13-02R).

Baker, D. (2014). *The schooled society: The educational transformation of global culture*. Stanford University Press.

Barefoot, B. (2004). Higher educations revolving door: Confronting the problem of student dropout in US colleges and universities. *Open Learning*, *9*(1), 9–18. doi:10.1080/0268051042000177818

Barefoot, B., Fidler, P., & South Carolina Univ, C. (1992, January 1). *National Survey of Freshman Seminar Programming, 1991. Helping First Year College Students Climb the Academic Ladder. The Freshman Year Experience: Monograph Series Number 10*. (ERIC Document Reproduction Service No. ED354842). Retrieved July 26, 2007, from ERIC database.

Bean, J. P., & Metzner, B. S. (1985). A conceptual model of nontraditional undergraduate student attrition. *Review of Educational Research*, *55*(4), 485–540. doi:10.3102/00346543055004485

Berger, J. B., & Braxton, J. M. (1998). Revising Tintos interactionalist theory of student departure through theory elaboration: Examining the role of organizational attributes in the persistence process. *Research in Higher Education*, *39*(2), 103–119. doi:10.1023/A:1018760513769

Browne, B., Kaldenberg, D., Brown, W., & Brown, D. (1998). Student as customer: Factors affecting satisfaction and assessment of institutional quality. *Journal of Marketing for Higher Education*, *8*(3), 1–14. doi:10.1300/J050v08n03_01

Cabrera, A. F., Nora, A., & Castaneda, M. B. (1993). College persistence: Structural equations modeling test of an integrated model of student retention. *The Journal of Higher Education*, *64*(2), 123–139. doi:10.2307/2960026

DeWitz, S., & Walsh, W. (2002). Self-efficacy and college student satisfaction. *Journal of Career Assessment*, *10*(3), 315–326. doi:10.1177/1067270201000300З

Elliott, K. (2003). Key determinants of student satisfaction. *Journal of College Student Retention*, *4*(3), 271–279. doi:10.2190/B2V7-R91M-6WXR-KCCR

Evans, A., Evans, V., & Evans, A. (2002, Fall). Historically Black colleges and universities (HBCUS). *Education*, *123*, 3–17.

Gandara, P. & Maxwell-Jolly, J. (1999, December). *Priming the pump: Strategies for increasing the achievement of underrepresented minority undergraduates.* Academic Press.

Higgins, M. J. & Glanville, M. (1983). *The impact of Act 101 programs in Pennsylvania.* (ERIC Document Reproduction Service No. ED243354)

Hout, M. (2012). Social and economic returns to college education in the United States. *Annual Review of Sociology*, *38*(1), 379–400. doi:10.1146/annurev.soc.012809.102503

Howard University. (2002). *Annual report.* Retrieved February 4, 2006, from http://www.ed.gov/about/reports/annual/ 2002report/edlite-howard.html

Jacobi, M. (1991). Mentoring and undergraduate academic success: A literature review. *Review of Educational Research*, *61*(4), 505–532. doi:10.3102/00346543061004505

Lenning, O. T., Sauer, K., & Beal, P. E. (1980). *Student retention strategies.* American Association for Higher Education.

Lincoln University. (2016). *2015 Fact Book.* Retrieved July 25, 2016, from http://www.lincoln.edu/sites/default/files/pdf/institutional-research/2015FactBook.pdf

Lotkowski, V., Robbins, S., & Noeth, R. (2004). The role of academic and non-academic factors in improving college retention. *ACT Policy Report.* Retrieved January 12, 2006, from http://www.act.org/research/policy/index.html

Lunneborg, P. W. (1975). *Longitudinal Criteria of College Achievement Predicted from Aptitude, Achievement, and Interest Measures.* Seattle, WA: Educational Assessment Center, University of Washington.

National Science Foundation. (n.d.). *Historically Black Colleges and Universities Undergraduate Program.* NSF 02-160. Retrieved June 19, 2007, from http://www.nsf.gov/pubs/2002/nsf02160/nsf02160.pdf

NCHEMS. (2016). *National Center for Higher Education Management Systems. Retention Rates: First-Time College Freshman Returning their Second Year*. Retrieved August 9, 2016, from http://www.higheredinfo.org/dbrowser/index.php?submeasure=223&year=2010&level=nation&mode=data&state=0

Nettles, M., Wagener, U., Millett, C., & Killenbeck, A. (1999, Winter). Student retention and progression: A special challenge for private historically black colleges and universities. *New Directions for Higher Education*, (108), 51-67. Retrieved June 2, 2007, from Academic Search Premier database.

Nichols, J. (2004). Unique characteristics, leadership styles, and management of historically Black colleges and universities. *Innovative Higher Education*, *28*(3), 219–229. doi:10.1023/B:IHIE.0000015109.49156.fb

Nicholson, E. (1973). *Predictors of Graduation from College*. Iowa City, IA: American College Testing Program, Research and Development Division.

Noel, L., & Levitz, P. (2006). *National Research Report: How students rate the quality service climate on campus*. Retrieved June 22, 2007, from http://www.noellevitz.com

Noel, L., & Levitz, P. (n.d.). Satisfaction-Priorities Surveys Interpretive Guide. Cedar Rapids, IA. *Ruffalo Noel Levitz, LLC.*

Noel-Levitz Student Satisfaction Inventory. (n.d.). Retrieved November 18, 2008 from www.paulsmiths.edu/offices/president/SSI%20InterpretiveGuide%202008-09.pdf

Oreopoulos, P., & Petronijevic, U. (2013). Making college worth it: A review of the returns to higher education. *The Future of Children*, *23*(1), 41–65. doi:10.1353/foc.2013.0001 PMID:25522645

Pew Charitable Trusts. (n.d.). Retrieved June 23, 2007, from http://www.pewtrusts.org/about_us

Strauss, L., & Volkwein, J. F. (2004). Predictors of student commitment at two-year and four-year institutions. *The Journal of Higher Education*, *75*(2), 203–226. doi:10.1353/jhe.2004.0007

Thomas, E., & Galambos, N. (2004). What satisfies students?. *Research in Higher Education*, *45*(3), 251–269. RetrievedJune22007 doi:10.1023/B:RIHE.0000019589.79439.6e

Tinto, V. (1975). Dropout from Higher Education: A Theoretical Synthesis of Recent Research. *Review of Educational Research*, *45*(1), 89–125. doi:10.3102/00346543045001089

Tinto, V. (2012). *Completing college: Rethinking institutional action*. University of Chicago Press. doi:10.7208/chicago/9780226804545.001.0001

Tobolowsky, B. F. (2005). *The 2003 national survey on first-year seminars: Continuing innovations in the collegiate curriculum* (Monograph No. 41). Columbia, SC: University of South Carolina, National Resource Center for The First-Year Experience and Students in Transition.

Turnbo, D. (2002). *A study of the impact on a new student intrusive advising model on undergraduate student attrition at a private higher education institution in Delaware* (Doctoral dissertation). Wilmington College.

Wilcox, P., Winn, S., & Fyvie-Gauld, M. (2005). The role of social support in the first year experience. Studies in Higher Education, 30, 707-722.

KEY TERMS AND DEFINITIONS

At-Risk: This term refers to any student who encounters or has encountered major obstacles to successful completion of college. These obstacles may include (but are not limited to) low socioeconomic status, low high school rank and GPA, low SAT/ACT test scores, and the need to take one or more developmental courses in college.

Attrition: This term refers to the reduction of enrolled students at an institution of higher education, prior to graduation, resulting from academic failure and/or personal decision. In this study it is used interchangeably with the words withdrawal and dropout. This study focuses specifically on attrition during or at the end of the freshman year.

Institutional Factors: This term refers to the professional, social and cultural elements to which the student is exposed on campus, within and outside of the classroom that may cause him/her to feel more or less satisfied with his/her college choice and/or ability to perform successfully. These elements include: academic advising; financial aid; registration; campus life; campus climate; student support services; classroom instruction; safety and security; service excellence; concern for the individual; and student centeredness (Noel & Levitz, 2006).

Retention: These terms refer to a student's ability to remain enrolled in the institution of higher education under study by way of internal/external motivation and meeting or exceeding the institutions minimum academic requirements for continuance through to graduation. In this study it is used interchangeably with retention.

Chapter 2
Minority Recruitment and Retention Among Gifted Students

Chandra A. Stallworth
Independent Researcher, USA

Ken D. Thomas
Auburn University, USA

ABSTRACT

Consistent with the national goal implemented by our current government, Auburn University is also working to recruit and retain underrepresented minorities in higher education. The rationale for this is simple, that is to allow a greater advantage when competing against others. One of the ways to foster this competition is by nurturing our gifted underrepresented minority students. In the 2010-2011 school year, the Honors College, which serves as a gateway for underrepresented minority students, developed a distinct focus on helping our students reach and their educational/academic goals. Within this paper we will go over some of the steps we have begun to take to reach our goal, in addition to future plans we have to continue these efforts.

DOI: 10.4018/978-1-5225-2212-6.ch002

BACKGROUND

Historically there has been an inequality that exists among minority students when education is examined. Previous issues have been in the form of tracking which has unfortunate consequences because it puts underrepresented students at a disadvantage. More recently, traditional curriculum tracking is on the decline, the content students learn and achievement outcomes are still differentiated by race and class (Gamoran, 2001; Lucas & Berends, 2002; Oakes & Guiton, 1995). Unfortunately this differentiation follows students into higher education and the labor market, influencing the choices they make. Lucas and Berends (2002) have documented de facto tracking where instead of division according to academic, general and vocational tracks, schools offer multiple levels of academic courses (such as mathematics) with some options (e.g. Calculus), being more rigorous than others (e.g. Consumer Mathematics), and with minority—especially Black—students being disproportionately more likely to be enrolled in the latter than the former. Regrettably there is a trend of minority students becoming the victims of misperceptions about achievement levels, being subject to negative attitudes, and getting less encouragement than other students (Diamond, Randolph, & Spillane, 2004; Good, 1981; Payne, 1994; Roscigno & Ainsworth-Darnell, 1999). Unfortunately schools contribute to the imbalance of power in society by communicating society's economic, political and cultural knowledge to students (Apple, 2004).

The importance of schools is more prevalent in regards to economic issues because the United States (U.S.) is becoming more diverse (Phinney & Alipura, 1996) and globalization has made it disadvantageous to continue to foster inequality of educational opportunity along ethnic lines. The National Academy of Engineers (2004), an organization that advises the government on issues concerning engineering, stated that if the U.S. is to maintain economic leadership and be able to sustain its share of high-technology jobs, it must prepare for a new wave of change.

A key population that will help in maintaining economic leadership among minority students are those who participate in Honors Programs or classes. Since the population of minority students in these classes and programs are so few, particularly at Predominately White Institutions (PWI) such as Auburn University, these students need to be treated with sensitivity. Ways that have proven to help with this are clubs, support groups, and mentors. Within school settings since the climate of a school has been proven to make a difference in school achievement (Brookover, Schweitzer, Schneider, & Beady, 1978) the academic norms within the school system, such as expectations and beliefs must be catered to in a variety of ways. Mentors in particular have been shown to correlate with gifted students performance (Bell, 1970; Gibson, 2004; Linnehan, Weer, & Uhl, 2005; Zeldin & Pajares, 2000). Tarek and Ford (2003) note that specifically in relation to African American students, in gifted programs that mentors are important because they help empower students by causing them to improve their decision-making skills and their ability to clarify goals for the future in a way that provides a sense of direction and purpose. Role models and mentors are believed to be important because they help strengthen individual growth. Zeldin & Pajares (2000) found that when women were asked about significant factors in their lives, they consistently recalled an influential person who helped them develop their beliefs about their capabilities.

In early America, unlike the fields of medical and legal, engineering education was never under the exclusive domination of a professional group, the curriculum was created strictly by educators (Grayson, 1980). Therefore those who were not engineers and knew little to nothing of the content it entailed had enormous input to shaping the field of engineering. As a result, the early engineering curriculum contained little technique of engineering practice. Prior to World War I few opportunities existed for African Americans to work in engineering fields. But some Historically Black Colleges and Universities (HBCUs) (such as Howard University, North Carolina Agricultural & Technical State University, and Hampton University) began offering engineering majors. These schools continue to be at the forefront of the education of African American engineers. Despite this fact, several theorists of higher education purport that HBCUs are generally thought of as having mediocre education standards and that their students attain substandard degrees (Harper, 2009).

The importance of educating all Americans including African Americans was identified in the 1930's and is still being restated here in the 21st century. The US has lost some technological ground and will continue to do so if everyone is not provided an equal opportunity to advance in all areas. Some have gone so far as to say that African Americans are not well educated in the field of engineering because the information they are obtaining is not relevant. Woodson (1933) concurs by stating that since African Americans were told what to learn by another race for so long, they must now be taught to think and develop something for themselves. Or some may go further back to say that during slavery African Americans developed a negative identity and now it is their responsibility to interpret, and come to terms with, their collective traumatic past and their relationship to that past (Eyerman, 2001). In doing this they can make strides to change their present identity. Regardless of what the source of this disparity in education all would agree that it is something that needs to be reconciled.

INTRODUCTION

There are a few organizations that attempt to make the Predominantly White Institution (PWI) that is Auburn University a lot more welcoming to all minority students at Auburn University. One of the largest student organizations with this agenda is the Black Student Union (BSU). The core mission of BSU is "To represent the interest and concerns of Black Students at Auburn University and to bring together all aspects of Black Student life for the purpose of improving the campus environment and encourage involvement of Black students in ALL campus activities. Discourage and abate institutional and individual acts and symbols of racism and promote the Auburn Spirit and tradition of Excellence and Heighten awareness and friendship between races." Much of the great works of BSU can be lumped into being a social organization where a few of their signature events include the Back to School Luau, Mr. BSU Pageant, and Jazz & Poetry Night (AUinvolve, n.d.a.)

Another important student organization that works at Auburn University with minority students is the Harold A. Franklin Society. The Franklin Society's mission "is to serve underrepresented male students at Auburn University, working to foster their growth in academic, social and professional areas. Through these avenues, we will raise the retention rate for minority male students." This society targets students that are an ethnic minority and identify as a male but there is a formal application for membership that assesses academic performance and leadership potential (Auburn University, n.d.). This means that incoming freshmen are not eligible for membership into this society.

The International Student Organization (ISO) is of critical importance to ensuring international students are adjusted and welcome into the Auburn University community. ISO's mission "is to improve multinational understanding and promote relationships between people of different cultures. The ISO also helps ease the process of adaptation for international students. We actively promote diversity and beneficial interaction between the international students and American Students." ISO is generally of importance to graduate students since there are very few undergraduate international students at Auburn University (AUinvolve, n.d.b.). However, this is changing with a most recent Pathway Program developed at Auburn University and a spike in international athletes coming to world class athletic training. The Honors College is practically devoid of international students but with the current Pathway Program in place, it is one area for diversity recruitment and retention efforts in the near future.

The aforementioned student organizations were all supported by the Office of Diversity and Multicultural Affairs (recently renamed the Office of Inclusion and Diversity) which is led by an Associate Provost. One of the chief roles of this office is also to award the Provost Leadership Undergraduate Scholarship (PLUS) to incoming freshmen. This scholarship is renewable and initial selection is not based on standardized test scores. The Office of Diversity and Multicultural Affairs included a Women's Resource Center as well as a Multicultural Center for support of all faculty, staff and students at Auburn University. It should be noted that none of this is particularly unique or novel to Auburn University or PWIs in general.

At Auburn University the enrollment of ethnic minority students who enter the Honors College has gradually increased in the past 15 years. Although this number is the highest it has ever been there is still a huge disparity between ethnic minority students and their white counterparts who also enter the Honors Program each year. Even greater, their exist a problem with retaining these same students as each year there is a drastic decrease further putting the Honors Program at a disadvantage when it comes to graduating minority students with Honors degrees.

Membership in the Auburn University Honors College has generally remained the same from 2010 to present. To be accepted in the Honors College, all students must have a 29 on the ACT (or equivalent SAT) as well as a high school GPA of 3.75. The GPA requirement was changed in 2016 to 3.85. In 2010 there were about 2000 students in the Honors College and in Fall 2010 approximately 1000 freshmen were brought into the Honors College. Currently, only about 550 freshmen are let into the Honors College and the total number in the college stands at around 1700. Percentage wise, the same number of minorities are allowed entry into the Honors College from 2010 to present (10% - 13%) and this percentage is on par with the percentage of minorities at Auburn University. To remain in the Honors College, students must show progress in the Honors curriculum and maintain at least a 3.2 Auburn University GPA. From 2010 to present over 70% of all Honors College students are in a STEM major and over 50% of them are pursuing a major in the Samuel Ginn College of Engineering. The Samuel Ginn College of Engineering at Auburn University is currently ranked as the top 31st College of Engineering among all public universities in the U.S. for undergraduate programs (U.S. News & World Report ranking). The Ginn College of Engineering is ranked 20th nationally in engineering degrees awarded to African-Americans (Diverse Issues in Higher Education ranking) among a few other accolades. With these types of college rankings as well as the Forbes ranking of Auburn University as the top university in the state of Alabama, there is no surprise that Honors students are increasingly attracted to the Samuel Ginn College of Engineering. It should be noted that it is the largest college – enrollment wise – at Auburn University.

The Samuel Ginn College of Engineering has put in place several things to attract and retain ethnic minorities in their undergraduate programs. Paramount among these was ensuring that the student chapters of minority professional engineering organizations were developed and housed in the college. Two prominent examples of these are the National Society of Black Engineers (NSBE) and the Society of

Hispanic Professional Engineers (SHPE). The administration of the college has taken strides to put things in place to improve the academic performance of ethnic minorities. These efforts include a learning community of African American students as well as the creation of the Alabama Power Academic Excellence Program. "The Alabama Power Academic Excellence Program (AEP) enhances the recruitment and retention of underrepresented students at Auburn University's Samuel Ginn College of Engineering. AEP supports participants from pre-college through graduation by focusing on expanding academic preparation, professional readiness and exploration of career paths. The preparation, readiness and exploration is facilitated through a wide range of programs and activities to support participants. This nationally recognized programs has been in existence since 1996, and has been the catalyst for graduating some of the nation's brightest underrepresented engineering students." (Auburn University, n.d.b.) At Auburn University students that intend to major in engineering are placed into Pre-Engineering and not directly into an engineering department. Some ethnic minority Honors students have reported that this has been a deterrent from participating in minority specific engineering organizations or department specific engineering student organizations as a freshman or even a sophomore.

In efforts to not only increase minority student participation but retain those students to graduate with Honors degrees, faculty within the Honors College are exploring ways to provide a support system/ intervention to these students. To help develop and implement this system faculty have begun Interviews with current upperclassman students within the Honors College. In these interviews, we have and will continue to ask questions pertaining to: how they came to be a member of the Honors College, what type of support they received, what type of support they think would have been beneficial, and brainstorm ideas of why more minorities don't participate in the Honors Program. Faculty members also met with administration at the College and University level. From these interviews and meetings, the faculty began to implement a new strategy to help with recruitment and retention.

This issue is viewed as important for a number of reasons but just to name a few it will provide ease of access to career and graduate school opportunities, create networking opportunities, expose students to prestigious scholarships, and grant a distinguished degree. Another key reason minority entrance and retention is important is the high correlation between Honors students and STEM majors. At Auburn University there is a very high number of Honors students in a STEM field. By increasing the minority population in these fields, the opportunities mentioned previously will help with the U.S.'s initiative to get more minorities in these fields.

EXCELLENCE IN HONORS

In the 2010-2011 school year the Honors College developed its first minority support group entitled "Excellence in Honors" in efforts to provide a resource for its students. Figure 1 shows the core elements that went into the creation of Excellence in Honors.

This college level commitment to minority and underrepresented students is consistent with not only Auburn University's strategic plan but the efforts align with that of other schools and colleges on the university campus. In Fall 2010 the initial meeting of Excellence in Honors was held to introduce the group to our minority Honors students. The demographics from that meeting are below in Table 1.

Students have had the opportunity to participate in two "Meet and Greet" informal meetings where faculty discussed the vision for the group. In the first meeting students were asked to a series of questions in order for faculty to learn about their views on the Honors College and their opinion regarding areas of improvement. All of the surveys were kept anonymous. The first question related to their motivation

Figure 1. Key elements involved in the creation of the minority support group "Excellence in Honors"

Table 1. Demographics of student attendees at initial Excellence in Honors meeting (Fall 2010)

Year in School		Major	Count
Freshman	14	Bio- Chemistry	1
Sophomore	1	Biomedical Science	1
Junior	1	Pre-Medicine	2
Senior	0	Pre-Engineering	1
		International Business	1
Ethnicity		Computer Science	1
African American	8	Pre- Veterinarian	1
Asian	4	Chemistry	1
Hispanic	3	Pre-Architecture	1
Cuban	1	Finance	1
		Chemical Engineering	3
		Psychology	1
		Undeclared	1

for joining the Honors College. From this question two key and consistent responses surfaced: smaller classes and scholarship opportunities. One student stated:

Smaller capacity in Honors classes provide a more personal, discussion-based environment in my classes. These allow me to retain more information and ultimately excel at a quicker pace.

When students were asked, how they felt the Honors College provided specific support to minority students, a number of them did not know. But for some, once again scholarships arose as their answer.

This meeting is a start. Perhaps inform us of minority-specific scholarships such as the HSF. Send us emails about those things.

The lack of knowledge regarding resources by nearly half of the students is a strong indication that our minority students are not being exposed to opportunities as much as they should be. From this information the faculty members within the Honors College developed a series of workshops to mediate the gap in knowledge. The planned workshop sequence is detailed in Figure 2 below.

Figure 2. Excellence in Honors workshops held based on student needs

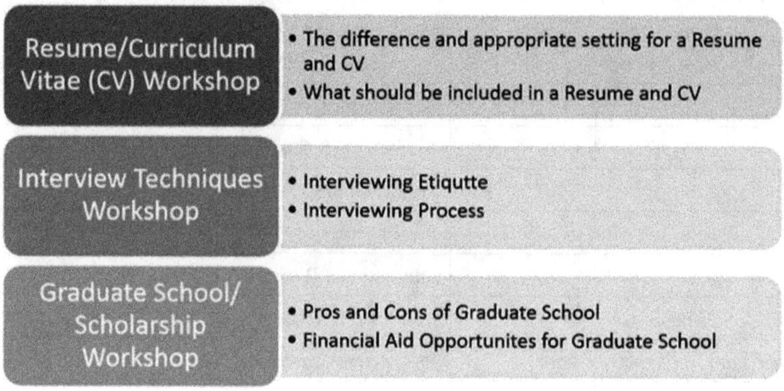

The first workshop was a "Resume/Curriculum Vitae workshop". During this workshop the students learned the difference between a resume and curriculum vitae, which they were previously unfamiliar with. Along with faculty they also critiqued and edited examples of bad resumes. During an activity entitled "Think like a hiring manager", students were assigned the task of examining resumes and hire someone based on criteria they thought would be acceptable if they had the responsibility of hiring. Overall students found the workshop extremely enjoyable. After the workshop a student came up to one of the faculty members and said:

I wanted to tell you and Dr. X thank you for holding the workshop. I have a father that is a Hiring Manager and an aunt who works in HR so I knew some of the stuff. But it was nice to hear another perspective on what companies are looking for...

In everything that we do, we are constantly requesting feedback from our students. It is important for the student to feel like they not only have a safe place to turn but that they feel comfortable with the faculty so in the event that something is needed they will not hesitate to ask.

As highlighted in Figure 2 above there are two additional workshops were held in the Spring 2011 semester for the students. These workshops were based on the themes of: Graduate school options and opportunities, and interviewing techniques. Faculty also worked to bring speakers to campus, in the Spring semester inclusive of an Associate Dean for Graduate Studies from a Research One University who came to speak to our students in the arena of graduate school and research opportunities for minority and underrepresented students. Aligning with our mission within the Honors College to make our graduates attractive to advanced programs, we provide them with an intimate opportunity to interact with faculty from graduate schools and employers from different industries. As we transitioned into the next school year we continued to grow Excellence in Honors and reach out to incoming and current students so that we could make sure to meet their needs. Our mission in the Honors College is to make sure the minority students receive a plethora of support throughout their tenure at Auburn University on their way to graduating at Honors Scholars. In consideration of meeting the needs of your minority students, we moved from a faculty-led "Excellence in Honors" program to a chartered student organization called "Diversity in Honors" in 2012.

DIVERSITY IN HONORS

Transitioning from "Excellence in Honors" to "Diversity in Honors" was one that was driven by the student members. A major motivation for this was the fact that many of these minority students felt that they did not have a fair chance at attaining leadership experience among the countless student organizations at Auburn University. Their reasons for this opinion are not important to this dialogue and so they will not be dissected here. The faculty advisors for the organization thought that there was great merit to providing minority Honors students with more leadership opportunities and worked with the first class of leadership to get "Diversity in Honors" (or DiH) chartered.

The structure of leadership for DiH took on that of a typical student organization with the conventional roles of President, Vice President, Secretary, Treasurer, through several Chair positions. To date, DiH had provided valuable and substantive leadership opportunities for twenty one (21) minority Honors College students. Naturally these minority student leaders become mentors and role models to the DiH members. It is no surprise that over 97% of all DiH members graduate from Auburn University with the Honors College's distinction of Honors Scholar or University Honors Scholar.

Under the direction of the students, the meetings took on a very similar format to when they were done under the "Excellence in Honors" umbrella. The major change that they inculcated was having a second meeting per month that was dedicated to just being social and enjoying the company of members with no fixed agenda. This addition was met with great reviews by members and some of them have informally suggested that their retention in the Honors College is hinged on the fact that to be part of DiH you must be a member of the Honors College.

The faculty advisors capitalized on this knowledge by continuing to push minority students towards excellence. At Auburn University, like most other U.S. college and universities, national recognition is sought through winning nationally competitive and prestigious fellowships and scholarships. These include, but are not limited to, the Rhodes Scholarship, Marshall Scholarship, and the National Science Foundation Graduate Research Fellowship. To date, DiH members have had success in winning some of these awards. Prior to "Excellence in Honors" and/or DiH, virtually no ethnic minorities won these top level awards at Auburn University. A brief summary of the impact of DiH on bringing national recognition to Auburn University is given in Table 2.

Table 2. Major prestigious awards won by Diversity in Honors (DiH) members through 2016

Award	Award Description	Award Level	DiH Member Winners	DiH Member Finalists/ Honorable Mentions
Rhodes Scholarship	"The Rhodes Scholarships are postgraduate awards supporting exceptional students from around the world to study at the University of Oxford. The Rhodes is the oldest and perhaps the most prestigious international scholarship programme in the world, which aims to nurture public-spirited leaders for the world's future." (Rhodes Trust, n.d.)	International	0	2 (1 African American; 1 Asian American; both males; both STEM majors)
Marshall Scholarship	"Marshall Scholarships finance young Americans of high ability to study for a graduate degree in the United Kingdom. Up to forty Scholars are selected each year to study at graduate level at an UK institution in any field of study." (Marshall Aid Commemoration Commission, n.d.)	International	1 (African American male; STEM major)	0
Mitchell Scholarship	"The George J. Mitchell Scholarship Program is a national, competitive scholarship sponsored by the US-Ireland Alliance. The Mitchell Scholarship Program, named to honor former US Senator George Mitchell's pivotal contribution to the Northern Ireland peace process, is designed to introduce and connect generations of future American leaders to the island of Ireland, while recognizing and fostering intellectual achievement, leadership, and a commitment to community and public service.....are chosen annually for one academic year of postgraduate study in any discipline offered by institutions of higher learning in Ireland and Northern Ireland." (US-Ireland Alliance, n.d.)	International	1 (African American female; non-STEM major)	0

continued on next page

Table 2. Continued

Award	Award Description	Award Level	DiH Member Winners	DiH Member Finalists/ Honorable Mentions
Fulbright Scholarship	"The Fulbright U.S. Student Program provides grants for individually designed study/research projects or for English Teaching Assistant Programs......to take place during one academic year in a participating country outside the U.S. During their grants, Fulbrighters will meet, work, live with and learn from the people of the host country, sharing daily experiences. The program facilitates cultural exchange through direct interaction on an individual basis in the classroom, field, home, and in routine tasks, allowing the grantee to gain an appreciation of others' viewpoints and beliefs, the way they do things, and the way they think." (Institute of International Education, n.d.)	International	5 (1 African American female; 2 African American males; 1 Hispanic American female; 1 Asian American females; 3 STEM majors [1 Engineering])	0
Truman Scholarship	This award had the mission of "supporting the potential of terrific young people from across the United States committed to public service, more than 3,000 so far." (The Harry S. Truman Scholarship Foundation, n.d.)	National	1 (Asian American; male; STEM major)	0
National Science Foundation Graduate Research Fellowship	"The NSF Graduate Research Fellowship Program recognizes and supports outstanding graduate students in NSF-supported science, technology, engineering, and mathematics disciplines who are pursuing research-based Master's and doctoral degrees at accredited United States institutions." (National Science Foundation Graduate Research Fellowship Program, n.d.)	National	3 (2 Hispanic American males; 1 African American female; 3 STEM majors [2 Engineering])	1 (Middle Eastern American male; STEM major [Engineering])
Goldwater Scholarship	"By providing scholarships to college sophomores and juniors who intend to pursue research careers in the natural sciences, mathematics and engineering, the Goldwater Foundation is helping ensure that the U.S. is producing the number of highly-qualified professionals the Nation needs in these critical fields." (Barry Goldwater Scholarship and Excellence in Education Foundation, n.d.)	National	1 (Asian American female; STEM major)	1 (Asian American female; STEM major)

Of paramount importance of having DiH members, ethnic minorities, apply for and win these prestigious awards is the mentorship and role modeling that it provides for younger and current DiH members. Though the impact has never been assessed through surveying the members, winners have shared that seeing the faces of past DiH members certainly encouraged them to know that winning was even possible and pique their interest to even apply. This thinking does not just apply to these major awards but also to internal fellowships like the Undergraduate Research Fellowship.

At the end of each academic year the faculty advisors for DiH survey the members to better plan for the next academic year. In Spring 2016 the focus was on trying to determine the intrinsic reasons as to why DiH members made certain choices to come to Auburn University, be a member of the Honors College, as well as to become a member of DiH. Only members that attended at least three (3) non-social meetings were invited to participate in the anonymous computer-based survey. The DiH student leadership was not eligible for participation in the survey nor were they involved in the preparation of the instrument to eliminate any notion of unfairness and bias. The survey and results are presented in Table 3. Only the top

Table 3. Spring 2016 survey of DiH members and the results

Number of participants = 12 (9 STEM; 4 non-STEM) (of 9 STEM, 7 Engineering) Freshmen = 10 Sophomores = 2 (11 in state; 1 out of state)
Why did you choose to study at Auburn University? 1. Best scholarship offer 2. Close to home 3. Strength of program/major/Auburn University
Why did you choose to become a member of the Auburn University Honors College? 1. Expects it will help with future career goals 2. Wanted smaller classes 3. Honors housing
Why did you choose to become involved with DiH? 1. Thought it would be a great way to meet minorities in Honors College 2. Being the only minority in every one of my classes, this would be a good way to hang out with other minorities 3. Parents thought this will be a good support group to be a part of
Did you consider attending a HBCU? Yes - 7 No - 5
Any comments or suggestions: "I love having this family support group that is so intimate…" "…all my true friends at AU are all members of this group…" "…compromise comfort in coming to Auburn over an HBCU for a good degree that will open doors…"

three responses for each question is given in the interest of focusing on responses with greatest frequency.

Just like in the early days of "Excellence in Honors," DiH serves a high number of freshmen and is a support group for them; just as the intent for "Excellence in Honors" was at its inception in 2010. DiH most certainly continues to have a place in recruiting and retaining gifted ethnic minorities in the Auburn University Honors College.

CONCLUSION AND FUTURE DIRECTIONS

Retaining ethnic minorities in an Honors College or Program at a PWI takes a concerted effort especially when the majority of them are in STEM fields. With administrative support, inclusive of funding, the pathway described above can be used with success to retain gifted minority undergraduates. There are a few interesting things that were highlighted in the Spring 2016 survey that warrant further research and investigation. This notion of HBCUs not having the ability to provide "a good degree that will open doors" is one that had been made commonplace decades ago and it is interesting that young gifted minority students have this perception though the majority of them seriously considered attending an HBCU over a PWI. Understanding why students of today are of this view will be able to inform student organizations like DiH of how best to give ethnic minorities what appealed to them at HBCUs. Another aspect of the survey that warrants further investigation is that some gifted ethnic minorities feel that attending a PWI is a "compromise [of] comfort" for obtaining a degree from a well-known PWI. The extent of discomfort that gifted ethnic minorities face at a PWI, especially in a state that has been a traditional Republican stronghold (like Alabama is), is very interesting to study and is a hole in the literature worth investigating.

REFERENCES

Apple, M. W. (2004). *Ideology and Curriculum*. New York, NY: Routledge Falmer.

Auburn University. (n.d.a). *Harold A. Franklin Society: Guidance Through Unity*. Retrieved May 20, 2016, from http://www.auburn.edu/student_info/hafs/mission.html

Auburn University. (n.d.b). *Alabama Power Academic Excellence Program*. Retrieved May 20, 2016, from http://www.eng.auburn.edu/admin/aep/

AUinvolve. (n.d.a). *Black Student Union*. Retrieved May 20, 2016, from https://auburn.collegiatelink.net/organization/bsu/about

AUinvolve. (n.d.b). *International Student Organization*. Retrieved May 20, 2016, from https://auburn.collegiatelink.net/organization/iso

Barry Goldwater Scholarship and Excellence in Education Foundation. (n.d.). *The Barry Goldwater Scholarship and Excellence in Education Foundation*. Retrieved September 2, 2016, from https://goldwater.scholarsapply.org/

Bell, A. P. (1970). Role modelship and interaction in adolescence and young adulthood. *Developmental Psychology*, *2*(1), 123–128. doi:10.1037/h0028613

Brookover, W. B., Schweitzer, J. H., Schneider, J. M., Beady, C. H., Flood, P. K., & Wisenbaker, J. M. (1978). Elementary school social climate and school achievement. *American Educational Research Journal*, *15*(2), 301–318. doi:10.3102/00028312015002301

Diamond, J. B., Randolph, A., & Spillane, J. P. (2004). Teachers Expectations and sense of responsibility for student learning: The importance of race, class, and organizational habitus. *Anthropology & Education Quarterly*, *35*(1), 75–89. doi:10.1525/aeq.2004.35.1.75

Eyerman, R. (2001). *Cultural trauma: Slavery and the formation of African American identity*. Cambridge Press. doi:10.1017/CBO9780511488788

Gamoran, A. (2001). American Schooling and educational inequality: A forecast for the 21st century. *Sociology of Education*, *74*, 135–153. doi:10.2307/2673258

Gibson, D. (2004). Role models in career development: New directions for theory and research. *Journal of Vocational Behavior*, *65*(1), 134–156. doi:10.1016/S0001-8791(03)00051-4

Good, T. L. (1981). Teacher expectations and student perceptions: A decade of research. *Educational Leadership*, *38*, 415–422.

Grayson, L. P. (1980). A brief history of engineering education in the United States. *IEEE Transactions on Aerospace and Electronic Systems*, *16*(3), 373–392. doi:10.1109/TAES.1980.308907

Harper, S. R. (2009). Niggers no more: A critical race counternarrative on Black male student achievement at predominantly White colleges and universities. *International Journal of Qualitative Studies in Education*, *22*(6), 697–712. doi:10.1080/09518390903333889

Linnehan, F., Weer, C., & Uhl, J. (2005). African-American students early trust beliefs in work-based mentors. *Journal of Vocational Behavior*, *66*(3), 501–515. doi:10.1016/j.jvb.2004.02.003

Lucas, S. R., & Berends, M. (2002). Sociodemographic diversity, correlated achievement, and De Facto tracking. *Sociology of Education*, *75*(4), 328–348. doi:10.2307/3090282

Marshall Aid Commemoration Commission. (n.d.). *Marshall Scholarships*. Retrieved September 2, 2016, from http://www.marshallscholarship.org/

National Science Foundation Graduate Research Fellowship Program. (n.d.). *About*. Retrieved September 2, 2016, from https://www.nsfgrfp.org/general_resources/about

Oakes, J., & Guiton, G. (1995). Matchmaking: The dynamics of high school tracking decisions. *American Educational Research Journal*, *32*(1), 3–33. doi:10.3102/00028312032001003

Payne, R. S. (1994). The relationship between teachers beliefs and sense of efficacy and their significance to urban LSES minority students. *The Journal of Negro Education*, *63*(2), 181–196. doi:10.2307/2967382

Phinney, J. S., & Alpuria, L. L. (1996). At the interface of cultures: Multiethnic/multiracial high school and college students. *The Journal of Social Psychology*, *136*(2), 139–158. doi:10.1080/00224545.1996.9713988 PMID:8691825

Rhodes Trust. (n.d.). *Rhodes House - Home of The Rhodes Scholarships*. Retrieved September 2, 2016, from www.rhodeshouse.ox.ac.uk/

Roscigno, V. J., & Ainsworth-Darnell, J. W. (1999). Race, cultural capital, and educational resources: Persistent inequalities and achievement returns. *Sociology of Education*, *72*(3), 158–178. doi:10.2307/2673227

The Harry S. Truman Scholarship Foundation. (n.d.). *A Living Memorial & the Future of Public Service Leadership*. Retrieved September 2, 2016, from https://www.truman.gov/our-mission/who-we-are

US-Ireland Alliance. (n.d.). *The George J. Mitchell Scholarship*. Retrieved September 2, 2016, from http://www.us-irelandalliance.org/content/3/en/George%20Mitchell%20Scholarship%20Program%20%7C%20US-Ireland%20Alliance.html

Woodson, C. G. (1933). *The mis-education of the negro*. Trenton, NJ: African World Press, Inc.

Chapter 3
Engineering Study Abroad:
High Impact Strategy for Increasing Access

Monica Gray
Lincoln University, USA

Connie Lundy
Lincoln University, USA

ABSTRACT

Successful engineers must be technically savvy, self-confident as well as culturally competent. Cultural competence is the ability to tolerate ambiguity and empathize with the socio-cultural nuances of different people groups. This calls for a diverse engineering workforce especially in today's increasingly global economy. In the United States, Minorities and Females constitute only 4% and 15% of the engineering workforce respectively. Research shows that women and students of color, dropout due to feelings of not belonging coupled with low self-efficacy. To change the profession's diversity portfolio requires a plethora of high impact approaches. Common among successful retention strategies is the provision of structured opportunities for all students to develop self-directing competencies in both the cognitive and affective learning domains. This chapter demonstrates that the study abroad experience engenders, facilitates and fosters these very aptitudes as well as cultural literacy, and advocates for its inclusion in discussions on increasing under-represented participations in engineering.

DOI: 10.4018/978-1-5225-2212-6.ch003

INTRODUCTION

Engineering is fundamentally about design, an iterative process at the nexus of art and science, creativity and competing criteria, and ambiguity and application. Successful engineers are of course technically savvy, but must be self-confident as well as culturally competent. In today's increasingly global economy, there is therefore an urgent need for a diverse engineering workforce. These mandates are codified in ABET's Student Outcome 3(h) "*the broad education necessary to understand the impact of engineering solutions in a global, economic, environmental, and societal context*" and Student Outcome 3(i) "*a recognition of the need for, and an ability to engage in lifelong learning*". Educating engineers, therefore, involves preparing culturally literate individuals as well as providing increased access to members of under-represented, under-served and minority (UUM) groups. According to the American Society for Engineering Education's (ASEE), *Going the Distance* report, 53.6% Hispanics, 61.4% Native Americans, 61.7% African Americans and 49% Female students who enter engineering programs do not graduate in this major (ASEE, 2012). This translates to an engineering workforce that comprise of about 6% Hispanics, 0.3% Native American, 4% African Americans and 15% females according to the latest National Science Foundation's report (NSF, 2015). Many strategies are currently being employed to reduce this disparity. These efforts include first-year seminars, internships, learning communities, and capstone projects (ASEE, 2012; Freeman, Sharon, & Duvon, 2008).

Study abroad, however, has not been seriously considered as a high impact retention activity. The research shows that UUMs struggle in mathematics and science intensive programs such as engineering, not simply because of academic under-preparation *per se*, but more importantly, due to feelings of not belonging and lack of confidence in their ability to learn coupled with low self-efficacy and self-direction (Felder & Brent, 2005; Gray & Lundy, 2016; Matthews, 2012). In fact, the common theme among successful high impact retention strategies is that they provide access to structured opportunities for all students, including UUMs, to develop self-directing competencies in both the cognitive and affective learning domains (Brainard & Carlin, 1998; Chubin, May, & Babco, 2005). Therefore, facilitating skill sets that are *sine qua non* both to successfully learn and complete engineering degree programs. Studying abroad inherently impact students' personal and academic development (Yang, Webster, & Prosser, 2011). Study abroad exercises students' cognitive analysis ability (students are able to deal with ambiguity in new, ill-defined situations by formulating models and seeing relationships), affective organizational skills (students are able to balance their responsibilities and formulate a cohesive and systematic approach to learning) as well as self-efficacy, which is a proximal predictor of proficiency in both these domains (Milstein, 2005).

The objectives of this chapter are to explore the reasons for high attrition from engineering programs among UUMs, to make the case for study abroad as a high impact strategy to increase retention and to discuss challenges in implementation.

BACKGROUND

Overall, the dropout rate for engineering programs is high. Between 40% to 50% of students switch to other majors within the first two years (Felder & Brent, 2005; Matthews, 2012). However, there exist demographic disparities, and minority students quit at significantly higher rates (ASEE, 2012). The problem of minority under-representation in engineering correlates to several factors. These include; under-preparation, poverty level and low self-efficacy. Inadequate pre-college preparation in basic Mathematics and Sciences create difficulty for students transitioning to a highly sequenced and vertically organized curriculum such as engineering (Lassibille & Navarro Gómez, 2008). Socio-economic status is a strong predictor of college preparedness. For example, there are higher proportions of minority students, under age 18, living in poverty (e.g., Whites (13%), Black (39%), and Hispanics (33%)) (NCES, 2014a). Large proportion of minority students attend urban high-poverty schools. For instance, 34% of students in urban/city, compared to 13% of suburban areas were enrolled in high-poverty schools. In contrast 35% of students in suburban areas were enrolled in low-poverty schools compared to 15% in urban/city schools (NCES, 2014b). High poverty schools tend to lack academic and financial resources and tend to have inadequately trained teachers. Thus, predominantly minority schools tend to have underqualified teachers. The percentage of teachers with a college major and standard certification in their main teaching assignment for schools with more than 50% Whites, Blacks and Hispanics enrollment are Mathematics (65.2%, 48.6%, 44.8%) and Science (73.3%, 57.3%, 64.7%) respectively (NCES, 2014c). In high school, minority students tend to be placed in general academic tracks, do not have access to college preparatory classes and are less likely to do well on standardized test for gateway courses such as Calculus. For example, Blacks are three times less likely to be enrolled in Calculus compared to their White peers (e.g., Hispanics 9%, Blacks, 6%, Whites, 18%) (NAEP, 2014). For the National Assessment of Educational Progress 12th-grade Calculus exam, students in high-poverty schools scored less (163) compared to students in low-poverty schools (199) (NAEP, 2014). According to Haycock (2001), 12th grade African American and Latino students had skill levels in Math and Reading similar to 8th grade White students, a disparity that persists into postsecondary education. Therefore, minority students tend to be academically under-prepared for highly sequenced majors such as engineering that demands specific levels of competencies.

While academic under-preparation certainly plays an important role in whether or not students persist in engineering, some interesting research findings indicate that women and students of color more often than not, dropout due to a loss in confidence and a feeling of "*lack of belonging*" within programs or the profession (Brainard & Carlin, 1997; Felder & Brent, 2005; Matthews, 2012). That is, women and students of color entering a predominantly white and male field are often times unable to see themselves as or being able to be engineering professionals. In fact, researchers corroborate that differences in learning abilities and rates among students only slightly correlates to natural ability (Bonaccio, Reeve, & Winford, 2012; Lashley & Best, 2003). In addition, studies into how and why people learn, have indicated that students' attitudes (affective development) was the best predictor of intellectual achievement (cognitive development) and subsequent academic buoyancy (Freeman et al., 2008; Martin & Marsh, 2008; Veloo, Ali, & Krishnasamy, 2014). When racial factors were controlled for, it was found that anxiety, which is strongly influenced by ethnicity and gender caused low academic performance and women were more negatively affected (Adigwe, 1997; Erdem, 2007; Nwachukwu, 2014; Osborne, Simon, & Collins, 2003; Tobias, 1985; Tocci & Engelhard Jr, 1991; Tryon, 1980). Of particular note is the fact that the preceding citations referred to national and international Mathematics and Science studies which are *sin qua non* for gaining admission to and advancing through engineering programs. These findings provide strong support for the hypothesis that students, especially women and minorities tend to leave engineering, not because of lack of competence, but loss of self-efficacy or confidence in their ability to learn in or to complete a degree program as well as continue their professional development outside the formal educational structure.

This research postulates that ABET's Student Outcome 3(h) "the broad education necessary to understand the impact of engineering solutions in a global, economic, environmental, and societal context" and Student Outcome 3(i) "a recognition of the need for, and an ability to engage in lifelong learning" are not mutually exclusive but rather interdependent and mutualistic in nature. Outcomes by their very nature describes what students should know or can do by the time of graduation. The implication is therefore, that lifelong learning and a global perspective must originate within the 4-year engineering curriculum/program. The proposition being advanced here is that the mutualism emerges from the commonality of the underlying skill sets required for success in achieving these two outcomes. That is, the skills required to persist and arrive at the end of one's undergraduate engineering program having acquired a "broad education..." are consistent with those that drive engagement in lifelong learning. It follows therefore, ceteris paribus, providing structured opportunities for minority and under-represented students to develop self-directing competencies facilitate the skill sets they need to both successfully learn and complete engineering degree programs, as well as participate in future lifelong learning.

Further, the case is also being made, that the study abroad experience engenders, facilitates and fosters these very competencies as well as cultural competence and must be seriously considered among high impact practices currently being employed to increase UUM participation in engineering programs.

PERSISTENCE, LIFELONG LEARNING AND STUDY ABROAD

Lifelong Learning: Precursor for Academic Buoyancy

In his seminal work, Philip Candy conveyed,

lifelong learning takes, as one of its principal aims, equipping people with skills and competencies required to continue their own self-education beyond the end of formal schooling (Candy, 1991).

Taking responsibility for one's education is more important today than ever before as the "*half-life*" (i.e., time before half of what is known becomes obsolete) of an engineer's undergraduate technical skills rapidly decreases, which for some disciplines can be less than 2.5 years (Smerdon, 1996). What therefore are the mental characteristics as well as attitudes of the individual who will continue their self-education beyond graduation?

To propose traits of lifelong learners, research in this area typically utilize the domains of learning framework which was developed to identify, measure and address how and why people learn. These constructs theorize that learning occurs on a continuum whereby growth and changes to the brain's architecture results from the ways the learner acquires information, make connections and apply what they know (Kraiger, Ford, & Salas, 1993). It categories learning outcomes into learning domains such as cognitive (mental skills/knowledge) and affective domains (attitude/value). According to Bloom's taxonomy of educational objectives, the cognitive domain consists of six progressively complex levels of competencies (i.e., knowledge, comprehension, application, analysis, synthesis and evaluation) (Bloom & Krathwohl, 1956). The affective domain describes the learner's personal motivation and interests and, consists of five proficiency levels (i.e., receiving, responding, valuing, organization and characterization) (Krathwohl, Bloom, & Masia, 1964). The consensus is, to subsequently be a lifelong learner, a graduating senior must, at minimum, demonstrate proficiency at the analysis and organizational levels in the cognitive and affective domains respectively (Mouros, 2003; Somerville, 2008). Here cognitive analysis is defined as *students are able to deal with ambiguity in new, ill-defined situations by formulating models and seeing relationships, while affective organization proposes that students are able to balance their responsibilities and formulate a cohesive and systematic approach to learning (Mouros, 2003).*

These two proficiencies work in tandem to promote learning, confidence and persistence. For example, a new ill-defined situation or problem requires analysis wherein prior knowledge and/or skills are used to deconstruct concepts, to examine the interrelationships of the parts and determine their contribution to the whole. While organization employs compare and contrast strategies to resolve conflicts between prior knowledge and the new challenge in a systematic effort to create a consistent value system. Both skills are thus used iteratively to formulate new mental models that will facilitate inquiry, reflection and application of the newly acquired knowledge to future encounters. In their groundbreaking work, Felder & Silverman (1988) concurred, "most of what we learn on our own (as opposed to in class) originates in a real situation or problem that needs to be addressed and solved…". Mouros (2003) offered the following link between the learning framework and Student Outcome 3(i) "a recognition of the need for (affective - organization), and an ability to engage in lifelong learning (cognitive - analysis)". Thus, the potential lifelong learner must at some point in their academic career develop value for information that pertains his or her discipline and has a strong enough sense of self-efficacy to be intrinsically motivation to independently learn. Concomitantly, in exploring his or her discipline, the student will face new, ill-defined and challenging tasks which require concerted, systematic and extended efforts in order to succeed and subsequently graduate. These principles are similarly applicable to Student Outcome 3(h) "the broad education necessary to understand the impact (cognitive – analysis) of engineering solutions in a global, economic, environmental, and societal context (affective - organization)". Given the rapid pace of new engineering solutions, subsequent impacts will similarly change. Thus, at the very heart of this outcome is the need for in-the-moment learning and future continual lifelong learning.

Study Abroad as a Structured Opportunity to Develop Self-Direction

Spatial mobilization of the engineering sector predicated by a global economy as well as engineering program accreditation requirements for students to have "the broad education necessary to understand the impact of engineering solutions in a global, economic, environmental, and societal context" have led to the market demand for more culturally savvy engineers. Studying abroad is a great way to satisfy these demands as well as to facilitate students' personal and academic development (Yang et al., 2011). Study abroad exercises students' cognitive analysis, affective organizational skills as well as self-efficacy, which is a proximal predictor of proficiency in both these domains (Milstein, 2005). Self-efficacy is, "an individual's judgments of his or her capabilities to perform given actions" (Schunk, 1991). Indeed, study abroad can be very challenging, as culture shock requires adjustment to the host culture, and may also elicit "feelings of not belonging" (Yang et al., 2011). However, culture shock provides students with opportunities to "deal with ambiguity in new, ill-defined situations". Felder & Brent (2005), describes the cognitive development trajectory of college students as a continuum of knowing everything, "ignorant certainty", recognition that context determines answer, "intelligent confusion" to a point where students begin to trust their own process of organizing their learning and judgment. A similar process occurs during the study abroad experience. According to self-efficacy framework and intercultural transformation theory, the study abroad participant leave their comfort zones feeling unconsciously certain of interpersonal interactions and societal expectations. The student's equilibrium is quickly disturbed when seemingly familiar interactions occur without familiar signs or social cues. This cognitive dissonance leads to "identity confusion" an analog to "intelligent confusion" (Zaharna, 1989). Each such interaction, however, leaves the student better equip to handle similar encounters because of "greater cognitive and affective capacity" (Milstein, 2005). Study abroad participants come away from the experience more confident in their ability to trust their own judgement and learning process.

The study abroad experience can potentially provide opportunities for UUMs to meet and interact with culturally different engineers which helps to change their world views of what it means to be different and/or to not belong. Firstly, the limiting societal reflection to African Americans are removed and unceremoniously replaced with a label normally reserve for the majority, they become, "*Americans*". Finally, depending on the cultural homogeneity of the society, minority students may have an opportunity to witness that a society, even those that are racially uniform, tend to exhibit elements of structural segregation, cultural alterity and baseless phenotypic bias. This has the potential to move the student from the culture shock of not belonging at home and in the host country to one of global solidarity and a redefinition of his/her group of peers.

Moreover, by its very nature, preparing to study abroad and actually going abroad requires setting and achieving a set of goals. From planning when to study to applying for a passport, each step provides opportunities for students to set goals and get feedback on how well they are doing. In self-efficacy theory, goal setting is an intervening or moderating process towards achieving self-directed learning (Bandura & Schunk, 1981). Students who perceive themselves as having low self-efficacy in engineering, will avoid engineering-related activities, employ minimal effort and will not persist (Schunk, 1990, 1991). However, the process of leaving one's country and surviving cultural differences in another, especially, if the student is the first in the family to do so, creates a deepened sense of belief in his or her abilities. Goals therefore mediate an individual's acquisition of learning and lifelong learning skills by rallying effort and persistence as wells encouraging the organization of action plans to creatively solve problems. Exercises in goal setting increase individuals' task performance, self-efficacy, and intrinsic interest in a specific task and related skills (Locke, Shaw, Saari, & Latham, 1981).

Engineering Study Abroad

The preceding discussion showcases the potential for study abroad to not only increase self-efficacy, but to assist in building learning and lifelong learning capacity. The leaving of one's country, independently learning how to assimilate into the host country's culture without direct support from one's social network is analogous to independent pursuit of learning without formal institutional support or affiliation. Studying abroad therefore represents a pivotal moment in a young person's life and informs one's outlook for years after the event has ended (Salisbury, Paulsen, & Pascarella, 2011). The experience provides opportunities to reflect on one's social norms as well as biases, thereby facilitating self-awareness and self-confidence. This immersion experience also fast tracks the language and intercultural development acquisition process. Increasingly, studies are showing that students who study abroad, are retained and graduate within four years at higher rates with higher GPA scores, than their non-study abroad peers (Hamir, 2011). Traditionally, engineers and minorities typically do not study abroad citing strictly structured curriculum or financial constraints. However, this perception is changing. For example, according to the latest Open Doors data the number of engineers who studied abroad in 2013 grew by 6% over the previous year. Finally, study abroad already have some built leverage. For example, more women study abroad than men, a strength that can be employed to encourage women to persist in engineering (Petzold & Peter, 2014).

SOLUTIONS AND RECOMMENDATIONS

Studying Abroad: Benefits, Challenges and Opportunities

According to the Institute of International Education's (IIE) 2015 Open Doors report, over the last two decades, the number of U.S. students studying abroad has tripled to a about 300,000. This indicates that students, faculty and administrators are recognizing the benefits of the study abroad experience and are availing themselves of the opportunities. For example, research shows that students who study abroad have higher GPAs, are more likely to graduate on time, and are more attractive to employers that seek to hire candidates with intercultural competencies (Hamir, 2011). African American students account for 15% of the overall U.S. undergraduate population but only represent 5.6% of the over 300,000 U.S. students studying abroad. There are several reasons for this disparity: finances, faculty leadership, attrition rates, choice of major, family support, anxiety resulting from travel inexperience, and lack of insight of the correlations between global cultural competence and career goals (Acquaye & Edmonds Crewe, 2012; Shih, 2009a). It is interesting to note that high attrition from degree programs has been listed here also as a contributing factor for under-participation. Traditionally, study abroad has been marketed as an opportunity to experience other cultures. For minority students, this angle has not been effective as they interact across culture on a daily basis (Salisbury et al., 2011). However, based on the research into the causative factors for under-representation, the subsequent realization that women and minorities leave the profession because they are unable to identify with the attributes of the successful professional and the drive for more culturally savvy engineer, a better approach to encourage under-served minorities to study abroad to acquire in demand skills and be more marketable to employers.

While STEM is now the top sending group accounting for about 23% of the study abroad cohort, engineering represents only 5% of that total. By and large, STEM fields, have very restrictive, highly sequenced and vertical curriculums, creating an additional barrier that prevents their majors from partaking of the transformative experiences gained from studying abroad. This still continues today especially in engineering. Students are fearful of not being able to find equivalent courses which may result in study abroad credits not transferring back to home institutions, falling out of sequence, getting behind, and subsequent delayed graduation and increased educational cost. Faculty are equally concerned with the quality of the academic experience overseas, their lack of control over the curriculum, and possible accreditation implications. Faculty leadership in encouraging study abroad participation under these conditions is, as would be expectedly, low.

There are several factors that are driving demand for more engineers to study abroad: ABET accreditation, globalized economy, and federal funding agencies. In addition, as the number of STEM majors studying abroad increases, this may assuage concerns of both students and faculty about the ability of engineers to study abroad and still graduate on time. Further, an opportunity may exist in the fact that females constitute about 65% of the study abroad participants. This is an inherent strength that must be capitalized upon, given the high female attrition rates from engineering programs.

Broadening UUM Engineering Majors' Participation in Study Abroad

The preceding discussion advances study abroad not only as a high impact strategic retention tool to ensure that more UUMs complete engineering programs but also a way to position this cohort of students for successful careers. However, to successfully encourage African American and other underserved engineering students to participate in study abroad, the experience must be adopted as part of an institution's overall culture, is not simply an additional item to an already packed curriculum. Internationalizing the curriculum requires strategic direction and system-wide buy-in. Successful strategies include cooperative agreements among U.S.-based and global institutions, development of U.S.-based consortia, curriculum integration, course matching, and scholarship funding, as well as a robust faculty-led program.

Cooperative Agreements

Cooperative agreements or memoranda of understanding (MOU) are an effective way for U.S.-based and international institutions with mutual internationalization missions and goals to broker relationships that advance the individual and collective vision of all signatories. MOUs that tend to work and are productive in the long run are characterized by up front, explicitly negotiated needs and strong bilateral support from individual parties. Some items that are typically included in internationalization agreements are scholarly exchanges, of both faculty and students, curriculum and course integration, formal agreements with individual departments, and English as the language of instruction.

For administrators who are concerned about ABET accreditation issues, a conservative option might include finding a partner university in a country that is a signatory of the Washington Accord and ensuring that the perspective institution is accredited by that country's accreditation body. The Washington Accord, signed in 1989, is an international mobility agreement among engineering accrediting bodies in seventeen countries. Each signatory recognizes accredited engineering programs in others' jurisdiction as being substantially equivalent to its own. For example, South Africa is a signatory of the Washington Accord, and its accrediting body is the Engineering Council of South Africa (Alliance, 2016). If a U.S. institution has an ABET accredited Civil Engineering program and wishes to sign an MOU with the University of Pretoria (UP) in South Africa, which is accredited by the Engineering Council of South Africa, then both civil programs are recognized as being substantially equivalent to each other. In this example, the U.S. institution's administrations would be even happier to realize that UP is also ABET accredited which is not a very common occurrence outside the U.S.

Local and Regional Consortia

Capitalizing on the idea of strength in numbers, local and regional universities sometimes come together to form a consortium. These partnerships may include stakeholders from across each member institution. It is common to include both study abroad officials and faculty. Individual schools may be relatively small and often do not have the enrollment numbers to create a cost effective study abroad contingent. By coming together, a group of schools can create agreements among themselves as well as with the global partner(s). The consortium then sends a large group of students to their partner institution(s) reducing the overall cost per student as well as ensuring the integrity of the educational experience. Member schools sometimes sign off on course credits though a process of triangulation. For example, if Lincoln and University of Delaware (UD) are part of a consortium and UD gave credit for a Statics course taken at the University of the West Indies (UWI), then Lincoln, by virtue of its consortium membership, would be more inclined to also accept those credits for its students studying at UWI.

Curriculum Integration and Course-to-Course Matching

Curriculum integration involves the time and offering synchronization of typical, fundamental engineering courses. Both the global and U.S. partner institutions evaluate their individual course sequence to determine common core offerings and work to offer those courses in tandem, see Table 1. Typical core courses that lend themselves to such processes include Statics, Dynamics and Fluids. Inherent in this process is the analysis to determine the equivalency of the courses as well as removal of the guess work about when is the best time for engineering students to study abroad. Courses are therefore pre-approved for engineering credit; thus,

Table 1. Sample engineering course-to-course matching and curriculum integration

U.S. Institution's Progression Plan	Global Partner's Equivalents
ME252: Applied Mechanics Dynamics	**DEN 4108**: Engineering Mechanics: Dynamics
ME291: Thermodynamics 1	**DEN 107**: Thermodynamics 1
MAT239: Differential Equations	**MTH 5123**: Differential Equations
CENE253: Mechanics of Materials	**DEN 4102**: Engineering Mechanics: Statics OR **DEN 5102**: Solid Mechanics
CENE253L: Mechanics of Materials Lab	Included in **DEN 4102** & **DEN 5102**
LISTCOURSE: Liberal Studies Course	**e.g., HST 5302**: Architecture in London II 1837 - to the Present

students go abroad confident that they can make adequate progression towards completing their degrees on time.

Scholarship Funding Opportunities

Financial constraint is a major concern for all students even though it has become relatively easier to get study abroad scholarships for STEM majors compared to other majors. Many scholarships specifically target minority and other underserved student populations. The following is a discussion of a few:

- **Benjamin A. Gilman International Scholarship:** Funded by the U.S. Department of State and administered by IIE, this scholarship funds traditionally under-represented cohorts in study abroad and shows a preference for non-traditional destinations. Students can study, research or intern while they are abroad. Funding is typically up to $5000 but can be increased to $8000 if students study a critical needs language. STEM is considered a priority group.
- **Vira I. Heinz Women in Global Leadership Program:** Funded by the Vira I. Heinz Endowment and administered by University of Pittsburgh, this program awards scholarships to three women from each participating institution of at least $5000 with an additional $1000 if they study outside of Western Europe. STEM is considered a privileged group.
- **Third Party Study Abroad Providers:** Usually offer scholarships to participating students. Universities or consortium can negotiate better rates for their students especially for large cohorts, typically at least 10 students.

Faculty-Led Programs

Faculty-led programming is an essential component to campus internationalization plans aimed at making study abroad part of the academic culture and not the exception. One iteration many include programing where faculty take students abroad and delivering course(s) in the host country. The abroad experience can be a continuation of a course taught on campus and/or a team taught course where U.S.-based and the host foreign country's students are grouped to work on a project during the semester and then come together at the end. Whatever the method they take, faculty-led opportunities ease many traditional fears and significantly impact the internationalization process. First, faculty who take students abroad are more likely to send students abroad. Second, since the faculty member delivers the content, there is no question as to the quality of the experience. Finally, if the abroad experience is a continuation of an on-campus course, then the program fee can be embedded in the course cost and can thus make use of the students' financial aid packages that are already in place.

CONCLUSION

Finally, while the number of students studying abroad has been steadily increasing, African Americans and Hispanics account for only 5% and 8% respectively (Open Doors 2012 data) (Shih, 2009b). Many strategies are currently being employed to reduce minority and under-represented attrition from engineering programs. Study abroad has not been previously considered as a high impact activity. Traditionally, study abroad has been marketed as an opportunity to experience other cultures. For minority students, this angle might not have the same impact as they interact across culture on a daily basis (Salisbury et al., 2011). However, an opportunity to increase skill sets that can assist in persisting and doing well in their current engineering programs as well as prepare them for continual lifelong professional development might just work.

In addition, spatial mobilization of the engineering sector predicated by a global economy, as well as engineering program accreditation with a student learning outcome requirement to have

the broad education necessary to understand the impact of engineering solutions in a global, economic, environmental, and societal context,

have led to the market demand for more culturally savvy engineers. It follows therefore that an engineering educational experience abroad not only provides a structured opportunity for UUM students to increase their self-efficacy and self-direction skill sets which will help them persist and do well in their current engineering programs but also prepares them to meet the demands of the global market place. Study abroad has not been previously considered as a high impact activity, there is, however, a movement afoot to change that paradigm. The National Science Foundation in a recent Louis Stokes Alliance for Minority Participation (LSAMP) program solicitation cited researchers Bowman and Sage as follows,

Preparing a diverse, globally-engaged scientific and technological workforce necessitates strengthening international research opportunities for students under-represented in STEM fields (Bowman & Stage, 2002).

REFERENCES

Acquaye, L. A., & Edmonds Crewe, S. (2012). International Programs: Advancing Human Rights and Social Justice for African American Students. *Journal of Social Work Education, 48*(4), 763–784. doi:10.5175/JSWE.2012.201100130

Adigwe, J. C. (1997). Ethnicity, test anxiety and science achievement in Nigerian students. *International Journal of Science Education, 19*(7), 773–780. doi:10.1080/0950069970190703

Alliance. (2016). *Washington Accord Signatories*. Retrieved October 15, 2015, form http://www.ieagreements.org/Washington-Accord/signatories.cfm

ASEE. (2012). *Going the distance: best practices and strategies for retaining engineering, engineering technology and computing students*. Retrieved October 15, 2015, from https://www.asee.org/retention-project

Bandura, A., & Schunk, D. H. (1981). Cultivating competence, self-efficacy, and intrinsic interest through proximal self-motivation. *Journal of Personality and Social Psychology, 41*(3), 586–598. doi:10.1037/0022-3514.41.3.586

Bloom, B. S., & Krathwohl, D. R. (1956). *Taxonomy of educational objectives: The classification of educational goals. Handbook I: Cognitive domain*. New York: David McKay.

Bonaccio, S., Reeve, C. L., & Winford, E. C. (2012). Text anxiety on cognitive ability test can result in differential predictive validity of academic performance. *Personality and Individual Differences, 52*(4), 497–502. doi:10.1016/j.paid.2011.11.015

Bowman, M. H., & Stage, F. K. (2002). Personalizing the goals of undergraduate research. *Journal of College Science Teaching, 32*, 120.

Brainard, S. G., & Carlin, L. (1997, November). *A longitudinal study of undergraduate women in engineering and science*. Paper presented at the Frontiers in Education Conference, Pittsburgh, PA. doi:10.1109/FIE.1997.644826

Brainard, S. G., & Carlin, L. (1998). A Six-Year Longitudinal Study of Undergraduate Women in Engineering and Science. *The Journal of Engineering Education, 87*(4), 369–375. doi:10.1002/j.2168-9830.1998.tb00367.x

Candy, P. C. (1991). *Self-Direction for Lifelong Learning. A Comprehensive Guide to Theory and Practice*. San Francisco, CA: Jossey-Bass.

Chubin, D. E., May, G. S., & Babco, E. L. (2005). Diversifying the engineering workforce. *The Journal of Engineering Education, 94*(1), 73–86. doi:10.1002/j.2168-9830.2005.tb00830.x

Erdem, E. (2007). Study of the relationship between test anxiety and the epistemological and problem-solving beliefs of students on a general chemistry course. *World Applied Sciences Journal, 2*, 750–758.

Felder, R. M., & Brent, R. (2005). Understanding student differences. *The Journal of Engineering Education, 94*(1), 57–72. doi:10.1002/j.2168-9830.2005.tb00829.x

Felder, R. M., & Silverman, L. K. (1988). Learning and teaching styles in engineering education. *English Education, 78*, 674–681.

Freeman, K. E., Sharon, T. A., & Duvon, G. W. (2008). Do Learning Communities Enhance the Quality of Students' Learning and Motivation in STEM? *The Journal of Negro Education, 77*, 227–240.

Gray, M., & Lundy, C. (2016). *Implementing and Integrating International Research into the Engineering Curriculum at Lincoln University, Pennsylvania and the University of the West Indies, Trinidad.* Paper presented at the American Society of Engineering Education International Forum, New Orleans, LA.

Hamir, H. B. (2011). *Go Abroad and Graduate On-Time: Study Abroad Participation, Degree Completion, and Time-to-Degree* (Unpublished doctoral dissertation). University of Nebraska, Lincoln, NE.

Haycock, K. (2001). Closing the achievement gap. *Educational Leadership, 58*, 6–11.

Kraiger, K., Ford, J. K., & Salas, E. (1993). Application of cognitive, skill-based, and affective theories of learning outcomes to new methods of training evaluation. *The Journal of Applied Psychology, 78*(2), 311–328. doi:10.1037/0021-9010.78.2.311

Krathwohl, D. R., Bloom, B. S., & Masia, B. B. (1964). *Taxonomy of educational objectives, handbook ii: affective domain*. New York: David McKay Company.

Lashley, C., & Best, W. (2003). *12 Steps to Study Success*. Cengage Learning EMEA.

Lassibille, G., & Navarro Gómez, L. (2008). Why do higher education students drop out? Evidence from Spain. *Education Economics, 16*(1), 89–105. doi:10.1080/09645290701523267

Locke, E. A., Shaw, K. N., Saari, L. M., & Latham, G. P. (1981). Goal setting and task performance: 1969–1980. *Psychological Bulletin, 90*(1), 125–152. doi:10.1037/0033-2909.90.1.125

Martin, A. J., & Marsh, H. W. (2008). Academic buoyancy: Towards an understanding of students everyday academic resilience. *Journal of School Psychology*, *46*(1), 53–83. doi:10.1016/j.jsp.2007.01.002 PMID:19083351

Matthews, M. (2012). *Keeping students in engineering: A research-to-practice brief.* American Society of Engineering Education. Retrieved from http://www.asee.org/retention-project/keeping-students-in-engineering-a-research-guide-to-improving-retention

Milstein, T. (2005). Transformation abroad: Sojourning and the perceived enhancement of self-efficacy. *International Journal of Intercultural Relations*, *29*(2), 217–238. doi:10.1016/j.ijintrel.2005.05.005

Mouros, N. (2003). *Defining, teaching and assessing lifelong learning skills.* Paper presented at the Frontiers in Education, 2003. FIE 2003 33rd Annual.

NAEP. (2014). *High School Coursetaking. The Condition of Education.* Retrieved on October 15, 2015, from http://nces.ed.gov/programs/coe/indicator_cod.asp

NCES. (2014a). *Children Living in Poverty. The Condition of Education.* Retrieved on October 15, 2015, from http://nces.ed.gov/programs/coe/indicator_cce.asp

NCES. (2014b). *Concentration of Public School Students Eligible for Free or Reduced-Price Lunch. The Condition of Education.* Retrieved on October 15, 2015 from from http://nces.ed.gov/programs/coe/indicator_clb.asp

NCES. (2014c). *Percentage of public high school teachers with a college major and standard certification in their respective main teaching assignment in mathematics, English, and science, by race/ethnicity concentration of schools: 2007–08. Status and trends in the education of racial and ethnic minorities.* Retrieved on October 15, 2015, from http://nces.ed.gov/programs/coe/indicator_cod.asp

NSF. (2015). *Women, Minorities, and Persons with Disabilities in Science and Engineering*. Retrieved January 02, 2016, from http://www.nsf.gov/statistics/wmpd/

Nwachukwu, P. (2014). Influence of Gender and Socio-Economic Factors on Students Performance in Mathematics in Shomulu Local Government Area of Lagos State. *Journal of Education and Practice*, *5*, 69–72.

Osborne, J., Simon, S., & Collins, S. (2003). Attitudes towards science: A review of the literature and its implications. *International Journal of Science Education*, *25*(9), 1049–1079. doi:10.1080/0950069032000032199

Petzold, K., & Peter, T. (2014). The social norm to study abroad: Determinants and effects. *Higher Education, 69*(6), 885–900. doi:10.1007/s10734-014-9811-4

Salisbury, M., Paulsen, M., & Pascarella, E. (2011). Why do All the Study Abroad Students Look Alike? Applying an Integrated Student Choice Model to Explore Differences in the Factors that Influence White and Minority Students Intent to Study Abroad. *Research in Higher Education, 52*(2), 123–150. doi:10.1007/s11162-010-9191-2

Schunk, D. H. (1990). Goal setting and self-efficacy during self-regulated learning. *Educational Psychologist, 25*(1), 71–86. doi:10.1207/s15326985ep2501_6

Schunk, D. H. (1991). Self-efficacy and academic motivation. *Educational Psychologist, 26*(3-4), 207–231. doi:10.1080/00461520.1991.9653133

Shih, K. (2009a). Study abroad participation up, except among minority students. *Diverse Issues in Higher Education*, 16.

Shih, K. (2009b). Study abroad participation up, except among minority students. *Diverse Education, November, 16.*

Smerdon, E. T. (1996). Lifelong learning for engineers: Riding the whirlwind. *American Ceramic Society Bulletin, 75*, 51–52.

Somerville, M. (2008, August). *Collaborative design and assessment: learning 'with and for' users.* Paper presented at the Building effective, sustainable, practical assessment, Seattle, WA.

Tobias, S. (1985). Test anxiety: Interference, defective skills, and cognitive capacity. *Educational Psychologist, 20*(3), 135–142. doi:10.1207/s15326985ep2003_3

Tocci, C. M., & Engelhard, G. Jr. (1991). Achievement, Parental Support, and Gender Differences in Attitudes Toward Mathematics. *The Journal of Educational Research, 84*(5), 280–287. doi:10.1080/00220671.1991.10886028

Tryon, G. S. (1980). The measurement and treatment of test anxiety. *Review of Educational Research, 50*(2), 343–372. doi:10.3102/00346543050002343

Veloo, A., Ali, R. M., & Krishnasamy, H. N. (2014). Affective Determinants of Additional Mathematics Achievement in Malaysian Technical Secondary Schools. *Procedia: Social and Behavioral Sciences, 112*, 613–620. doi:10.1016/j.sbspro.2014.01.1208

Yang, M., Webster, B., & Prosser, M. (2011). Travelling a thousand miles: Hong Kong Chinese students study abroad experience. *International Journal of Intercultural Relations*, *35*(1), 69–78. doi:10.1016/j.ijintrel.2010.09.010

Zaharna, R. S. (1989). Self-shock: The double-binding challenge of identity. *International Journal of Intercultural Relations*, *13*(4), 501–525. doi:10.1016/0147-1767(89)90026-6

KEY TERMS AND DEFINITIONS

ABET: Formerly, The Accreditation Board of Engineering and Technology, this organization evaluates the quality of engineering programs mainly in the United States for accreditation.

Academic Buoyancy or Resilience: The ability of students to persist and make progress through to degree completion.

Affective Domain: A construct in the domain of learning framework that categories student learning outcome based on the learner's personal motivation and interests. It consists of five progressive proficiency levels, namely, receiving, responding, valuing, organization and characterization.

Affective Organizational: The fourth proficiency level of the Affective Domain.

Cognitive Analysis: The fourth competency level of the Cognitive Domain.

Cognitive Domain: A construct in the domain of learning framework that categories student learning outcome based on the learner's mental skills. Most typically associated with Bloom's taxonomy of educational objectives. It consists of six progressively complex competency levels, namely, knowledge, comprehension, application, analysis, synthesis and evaluation.

Cultural Competence: Is the ability to tolerate ambiguity and empathize with the socio-cultural nuances of different people groups.

Lifelong Learning: Possessing the skill to acquire and use knowledge in both formal and informal settings. The notion of taking responsibility for one's self-education.

Self-Efficacy: A belief in one's ability to perform and deliver an assigned task.

Section 2
Micro-Community or Target Approaches to Diversifying Engineering

Chapter 4
Empowering Women in STEM: Embedding STEM in K-12 Education

Gretchen Dietz
University of Mount Union, USA

Terry Long
University of Mount Union, USA

Julie Hessedence
University of Mount Union, USA

Helen E. Muga
University of Mount Union, USA

ABSTRACT

This chapter covers a project that was part of an I3 (Ideas, Innovation, Invention) challenge within a global engineering, junior course at University of Mount Union. The course exposes engineering students to global societal challenges and their solutions. Of importance in this project is the need to boost numbers in the STEM (Science, Technology, Engineering, and Mathematics) and the need to increase diversity. With this in mind, the team developed and tested five different activities that involved different areas of STEM that were both fun and didactic in nature. Each activity had an associated module to allow for assessment of understanding. An orange buoyancy module, a strength of paper module, a soil erosion module, a simple circuit module, and a simple electric motor module were developed. They were tested on K-5 students at Washington Elementary School, Alliance, OH and on K-5 students at Ollas Arriba Elementary School, Panama City, Panama.

DOI: 10.4018/978-1-5225-2212-6.ch004

INTRODUCTION

The project conducted by the University of Mount Union engineering students in EGE 320 International Engineering Field Experience focused on diversifying engineering, specifically the empowerment of women to allow for a greater accessibility and availability of STEM-related careers. To meet the constantly evolving global economy and workforce demands within and between other nations, STEM is needed to further develop modern technology to prevent a country from falling behind in a technology-driven society. The idea behind this research was to develop methods and approaches that introduce students to STEM at a young age. Engagement and exposure to STEM at a younger age as well as maintaining a student's interest is critical in whether a student goes into STEM fields and careers or drops out completely.

Careers in this field are priority because too few college students are pursuing degrees in these fields. It is argued that with the creation of STEM-related programs, innovation and the establishment of more jobs will follow, and consequently, make a country more competitive in the global economy. An important objective of this project involved the creation and implementation of gender-neutral engineering-related activities used in a classroom setting. These activities were designed to be enjoyable and engaging for many students with the expectation of exposing each student to the world of STEM-related fields. Since a general interest in STEM related fields will largely contribute to a better economy and society for all, it was favorable that this project was feasible and easily implemented into primary and secondary levels of education throughout the world.

In order to assess the empowerment of diversity within engineering, kits were created by the team that were implemented within classrooms. The kit that was designed featured five different lessons taught students ideas from electrical engineering, structural engineering, environmental engineering, mechanical engineering, fluid mechanics, and physics. It featured products that were easily accessible, inexpensive, safe, and user friendly. The kit explained to the adult the background behind the lesson, the procedure the students should follow, and the concept that should be explained to the students following the completion of the experiment. These short, engaging, STEM based activities were a fun way to engage kids of various ages. Throughout this chapter, the five modules used will be explained more in depth along with the results found, and background research done to get to the conclusions.

BACKGROUND

As the world progresses into the 21st century, so too does the need for education embedded with modern technology and its implementation in developing countries. Within the past decade, STEM Education has become an international topic of interest because of its influence in a changing economy and workforce demands.

As a result of these global pressures, there has been a shortage of STEM-prepared workers and educators in many countries around the world (Kennedy & Odell, 2014). The need for STEM educated persons is vital for the progression and implementation of modern technology as STEM related fields explicitly involve the use of science, technology, engineering, and math. Without these fields or the further education of personnel, many countries and its citizens will fall behind in their global competitiveness.

STEM-related fields have been a focus since the late 1990s to the point where it has drawn adequate funding in the US, UK and Australia ever since. In order to understand the relation between STEM related education and developing countries, the first step would be to understand the deep history and progression of STEM itself. Each country contains its own distinct cultural background, full of unique social norms and customs. Because of this, the topics of education and schooling vary discretely from one country to another. These topics are in a constant fixation on new and evolving political, social, economic, and global connections. Drawing from these connections one can begin to see the possibilities of evolution and integration in a different context. To synthesize a new educational system in a developing society, we can first examine different areas in the world and how they incorporate an effective innovative system to achieve an effective lead on the global economy.

In the East, many Asian countries with high performing education systems and growing economies such as Korea, Japan, China, and Taiwan have established national policies revolving around science and technology more broadly with university and industry-driven research and development (Blackley & Howell, 2015). The Western STEM agenda on the other hand comprises of vocational and economic goals that are being backed up and funded by politicians to further enhance its importance and shifts in the workforce and economy (Blackley & Howell, 2015). As a result, the United States is generally believed to perform poorly in STEM education, where scholars suggest that this negative outlook on the US may largely be due to the complex nature of the U.S. educational system (Kennedy & Odell, 2014). This 'STEM crisis' resides mostly in English-speaking countries (with the exception of Canada) as quantitative indicators show that performance in international comparisons of achievement is declining and placing the nation at a lower status it should occupy. In the future, however, statistics predict that projections for Fall 2021 enrollments in grades 9-12 will show the highest increase in student population contributed largely by Hispanic/Latino students, indicating that additional bilingual fluency will need to be implemented to meet the future needs of U.S. classrooms (Kennedy & Odell, 2014). In spite of the differences, several international organizations place significant attention to STEM issues to help maintain a global focus, such as the Organization for Economic Cooperation and Development (OECD), the World Bank, United Nations Science, Education and Cultural Organization (UNESCO), the European Union (EU), and the International Association of the Evaluation of Educational Achievement (IEA) (Blackley & Howell, 2015).

As previously mentioned, the emergence of STEM was largely contributed by the political agenda based off of vocational and economic responsibilities. This push from governments to increase the number of students vouching for STEM subjects in senior level STEM courses has been a posing challenge for educators. The main challenge has been due to the act that teachers default to the notion of S.T.E.M. The focus should rather be on STEM. By this, it is meant that teachers focus on sciences, technologies, engineering and mathematics as individual subjects rather than the idea of incorporating them as a whole. Studies have found that teachers often ignore the importance of technology and engineering to focus on science and mathematics. This is largely due to the facts that science and mathematical teachings align with their traditional teachings and recent inclusion of technology without proper guidance. The educational system lacks a good guide in order to assist the teachers with their lessons. Blackey had stated that "In schools S.T.E.M. is enacted as S.t.e.M.; while outside schools (i.e. vocational settings) it is enacted as s.T.E.m." (Blackley & Howell, 2015). This suggests that the focus is indeed on science and mathematics within the traditional classroom. The disconnection between schools and vocational settings over the enactment of STEM and its definition represents a huge crisis point in the STEM curriculum. Another crisis point involves the skill level and preparation of teachers, where during initial teacher education programs, there are discrete subjects that specifically cover two of the subject areas: science and mathematics (engineering is excluded since there is no curriculum document to teach it, hence no obligation to teach it). These programs must meet national guidelines in order to achieve accreditation; as a result, only a minimum level of S.t.e.M. skills are being acquired by graduate teachers. To keep up with the standards and guidelines of STEM, a change in the accreditation processes and guidelines will need to outline new learning objectives. Another possible solution would to be reinvent the initial teacher education programs such as a move to postgraduate qualifications (Blackley & Howell, 2015).

The final process in STEM education is the move towards integration. The concept of integrated STEM education is the switch between the teaching of two or more STEM subjects or between a STEM subject and a non-STEM subject such as the arts throughout the class. This suggests that integrated STEM education could be based on two or more STEM subjects instead of having all four subjects be included. This application is more reflective of real-world scenarios that utilize STEM, and is also more manageable and logical from both a didactic and comfortable perspective. This concept was refined further by Moore and Smith (2014) who described two ways to integrate STEM education:

1. **Context Integration:** Where engineering design is recognized as a motivator to teach mathematics and science discipline content, and
2. **Content Integration:** Where engineering skills are part of the learning objectives and mathematics and science content are incidentally developed.

Since engineering skills generally are not included in the school curriculum learning objectives, the notion of context integration has been seen to be the most feasible method and likely to succeed in schools (Blackley & Howell, 2015). The benefits of having contextual integration involve new opportunities for students to develop the 21st century skills of: adaptability, complex communication, social skills, non-routine problem solving, self-management, self-development, and systems thinking (Blackley & Howell, 2015). This then poses the question of how can one include engineering in STEM education? This is one of many obstacles when referring to a STEM-driven curriculum, as it is difficult to be inclusive in addition to the disconnection between traditional schooling and vocational education and the initial teacher education programs.

With the framework of STEM being established, it provides a roadmap to achieving STEM in schools. From a global standpoint, there are many different approaches to STEM, which is largely contributed by the differences in culture and society. The four geo-social realms we can observe include:

1. English-speaking,
2. Western European countries,
3. Asian countries, and
4. Developing countries.

The English-speaking realm was discussed previously where Australia was used as a model for implementing STEM into schools via an integrated approach. Western-European countries can also follow along these same guidelines as English-speaking models. Many Western European countries, such as France and Germany, have already placed great emphasis on STEM for some time as part of the framing of national policy on education and industry (Blackley & Howell, 2015). For the Asian countries, national policies around science and technology have already been established more broadly and have been recognized for their excellence in the teaching of these subjects (Blackley & Howell, 2015). Developing countries can benefit greatly from the integrated STEM curriculum, where participation in basic education and the development of a qualified teaching workforce can be improved (Blackley & Howell, 2015).

We can see from the history of STEM that there are many challenges and obstacles to overcome for an integrated STEM education to be successful. As teachers and students are subject to rigorous standardized testing where the results greatly impact the school's funding, image, and teacher performance pay, priority will be allocated to the subject areas being tested the most: mainly mathematics and literacy (Blackley & Howell, 2015) This poses as a challenge because the embracement and implementation of STEM related education is not feasible at first glance largely due to the fact that integrated STEM education and preparation for standardized testing are not mutually exclusive (Blackley & Howell, 2015). High financial costs and personal commitment are also difficult factors to overcome as retraining teachers and adapting to new STEM integrated education programs will require great effort and time. Furthermore, there has been a small amount of evidence and concrete proof where the positive outcomes of STEM integration are not readily available for teachers to be encouraged to consider this form of education. Conceptually, it may take as long as 12 years after implementation—the time it takes for a student to complete a high school level of education, for results to be readily accessible (Blackley & Howell, 2015). Once a student completes their secondary-level of education, positive outcomes and success of STEM education will largely draw from the level of parental involvement and support, availability of bilingual education, culturally relevant pedagogy, early exposure to STEM fields, interest in STEM careers, self-efficacy in STEM subjects, and STEM-related educational opportunities and support programs (Kennedy & Odell, 2014). For STEM education to really show its benefits, an amendment to the traditional curriculum of K-12 education will need to be considered to further progress a nation into this competitive global economy.

MAIN FOCUS OF THE CHAPTER

The first step in the engineering design process was to identify the needs and constraints of the project. There was an obvious need to encourage critical thinking at a young age, especially in girls, in order to provide a pathway to a better future in engineering. The second step in the design process was to research the related issue. There were many articles that lead to the same conclusion of women in the engineering field was lacking, not just in Panama but also in the world as a whole. In order to help encourage children to think like engineers in a critical manner, different activities were researched. These activities were to be implemented on the ground in Panama for the school. They were engaging, scientific activities in order to assess the thought processes taken by the school students. After researching different activities, the team chose and created six lesson plans that focus on different aspects of engineering. These plans include an orange buoyancy experiment, testing the strength of paper, soil erosion exploration, simple circuit creations, assembly of a simple electric motor, and testing spatial skills with Legos. The fifth step in the engineering design process was to build a prototype. For this, the team created sample lesson plan activities to see the ease of creating them. Next, these plans had to be tested and evaluated. This was done at a local elementary school. The students were to complete the activities just as Panama students would, but the Team kept a close eye on the progress in order to assess the design. Errors found were improved upon with iteration and the process repeats.

When selecting the design criteria, the needs of the project must be taken into consideration. The needs include encouraging engineering at a young age by engaging critical thinking skills in a fun manner. Some constraints that come with the project include the language barrier between the team and the Panamanians, restricted funds to continue in Panama, and the project must be simple enough for young ages. With this in mind, different activities to promote thinking like an engineer were researched. These designed lesson plans were geared toward students in elementary to middle school. They included different aspects from differing fields of engineering including civil, mechanical, electrical, and geotechnical.

The following recommendations are intended for high school STEM educators. Each curriculum is required to engage students in STEM-based instructional activities that challenge students to innovate and invent. By doing so, the student has demonstrated their ability to apply science and mathematics-related knowledge they learn to an engineering problem and utilize technology in finding a solution. To accomplish this, the teachers need to be equipped with innovative instructional tools for promoting applied and collaborative learning. Technology also needs to be integrated into the culture, curriculum, teaching strategies, and daily operations of classrooms as well so that they can both enhance learning and provide relevance towards the students. For the team's project, a model curriculum for a high quality STEM education was found and credited to the International Journal for Science Education International (2014).

High quality STEM education programs and curricula should reflect the following features:

- Include rigorous mathematics and science curriculum and instruction;
- At a minimum, (if separate STEM courses are not available in all areas) integrate technology and engineering into the science and mathematics curriculum;

- Promote engineering design and problem solving—(scientific/engineering) the process of identifying a problem, solution innovation, prototype, evaluation, redesign —as a way to develop a practical understanding the designed world;
- Promote inquiry—the process of asking questions and conducting investigations—as a way to develop a deep understanding of nature and the designed world (NSTA, 2016);
- Be developed with grade-appropriate materials and encompass hands-on, minds-on, and collaborative approaches to learning;
- Address student outcomes and reflect the most current information and understandings in STEM fields;
- Provide opportunities to connect STEM educators and their students with the broader STEM community and workforce;
- Provide students with interdisciplinary, multicultural, and multi-perspective viewpoints to demonstrate how STEM transcends national boundaries providing students a global perspective;
- Use appropriate technologies such as modeling, simulation, and distance learning to enhance STEM education learning experiences and investigations;
- Be presented through both formal and informal learning experiences;
- Present a balance of STEM by offering a relevant context for learning and integrating STEM core content knowledge through strategies such as project-based learning.
- K–16 teachers of STEM, school and district leaders, community college and university STEM leaders, and other key stakeholders should embrace the following key points:
- Teachers of STEM should recognize the compelling and inherent opportunities of STEM to strengthen and support the teaching of STEM education, and, where possible, integrate STEM applications into the curriculum.
- Teachers of STEM should seek out and participate in quality professional development opportunities to enhance their knowledge of STEM and its application in meeting curricular requirements, and to gain exposure to practicing STEM professionals.
- Teachers of STEM should locate and use quality resources from STEM organizations to enhance and strengthen their curricula.
- School administrators and principals should support teachers in their efforts to integrate STEM within science and mathematics curricula.
- Collaborations among stakeholders in education, government, business, the community, and the media should be encouraged to coordinate the development and availability of STEM educational resources. STEM as an interdisci-

plinary discipline requires that pedagogical approaches must be altered from traditional approaches to support student learning.
- STEM educators must implement instructional strategies that integrate the teaching of STEM in a way that challenges students to innovate and invent.
- STEM educators must use problem-based and project-based learning with a set of specific learning outcomes to support student learning.
- STEM educators must create meaningful learning opportunities provided context learning is delivered using applied and collaborative learning.
- STEM educators must require students to demonstrate their understanding of these disciplines in an environment that models real world contexts for learning and work.
- STEM educators must provide students with interdisciplinary, multicultural, and multi-perspective viewpoints to demonstrate how STEM transcends national boundaries providing students a global perspective that links students with a broader STEM community and workforce.

From this model, the team was required to carefully follow this curriculum, unless any additions or changes were deemed necessary. On the ground, the team was only able to address some of these criteria as it is not possible to accomplish every provision of this curriculum or design in such a short period of time. The goal was to introduce these concepts in an effort to encourage young men and women to pursue interest in STEM related programs throughout primary and secondary schools.

SOLUTIONS AND RECOMMENDATIONS

The team's focus within this chapter was on diversity, gender equality, and the empowerment of women. In order to achieve this goal, STEM research was implemented within the design of the project. The overall idea created by the team was to incorporate more STEM related activities and lessons for children at a younger age in order to spark their interest within the related fields. With more critical thinking and difficult activities, it would stimulate them to work harder and think in a more complex manner. To achieve this, the team developed five different modules that involved different areas of science, technology, engineering and math. The modules included an orange buoyancy module, a strength of paper module, a soil erosion module, a simple circuit module, and a simple electric motor module. These modules were tested in a local elementary school, Washington Elementary, along with an elementary school in Panama. Due to the language barrier, statistical values were not assigned to the testing done in Panama, but instead observations were made within the classroom. From surveying done in Washington Elementary, of the male students 92% had a favorite course in STEM, while the females had 50% of STEM related courses as their favorite. This further proved the need to empower women and educate them within STEM related classes and fields. A recommendation would be to further improve the questionnaires and lessons in order to have more in depth results and meaning to the modules. The students tend to get overwhelmed, so it'd be best to make these as clear cut as possible. The lesson plans for the teacher along with student instructions for each module can be seen in the Appendix.

FUTURE RESEARCH DIRECTIONS

The possibility of more modules is ever growing. Within society, the team would like to promote critical thinking and higher education in the STEM fields. With this, many sustainable lessons would be taught and over time societies education and views would expand. In order to assess the impact made, in the future the country would have to be studied again. Students loved the fun and didactic activities and the continuation of this would be beneficial to the future generations that will someday run the world. Research is in the works on developing differing activities, ways to empower and encourage the fields of STEM in all ages. Top institutions including Purdue, the Ohio State University, and the University of Florida have specific funding focusing on developing research in education for engineering.

CONCLUSION

Five STEM modules were designed and tested within two schools by the Mount Union students. From surveying done in Washington Elementary, of the male students 92% had a favorite course in STEM, while the females had 50% of STEM related courses as their favorite. This further proved the need to empower women and educate them within STEM related classes and fields. With the rise in interest in an early age, and confidence that develops over time, the rates of diversity within engineering should increase. This is crucial in order to change the future for kids that will someday be the backbone of society.

REFERENCES

Blackley, S., & Howell, J. (2015). A STEM Narrative: 15 Years in the Making. *Australian Journal of Teacher Education, 40*(40), 102–112. doi:10.14221/ajte.2015v40n7.8

Georgia State University. (n.d.). *Pressure - Hyperphysics*. Retrieved May 21, 2016, from http://hyperphysics.phy-astr.gsu.edu/hbase/pbuoy.html

Henderson, T. (n.d.). *The Physics Classroom - What Is an Electric Circuit?* Retrieved May 21, 2016, from http://www.physicsclassroom.com/class/circuits/Lesson-2/What-is-an-Electric-Circuit

History Channel. (n.d.). *Dust Bowl*. Retrieved May 21, 2016, from http://www.history.com/topics/dust-bowl

Jenkins, S. (2015). *Electricity*. Retrieved May 21, 2016, from https://prezi.com/3f91mznpuew3/electricity/

Kirkwood, J. (2015). *STEM Activities for Kids: How strong is a piece of paper?* Retrieved March 2, 2016, from http://creeksidelearning.com/stem-activities-for-kids-how-strong-is-a-piece-of-paper/

Lappapadolce. (n.d.). *Science experiment on soil erosion*. Retrieved March 30, 2016, from http://www.lapappadolce.net/en/science-experiment-on-soil-erosion-2/5/

Moore, T. J., & Smith, K. A. (2014). Advancing the state of the art of STEM Integration. *Journal of STEM Education, 15*, 5–10.

National Geographic. (n.d.). *Deforestation Facts, Deforestation Information, Effects of Deforestation*. Retrieved March 31, 2016, from http://environment.nationalgeographic.com/environment/global-warming/deforestation-overview/

National Science Teachers Association (NSTA). (2016). *Principles of Professionalism for Science Educators*. Available from http://www.nsta.org/docs/PositionStatement_Professionalism.pdf

Plant and Soil Sciences eLibrary. (n.d.). *Water Erosion*. Retrieved May 21, 2016, from http://passel.unl.edu/pages/informationmodule.php?idinformationmodule=1086025423&topicorder=18&maxto=7

Woodford, C. (n.d.). *Electric motors*. Retrieved March 31, 2016, from http://www.explainthatstuff.com/electricmotors.html

APPENDIX

1.1 Orange Buoyancy: Teachers

The purpose of this lesson is to teach students about buoyancy. Buoyancy is defined by Wikipedia as "an upward force exerted by a fluid that opposes the weight of an immersed object" (Georgia State University, n.d.). In other words, it is the inclination of an object to sink or float in a fluid, such as water.

1. Collect the following materials: one orange, one clear large cup or bowl, and water to fill the container.
2. Ask students if they think an orange will sink or float and why.
3. Pour the water in the cup until it is about ¾ full.
4. Place the unpeeled orange in the water to find out if it sinks or floats.
5. Once the students determine that the unpeeled orange does not sink, have the pupils unpeel the orange.
6. Ask the students if the unpeeled orange will sink or float and why.
7. Place the peeled orange in the water to find out if it sinks or floats.
8. The pupils should determine that the peeled orange sank.

The unpeeled orange floats due to the peel surrounding the inside of the orange. The peel traps the air inside and when the orange is placed in the water, the air bubbles float to the top. The flesh orange inside the peel is denser than the water which would make you think that the orange would sink. However, the air bubbles inside the peel are lighter than the water and causes the orange to be buoyant Density means that the object weighs more for its size (Georgia State University, n.d.).

1.2 Orange Buoyancy: Student

What is buoyancy? (Circle Answer)

a. A fruit
b. The ability of something to sink or float in a fluid
c. An object used in rivers or lakes
d. A material used to make a cup

1. Collect the following materials: one orange, one clear large cup or bowl, and water to fill the container. Will the orange sink or float in the cup of water? (Circle Answer)
2. Pour the water into the cup until it is a little over halfway full
3. Place the orange into the water. Did the orange sink or float? (Circle Answer)
4. Unpeel the orange. Will the unpeeled orange sink or float? (Circle Answer)

5. Place peeled orange into water.
6. Did the peeled orange sink or float? (Circle Answer) What is buoyancy? (Circle Answer)
    e. A fruit that you eat
    f. Something that is in a lake or river
    g. The ability of something to sink or float in a fluid
    h. A material used to make a cup

1.3 Design Analysis of Orange Buoyancy Module

The buoyancy module tested the concept of the floatability or buoyancy of an orange when it was peeled and unpeeled. In Washington Elementary School before the module, it was found that 64% of the students thought the orange would float. When asked why, they did not have a true answer. Most students just said it was a guess. In Panama the majority of the students guessed the orange would sink. Once the unpeeled orange was placed into the water, all faces lit up with excitement. After the physical testing of the peeled orange was done, all learned the definition of buoyancy. For the peeled orange, 57% of the Washington students guessed it would sink. The 43% of the students that believed it would float gave reasons based on the fact that it had previously floated. The Panamanian students had the same results of guesses. After the orange was placed into the water, all students were taught that the orange floated due to air pockets within the peel and sank because of the density of the orange without the peel. They also learned that apples, pears, and other fruits with more air than liquid and the thinner peel will always float, even when cut in half.

2.1 How Strong is a Piece of Paper: Teacher

The purpose of this lesson is to teach students about how the shape of a column affects how much weight it can hold. This lesson will test the strength of paper columns in the shape of a rectangle, triangle, and circle by placing books on top. By loading the structure on top, it is similar to how buildings or floors are supported by columns.

1. Collect the following materials: tape and paper.
2. Ask students whether they think the rectangular, triangular or circular shaped column will hold the most number of books and why.
3. Have the students construct a rectangular shaped column using a piece of paper and tape.
4. Count the number of books that the rectangular shaped column can hold before collapsing.

5. Have the students construct a triangular shaped column using a piece of paper and tape.
6. Count the number of books that the triangular shaped column can hold before collapsing.
7. Have the students construct a circular shaped column using a piece of paper and tape.
8. Count the number of books that the circular shaped column can hold before collapsing.
9. The students should determine that the circular shaped column can hold the most number of books.

The circular shaped column can support more books than the triangular and rectangular shaped column because it has no edges. Sharp edges are not ideal because the weights of the books are concentrated at sharp corners. That is why the triangle and rectangle deform easily and collapse with less weight being applied. For the circular column, the weight of the books is evenly distributed, which is why it can hold the most books. The entire wall of the circular column is supporting the weight of the books (Kirkwood, 2015).

2.2 How Strong is a Piece of Paper: Student

1. Collect the following materials: tape and paper. Do you think the rectangular, triangular or circular shaped column will hold the most number of books? (Circle Answer)
2. Construct a rectangular shaped column using a piece of paper and tape.
3. Count the number of books that the rectangular shaped column held before collapsing.
4. Construct a triangular shaped column using a piece of paper and tape.
5. Count the number of books that the triangular shaped column held before collapsing.
6. Construct a circular shaped column using a piece of paper and tape.
7. Count the number of books that the circular shaped column held before collapsing. Which column held the most number of books? (Circle Answer) Rectangular Triangular Circular

The cylinder can support the most books because its walls do not have any edges. The force of the books cannot become concentrated in a particular area. The load is distributed evenly. In other words, all parts of the cylinder are sharing the weight of the books. All parts of the cylinder, therefore, contribute to its overall strength until, finally, it collapses. The square and triangle deform more easily. They shift the weight of the books to their edges and corners, which deform their walls and leads to a quick collapse.

2.3 Design Analysis of Structural Module

The structural module was a test to show the strengths of paper when designed into different shapes. This related to structural engineering and the design of buildings, supports and other column dependent designs. In Washington Elementary, 36% of the students guessed that the circular paper would be the strongest. One student had a reason being that it is what buildings have in the front (column posts). The other students had no reason for their choices. In Panama, the classroom of 3rd graders had a majority of circular paper being the strongest. They knew the reasoning why. As one student put it, "it holds more mass". These results were found to be overwhelming better than the United States students. In the younger grades, the concept wasn't as clear but the older all knew the correct results. They were able to stack 6 books onto the circular paper structure, 3 on the rectangular, and 2 on the triangular. This activity really engaged the students and they were thrilled to keep testing and destroying paper structures to see just how much they could get it to hold. The concept of the less points a structure had, the less failure would occur was taught. They related the concepts to the building of bridges, desk chairs, columns in front of a house, and other structural engineering designs. After testing, all students learned the concepts that were taught for this module.

3.1 Soil Erosion: Teacher

This lesson is effective because it visually shows the importance of vegetation and it also very simple to perform. It also demonstrates the effect that deforestation has on soil erosion. Removing vegetation and heavy rainfall have some very serious effects on the environment.

The materials needed for this lesson include: 6 identical water bottles, string, scissors, hot glue, flat wood, marker, and a plastic cup. In order to ensure students safety, pre-cut three of the water bottles by removing the upper portion of the bottle when it is laying horizontal. The following picture is an example:

Use hot glue or super glue to attach the three bottles to a small piece of wood. Leave the cap on the bottle and Put in each bottle the same amount of dirt and press hard to pack as much as possible. The ground must be below the level of the opening of the bottle. Cut the bottom of other three water bottles and make two holes for the string. These cups will be used to collect the excess water during the experiment, the water in excess. Place a plant (grass would work well) in one of the bottles and in another bottle place leaves and sticks. The third bottle will only contain soil (Lappapadolce, n.d.).

1. Collect the water bottles glued on to the wood and attach the water collection cups.
2. Ask students which soil sample they think will have the least amount of soil reduction and why.
3. Have a student pour the same amount of water into each bottle at the same point (preferably close to the end furthest from the opening). To ensure that the students pour an equal amount of water into each soil sample, make a mark using the marker inside a cup as a fill line.
4. The students should notice that the collection cup for the soil sample with the plant is the clearest compared to the other soil samples.

Deforestation is a very serious problem around the world. At the current rate of deforestation, the forests could be depleted in 100 years. The most devastating effect of deforestation is the destruction of animal habitats. This is a serious problem because about 70% of the animals that live on land, live in forests and if the forests are gone then many of these animals will be unable to survive. Deforestation also leads to erosion. Soil erosion can pollute water sources and contaminate clean water. This is harmful to the animals in the water and the people who get their water from the contaminated sources (National Geographic, n.d.).

Vegetation is also important because it holds the soil in place and prevents erosion (Plant and Soil Sciences eLibrary, n.d.). The Dust Bowl occurred in the 1930's because of harsh farming. Farmers removed the vegetation from the land and failed to take proper steps to prevent wind erosion. The Dust Bowl killed thousands of livestock and forced families to leave their homes (History Channel, n.d.). Erosion is still a big problem and engineers are innovating new ways to prevent erosion.

3.2 Soil Erosion: Student

Which soil sample do you think will have the least amount of soil reduction and why?

Pour the same amount of water into each bottle filled with soil at the same point (preferably close to the end furthest from the opening).

Which collection cup is the clearest? Why?

3.3 Design Analysis of Soil Erosion

Because of lack of time to properly complete this module, it was not tested at either of the collaborating schools. Future testing will be implemented to see the results of the students' intuition on the subject of soil erosion. This module was specifically chosen in order to teach international students about the dangerous effects of removal of trees and grasses for farming specifically in areas that are prone to landslides.

4.1 Simple Circuit: Teacher

This lesson is a great way to teach students the basics of electricity and how circuits work. The best part of this lesson is that there is no soldering, it uses materials that can be easily found, and it is not expensive. Students could take this lesson home and have fun with parents building their own circuit.

4.2 Simple Circuit: Student

1. Collect the following materials: Cardboard, white LED, blue LED, 3 Volt Button Cell Battery, 18 Metal Brads, 13 Paperclips, and electrical tape.
2. The cardboard already has holes in it for location of the items.
3. Place a brad in the bottom left hole.
4. Put the paperclip on the metal brad and place the metal brad in the bottom right hole as shown.

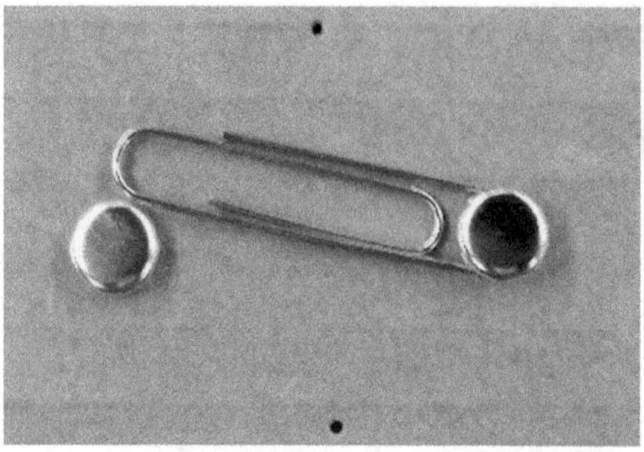

5. Flip the cardboard over and place a paperclip over the left and right brads. Flatten the wings on each brad to hold the paperclip and brads in place.

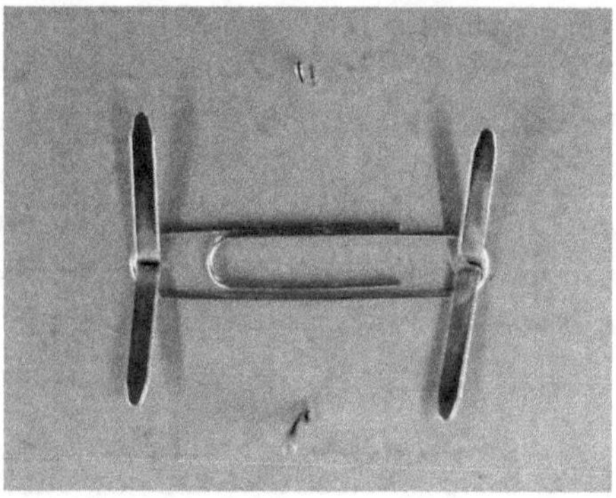

6. Tape the brads down as shown.

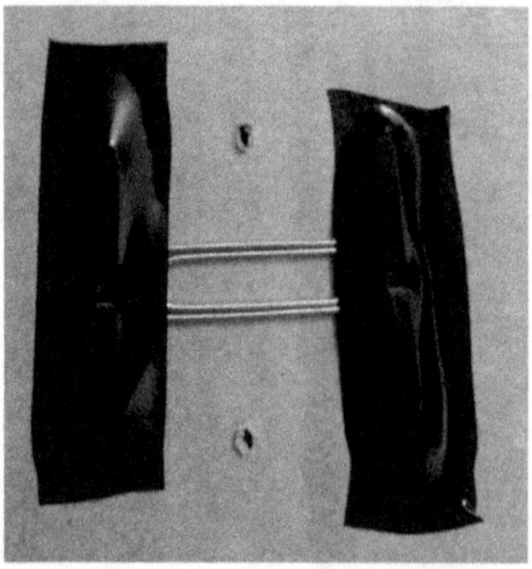

7. Flip the cardboard over and push a brad through the bottom. Place a brad through a paperclip and then through the top hole as shown.

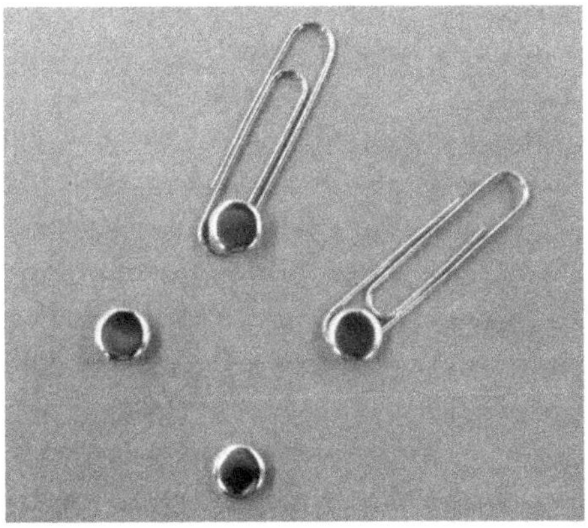

8. Flip the cardboard to the back and place the 3V battery on the paperclip with negative end facing down and positive facing up.
9. Place a paperclip over the top and bottom brads as shown.

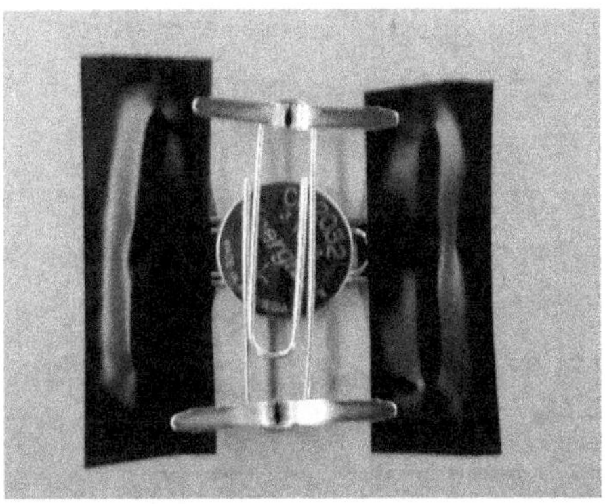

10. Tape the brads down as shown.

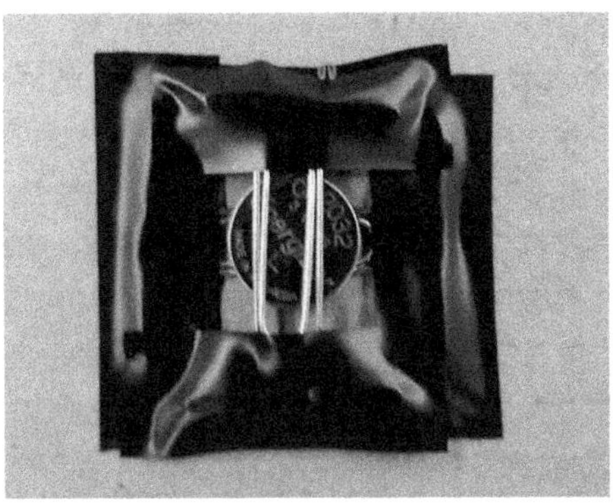

11. Next, place a brad through a paperclip and push it into the hole above the battery as shown.

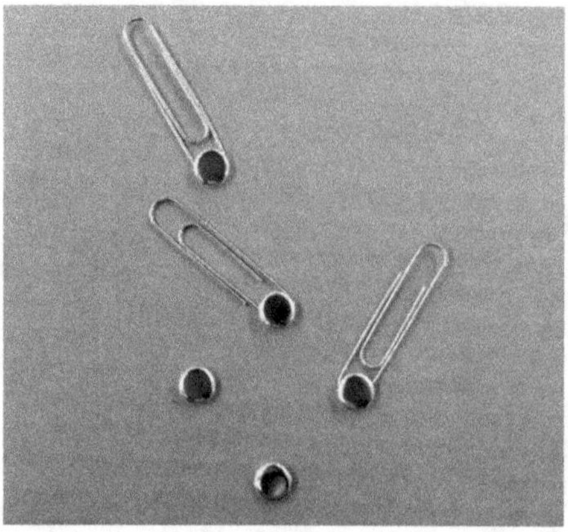

12. Place brads and paperclips in the cardboard as shown below. In the upper left corner, push the Light Dependent Resistor (LDR) through the hole.

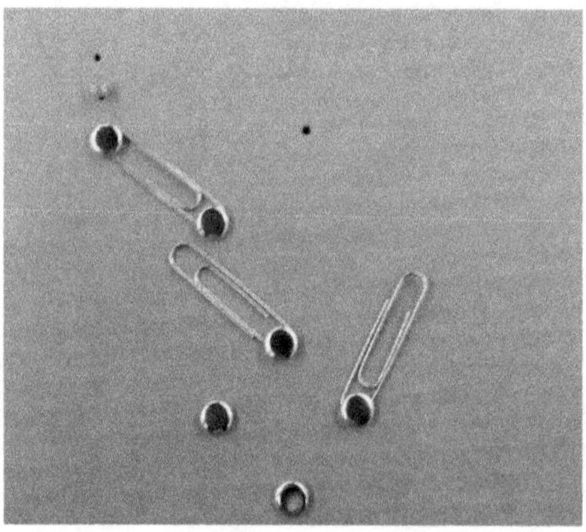

13. Turn the cardboard to the back and wrap the legs of the LDR around the flattened brad wings as shown.

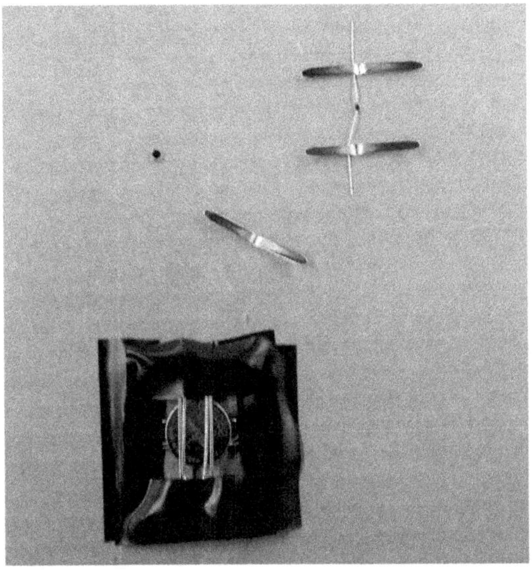

14. Flip the cardboard to the front and place the brads and paperclips as shown below.

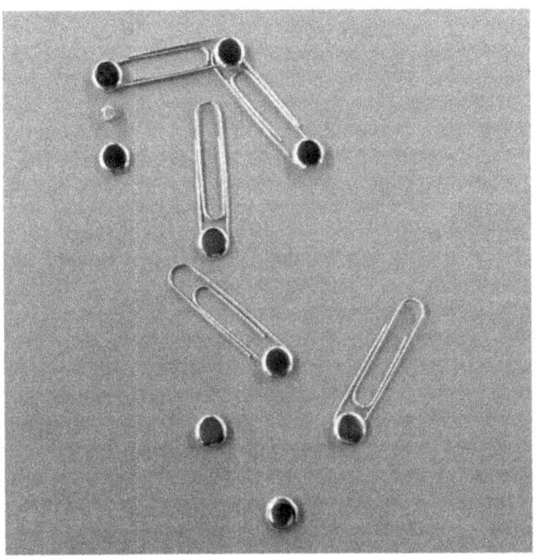

15. Flip the cardboard over and flatten the wings of the brads.
16. Flip the cardboard to the front and place the brads and paperclips as shown below.

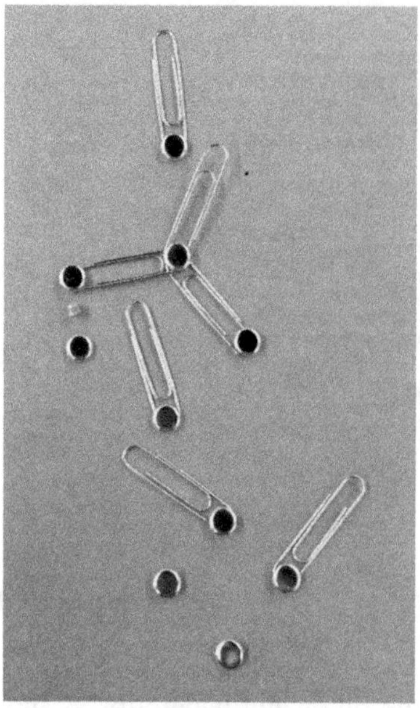

17. Flip the cardboard to the front and place the brads and paperclips as shown below.

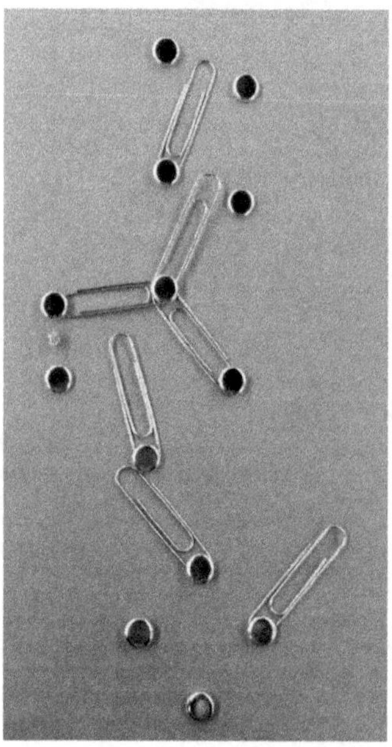

18. Push the LEDs through the hole shown below. The longer leg is the positive end and should be facing the left. And the short leg should face the right. Also, place the brads and paperclips through the holes as shown.

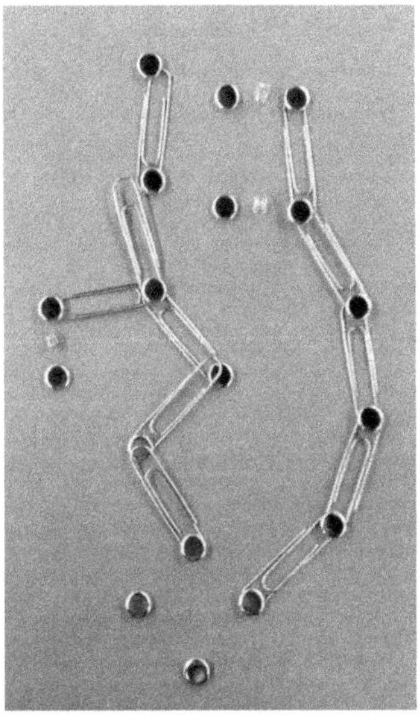

19. Flip the cardboard to the back. Be sure that the longer leg of the LED touches the brad to the right and the shorter leg of the LED touches the brad to the left. Flatten the brads as shown.

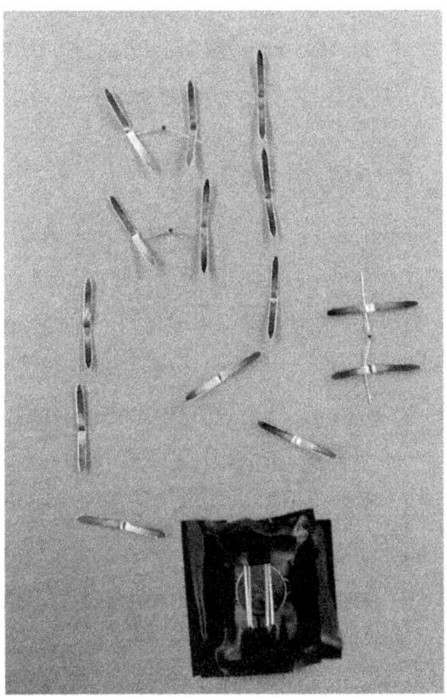

20. Test to see if your circuit works by setting the paperclips as shown. If the LED does not turn on ask your teacher for help.

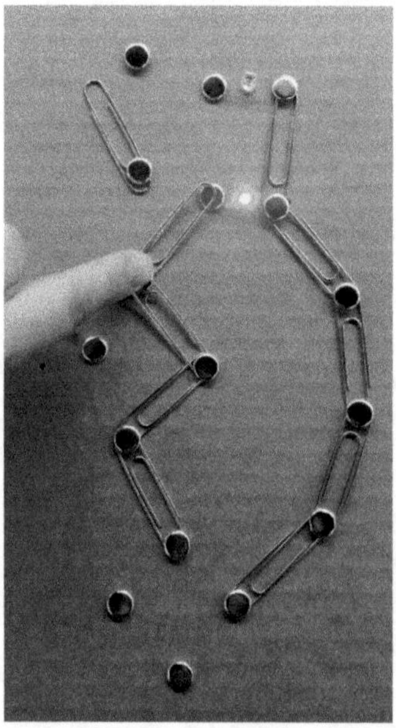

Electrical circuits are important because it is found in items we use every day (Jenkins, 2015). Circuits are in many of items such as cell phones, televisions, and all of other electrical devices. An important component to consider is the fact that the bulbs and buzzer need to be able to handle the voltage of the battery. If they are not rated for the voltage of the battery, then a resistor will be needed. A resistor only allows a certain amount of voltage to go through it. If the resistor is not in place, a bulb will burn out when too much voltage is run through it (Henderson, n.d.).

4.3 Design Analysis of Simple Electric Circuit

The simple electric circuit was tested in Panama and Washington Elementary. In Washington Elementary, the students were given the module packets and kits, and were allotted time to build the circuit as a team. This went well for the older kids, but the younger students needed much assistance. This testing aloud the team to improve upon the directions given to students in order for them to be able to complete the circuit with ease. The fourth and fifth grade students excelled at this module due to previous lessons taught in class. One particular student grabbed the battery and the light, connected them and created a current through the bulb. His classmates were amazed. In Panama, pre-made circuits were given so the students could experiment with them, without wasting too much classroom time. The concepts of negativity and positivity were taught.

5.1 Simple Electric Motor: Teacher

The kit that will be bought is The World's Simplest Motor™. Students will assemble this kit following the instructions provided. When assembled, the coil rotates freely above the magnet, which demonstrates a simple electric motor. This can be conceptual understood by allowing the students to play with two magnets. They can see how the magnets repel each other in some instances and how it sometimes is attracted. This is based on the north and south poles of the magnet, where the opposite poles attract while the same poles would repel. This is essentially what is happening in the simple electric motor. A current is passed through the copper wire by the battery which produces a magnetic field in the copper. The magnetic field acts like a magnet, with a north and south pole. The magnet is used to repel the copper "magnet" which causes the copper coil to spin away from the magnet or be pushed away. The copper coil continues to spin because the copper is scraped off the one side in order to prevent a current from flowing, which allows the coil to continue its spinning motion. Once the copper comes into contact with the metal again, the current is pushed through the wire and the magnet repels the coil. This process is repeated which is how the wire is able to continually spin similar to how a motor works (Woodford, n.d.).

5.2 Simple Electric Motor: Student

1. Collect the following materials:

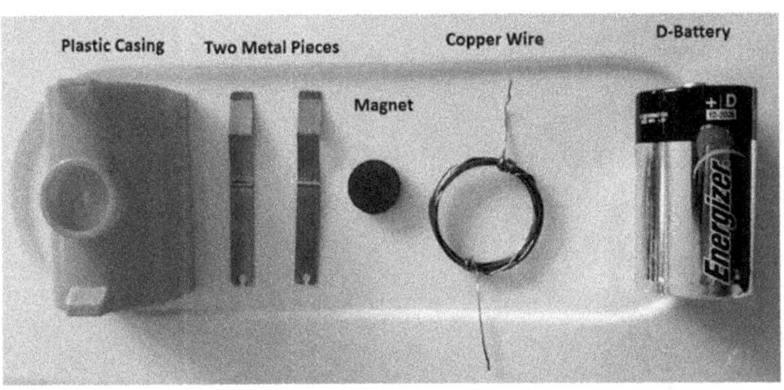

2. Slide the two metal pieces (U-end first) up through the slots in the plastic casing through the bottom. The bump in the metal pieces should face towards the battery as shown. Push the battery in the bottom so that the battery touches both metal supports as shown.

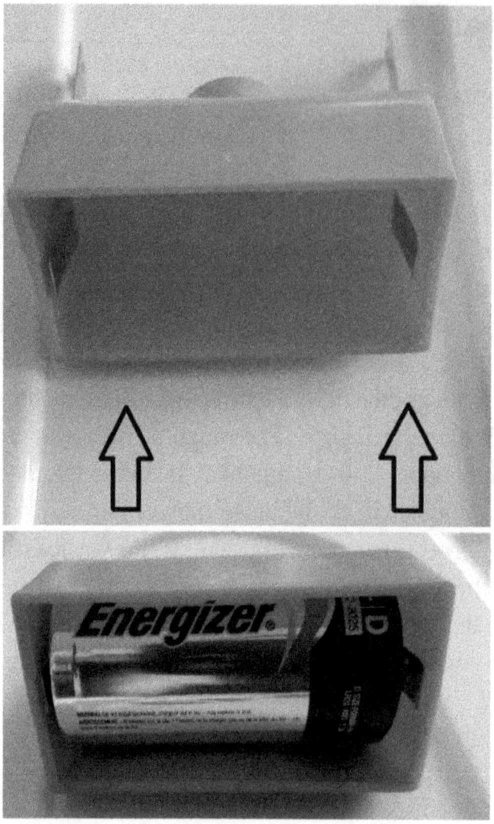

3. Put the magnet in the circle on the top of the plastic casing and place the copper wire on the metal supports as shown.

(Give the copper wire a gentle spin, if it does not continue spinning try flipping the coil the other direction or ask an assistant for help.)

5.3 Design Analysis of Simple Electric Motor

This design allowed the lesson of magnetism to be taught. No pre-questions were given during the experiment due to the complexity of a motor. After testing, the students were asked questions to assess how much they learned from the module on the subjects of magnetism and motors. 100% from the Panamanian and 79% Washington Elementary students were able to answer correctly. They enjoyed the experiment and were good at problem solving when the motor didn't spin properly.

Chapter 5
Developing a Learning Community of Engineers Through an Honors First-Year Seminar

Melissa L. Johnson
University of Florida, USA

Kristy Spear
University of Florida, USA

ABSTRACT

First-year engineering majors face a myriad of obstacles as they begin college. Taking challenging foundational coursework, navigating new expectations for performance and experience, and understanding the broader impact of their academic interests are just a few of those obstacles. In addition, female students sometimes face additional barriers to success, particularly as some question their own competence in the field. This chapter focuses on a first-year seminar for honors students that highlights the high impact practices (Kuh, 2008) students should participate in throughout their undergraduate career. These practices include global engagement, undergraduate research, and internships that are essential for early exposure to future career interests. By developing both formal and informal learning communities within the seminar (Johnson, Pasquini, & Rodems, 2013), first-year students are exposed to opportunities, mentoring, and support that help them make informed decisions about their major and career.

DOI: 10.4018/978-1-5225-2212-6.ch005

INTRODUCTION

The purpose of this chapter is to highlight a first-year seminar course designed for honors students. With a large proportion of first-year honors students majoring in engineering, the Honors Program has developed outreach opportunities adapted from their first-year professional development seminar. By creating specific course sections for STEM majors in general, as well as for engineers, first-year students are able to develop a learning community with other like-minded, high achieving individuals.

It is important to understand the challenges facing any first-year engineering major – major and career decision making, self-efficacy, and other aspects of career and professional development are all explored. In the case of the Honors Program, additional barriers have been discovered among female engineering majors. Data about first-year retention and persistence among honors first-year engineering majors as compared to the general undergraduate student population, are provided to highlight these barriers to success. Finally, components of the first-year seminar are described in detail, including exposure to relevant opportunities in the field which help students make more informed decisions about their future plans.

BACKGROUND

Major and Career Decision Making

The process of choosing a major and career can be difficult for many students. Even after choices are made, decisions have a tendency to waver. This challenging process is often influenced by a number of factors. External factors including familial pressure and the current the job market can impact choices as much as internal factors which may include interest, ability, knowledge of the career options, and self-efficacy. Several studies have tied the career decision-making process back to presumed financial reward, recognition, sense of social purpose, flexibility, and intellectual stimulation (Xu, 2013). Major and career decisions can also be influenced by a student's ability to feel that he or she is achieving a set of desired goals. These goals generally relate to academics or knowledge of the skills needed to be successful in a certain career. With all of these variables in play, major exploration and the uncertainty surrounding the real-world implications for career choices confound many undergraduate students.

Limited knowledge of career options leave many students feeling ambiguous about major choices. Even students who declare majors that seemingly correlate with careers often lack critical knowledge of the range and extensiveness of the career choices available. This particular issue presents often with engineering students, as they narrowly interpret the profession and the options that exist after graduation (Singer et al., 2015). Providing engineering students with resources, support, and information about potential careers in a community setting during the critical first-year is one way that colleges and universities can help address this uncertainty regarding major and career.

The transitional first year of college is difficult for many students, and the pressure to make an immediate and absolute decision about majors and careers can cause additional stress. For many students, the major decision-making process extends well beyond the first year. This exploratory process can be ongoing. For engineering students, additional support during the first year can build confidence in the choice of major and provide evidence of career options that may be a good fit. With a lack of qualified workforce and a major push by the United States government to increase the number of students graduating with degrees in science, technology, engineering and math (STEM), particularly women and racial and ethnic minorities (Olitsky, 2014), faculty and staff must work to reach out to these particular students as early as possible. Providing support and mentoring for first-year engineering students through community building and high impact practices can build confidence in major choice, supply resources that a student may access if he or she is struggling, and help to retain more students in the field and the profession.

Supporting and Educating Engineers

Despite repeated efforts to increase enrollment and graduation rates, the current workforce of engineers, both in industry and academia, is struggling (Jones et al., 2010). While the lack of suitable candidates remains a concern in the United States, and an issue many are trying to address, the diversity of those candidates' is even more alarming (Jones et al., 2013). Women, for example, who account for 55% of the population of bachelor's degree recipients, are still the minority in STEM fields (Xu, 2013). Racial and ethnic minorities are also severely underrepresented in the engineering majors and in the field. A number of federal and national commissions have fought to increase policies, funding, and initiatives to expand the workforce in science, technology, engineering and mathematics through educational investments that seek to reach a broader portion of the population interested in pursuing these fields (Byars-Winstron, 2013). To further combat this issue, a commitment to address this inadequacy is necessary at the institutional level. Colleges and universities must provide enhanced academic and enrichment opportunities for students interested in and pursuing an engineering education early in the student's academic career.

By large, the curriculum for a first-year engineering student focuses on the foundational math and science courses necessary to understand the upper-division engineering curriculum. The in-depth coursework that is relevant to the specific branch of engineering that a student is studying is not generally taken until after a solid foundation has been built and the student is further along in his or her undergraduate education. There have been several studies that have found the disconnect between curricular content and professional practice to be the primary factor for not attracting students into engineering majors (Singer et al., 2015); this is also an issue from a retention perspective. For many individuals, this delayed start in engineering-specific courses contributes to major and career uncertainty as well as confusion. This delay can be particularly perplexing for a student who is uncertain about the field or his or her ability to succeed in the required upper-division coursework. Evidence shows that students who no longer feel passionate about a field or successful in the required coursework face the challenge of whether or not to persist with that area of study and potentially the subsequent career.

Additionally, the highly structured and regimented curriculum required for an engineering student often leaves little room for exploration inside or outside of the classroom. As a result, younger majors are at risk for wavering in their decision to stick with engineering, or at, worse, leaving the field altogether. This issue is particularly challenging for the students in underrepresented populations, who may not have the support networks available to keep them focused on the big picture.

Providing engineering students with the tools necessary to explore career opportunities and develop the skills necessary to be successful in major coursework is one way colleges and universities can address this problem. Engaging engineering students in collaborative environments and high impact strategies provides an outlet for exposure and exploration, a means to promote the student's self-efficacy, and an opportunity to grow and retain engineering majors. There are several ways for colleges and universities to provide students with exposure to the field of engineering and promote self-efficacy early on. Suggestions for such opportunities will be provided later in this chapter. In exhibiting the usefulness and societal impact of engineering as early as possible, educators can help to attract students into engineering majors (Jones et al., 2010) and create meaningful experiences to engage and retain them in this field of study.

Self-Efficacy and Career Exploration

To attract more students to engineering majors, colleges and universities must embrace a sociocognitive approach. What people think, feel, and believe affect how they behave. Unless a student believes that he or she can be successful, there is little reason to proceed (Bandura, 1986). This idea holds true for many students that may be considering an engineering major. Without resources and guidance for engineering majors and prospective students, the lack of trained engineers in the workforce will be a perpetual problem. Providing informal and formal learning opportunities for students within the field of engineering can help develop the confidence and the critical skills needed to possess a strong sense of self-efficacy, and make an informed choice about major and career.

Social cognitive theory defines self-efficacy as a person's belief that he or she can perform a task within a specific domain (Bandura, 1986). Several studies have linked self-efficacy to persistence, achievement, and interest with engineering students. As such, a number of interventions have been created and executed to address self-efficacy with this population (Jones et al., 2010). Many new students struggle with self-efficacy as it pertains to the field of engineering because of the lack of exposure early on. Both curricular and co-curricular endeavors can help with exploration and the development of a deeper understanding of the field, thus increasing self-efficacy. According to Bandura (1986), self-efficacy stems from four sources: mastery experience, vicarious experience, verbal persuasion, and physiological states. By providing students with opportunities to explore the field, mentors and role models to guide the investigation, and actionable plans for moving forward, institutions can encourage students to confidently approach the major. Social Cognitive Career Theory is often used to understand interest and persistence within engineering; it identifies three aspects of academic and career development-how those interests develop, how choices are made, and what factors affect success (Miller et al., 2015).

From a retention standpoint, self-efficacy is critical for an engineering student. As such, this area for growth should be addressed institutionally. The choices students make often revolve around the types of activities they participate in, the amount of effort exerted, and their persistence and performance with those activities (Bandura, 1986). A student's beliefs can be tied to persistence. If the student believes that he or she will be successful in a major, the student is more likely to persist for longer.

The clear need for students to possess self- efficacy is relevant for attracting and retaining engineering students in both the academic and professional realms. Self-efficacy requires reflection and a certain amount of understanding of self-ability. These processes and the time they take to develop are not something often addressed in the engineering curriculum or classroom. A conscious effort has to be made if institutions seek to attract and retain diversity in engineering. The more awareness a student has of his or her ability, the more likely they are to make informed choices. Institutions need to engage students in high-impact practices early on to raise awareness of the field and increase a student's ability to be successful therein.

High Impact Practices

In a quest to identify effective educational practices that produce positive results for a diverse range of students, including those who have historically been underserved, it is critical for educators to engage students in meaningful experiences, both in and out of the classroom. Using data gathered from the National Survey of Student Engagement, Kuh (2008) identified several best practices for colleges and universities to engage students and increase student success; these active learning initiatives are also known as high-impact practices. If implemented properly, these high-impact practices have the ability to promote self-efficacy in students.

There are several commonalities among the type of activities deemed high-impact practices: they demand time and effort from students, engage students in out-of-class activities, facilitate interaction between peers and faculty, and provide opportunities for growth through feedback (Kuh, 2008). These utilitarian activities are known to deepen the students' investment and commitment to their education and chosen academic program. Depending on the student, the institution, and the way these activities are integrated into the curriculum, high-impact practices have the potential to enrich the student's academic experience. They can also provide opportunities for individual growth and development.

Kuh (2008) contends that to enhance student engagement and increase student success, colleges and universities must, "make it possible for every student to participate in at least two high-impact activities during his or her undergraduate program, one in the first year, and one taken later in relation to the major field" (p. 29). These high-impact practices include common intellectual experiences, writing-intensive courses, diversity/global learning, service learning and community based learning, collaborative assignments and projects, undergraduate research, internship, capstone-courses and projects, learning communities, and first-year seminars and experiences (p.16).

First Year Experience

As discussed, students transitioning from high school to college face a number of challenges. Adapting to the demands of a new academic environment, learning critical soft skills for success, and acquiring study habits that effectively meet the needs of college-level course work are a few of the academic-based challenges they face. From a social perspective, students are also trying to find groups (Nicol, 2009). Historically, first year seminars serve to create a classroom-based learning community (Johnson, Pasquini, & Rodems, 2013) and connect students to resources that can enable their success.

Research proves that it is important for higher education institutions to assist incoming students with the critical process of assimilating to this new environment (Nicol, 2009) to foster self-efficacy, and assist with major and career exploration and decision making. As such, many institutions develop first-year experience programs to teach students to navigate the transition, develop critical skills that will aid in achievement, and find the path towards success. These programs can include orientation programs, first year seminars, living learning communities, designated housing, and learning communities (Johnson, Pasquini, & Rodems, 2013).

Connecting engineering students with high-impact practices and opportunities to engage in community learning through a first-year experience is one way to invest in their continued success in the field. Jones et al., (2010) examined several constructs with first-year engineering students including self-efficacy, expectancy for success in engineering, engineering intrinsic interest value, engineering attainment value, engineering extrinsic utility value, identification with engineering, engineering achievement, and engineering career plans. Through their research, they found that first-year engineering students began their undergraduate education with high-expectations for their performance in engineering-related courses, they valued engineering, enjoyed it and believed it was important and useful, and felt that they would eventually chose a career pertaining to engineering (p. 325). As the students in the study approached the end of their first year, perceptions in all areas decreased. Based on their findings, the authors offer a suggestion- "It would be interesting to investigate how the curriculum and mentors could be used to affect students' values towards engineering" (p. 331). It seems that a first-year seminar course, designed to meet the specific challenges that face engineering students and the curriculum, offers a solution.

Informal Learning Communities

Formal education is only part of the learning that takes place in college and universities today. Much of an undergraduate's time is spent outside of the classroom and creating meaningful settings for informal learning can enhance the experience. Informal learning generally involves talking, observing, experiencing and collaborating, and is effective because it is personal (Cross, 2007). Informal learning communities exist in many settings and are prevalent in higher education (Johnson, Pasquini, & Rodems, 2013). Research validates the importance of informal learning environments, and some researchers assert that informal education has the potential to reform science, technology, engineering and math education (Denson et al., 2015). To be effective for engineering students, informal learning environments must

1. Promote awareness of the field,
2. Provide academic enrichment,
3. Be led by trained and competent instructors and
4. Have the support of the student's educational system (Chubin et al., 2005).

Denison et al. (2015) noted eight themes that emerged from their study on informal learning environments. In their experience, the benefits of this type of educational environment present the opportunity for students to build-

1. Informal mentoring relationship,
2. Fun learning experiences,
3. Time management skills,
4. Experiences to apply math and science,
5. Feelings of accomplishment,
6. Confidence,
7. Comradery, and
8. New opportunities (p.14).

Creating informal learning environments for first-year engineering students exposes them to opportunities, connects them with resources and mentors, and helps them make informed decisions about their major and career choices.

MAIN FOCUS OF THE CHAPTER

Issues, Controversies, Problems

The following sections provide an overview of the issues and challenges faced by first-year engineering majors in the Honors Program.

Honors Students at the University of Florida

Only half of all first-year applicants are admitted into the University of Florida Honors Program each year. Admission is highly competitive, with the average SAT scores exceeding 2100/2400 (math, verbal, and writing) in 2015. Students are selected for the program before or during their first year at the university and continue in the program throughout their undergraduate career. Admission is holistic and based on essay responses and indicators of academic preparation such has SAT / ACT scores and high school GPA. With strong test scores, and an institutional, state and national focus on STEM education, it is no surprise that close to one-third of all honors students select engineering as their major (1000/3200).

Honors students represent less than 10% of the total undergraduate population at the university. To complete the Honors Program, students must demonstrate involvement in the academic, co-curricular, and extracurricular life of the university. Academic activities include honors courses, undergraduate research, publications, and presentations. Co- and extracurricular activities include internships, study abroad, volunteer service, and leadership. In addition, students must graduate with honors in their major which typically includes attaining a certain GPA in their upper-division (i.e. junior and senior) major courses.

Honors Engineering Majors

Of the roughly 750 first-year students in the program, approximately one-third are declared engineering when they enter the university. According to the University of Florida Office of Institutional Research and Planning (OIRP), honors engineering students represent 18% of the total number of engineering majors, almost double their representation among the entire undergraduate population. OIRP does not separate enrollment data for retention and graduation rates by race / ethnicity, but it does by gender. Latest figures available (entering undergraduate class of 2013) show that males represented 69% of all honors engineering majors and 70% of all engineering majors. Honors status does not seem to impact interest in the major.

Academic preparation, as indicated by honors admission, may impact retention in and graduation from the major, however. OIRP provides retention data from freshman to sophomore year by major. For students who entered the university in 2013, 84.44% of all males and 82.64% of all females in engineering persisted to their second year. By contrast, 88.89% of honors males and 87.84% of honors females persisted. Of the students who persisted to the second year, incoming test scores and high school GPA's differ between honors and non-honors students. The average high school GPA of all males who persisted in engineering was 4.29 versus 4.42 for honors males. All females averaged a 4.33 GPA in high school versus 4.44 for honors females. Test scores provide a similar picture. The average SAT score for all males was 1341 versus 1429 for honors males. All females averaged 1308 versus 1413 for honors females. In both cases, honors females had high school GPA's and SAT scores that were closer to honors males than to all females.

In looking at their GPA's from the first year in college, honors students continue to outpace their peers. Of the students who persisted to their second year, all males and females both averaged a 3.48 GPA during their freshman year. Honors males averaged a 3.68 GPA, while honors females averaged a 3.72 GPA. Academic success in their foundational courses might have had a strong impact over whether or not students persisted. On the other hand, it is interesting to note that the honors females who did not persist to their sophomore year in engineering, they averaged a 3.37 GPA. In fact, the GPA for honors females who did not persist was closer to the GPA for all females who did persist (3.48) than the GPA for all females who did not persist (3.17). Considering the GPA for engineering majors to graduate with honors is a 3.3, something other than academic achievement might explain the reason for the honors females leaving the major.

Did these trends continue throughout the college experience? A look at 5-year graduation rates of the entering class of 2009 tells a similar story. Five-year graduation rates are selected, as opposed to 4-year graduation rates, because the engineering curriculum at the University of Florida extends to five years. Therefore, graduating in five years for a full-time student in engineering is considered "on time." Students who declared engineering as a first-year student graduated at much higher rates if they were in the Honors Program. Of all students who declared engineering as a major during their freshman year, 61.32% of males and 64.79% of females graduated in five years. Honors students graduated at a rate of 73.47% for males and 82.22% of females if they declared engineering as a freshman and continued with it through graduation.

Again, a look at the academic performance of those honors females who did not graduate in five years raises questions. During their first year in college, the average GPA of honors males who graduated in five years in engineering was a 3.68 compared to 3.65 for honors females. On the other hand, honors males who did not graduate in five years earned a 2.92 GPA during their first year. One could extrapolate that academic performance during the first year was a key indicator for honors males who did not persist. Honors females who did not graduate in five years, however, earned a 3.52 GPA during their first year. A 3.52 GPA, according to many academic advisors and college personnel, would be an indicator of academic success. So what, then, is impacting honors females who are leaving engineering?

Potential Causes of Honors Female Attrition in Engineering

Something is happening during the first year in college for honors females in engineering. Although the number of female honors students who do not persist to sophomore year or graduate in five years is small, understanding their experiences is important. In general, academic deficiencies are not the root cause of female honors students leaving the major. Both the female honors students who do not persist after their first-year and do not graduate in five years are still excelling academically according to their GPA's.

As observed in academic advising appointments with female honors students who are contemplating major changes outside of engineering, a number of other issues arise. As mentioned previously, the engineering curriculum is highly regimented and allows little room for exploring other interests. Some females lament the fact that they do not have room in their schedules to double major, minor, or even take elective coursework in their other areas of interest. Growing frustration with a singular focus on one field drives many to look into more flexible options.

Other females do not understand the long-term impacts of a career in engineering. During the first year when they are taking foundational courses, they face a disconnect between coursework and the practical implications of their field. Honors advisors frequently hear from female students that they want to "help people" or follow a caring profession. Some female students will move out of engineering and into a different STEM field, particularly ones that will steer them towards a pre-health profession where the societal impact is clearer and more tangible. These students can easily change their major, as all STEM fields tend to focus on the same foundational courses.

Particularly salient to this population is a poor sense of self-efficacy. Honors advisors frequently hear mentions of imposter syndrome from female engineers (and STEM majors in general) who fear that they cannot "cut it" even in their foundational courses during the first year despite all academic indications to the contrary. Convincing a female engineering student that she is competent, deserving, and capable of success can be a difficult task for many advisors and instructors. Even strong students will decide to leave engineering because they simply cannot see themselves succeeding, regardless of academic performance.

SOLUTIONS AND RECOMMENDATIONS

Developing a First-Year Course

Sensing a need for high achieving first-year students to form a learning community around their interests in topics beyond traditional orientation and transition, a move away from the university-wide first-year course was sought. In 2005, the Honors Program developed their own first-year seminar to expose high achieving students to a broad array of co- and extracurricular activities. Since that time, the course has been offered every fall and spring, with multiple sections offered in fall.

Co-taught by a faculty member in the Honors Program, many of whom are also academic advisors, and an involved honors student who serves as a peer leader, the 1-credit course meets in-person for 50-minutes weekly to showcase opportunities relevant to honors students' academic and professional goals and interests. In addition to the in-class resources, technology is often utilized beyond the classroom to reinforce lessons and provide additional opportunities for growth and exploration. Limited to 25 students, each class is designed to provide opportunities for this high-impact practice to provide guidance and personalized assistance tailored to the specific needs of each of the registered, high potential students.

The course is optional for first-year honors students, and approximately 15% of the first-year cohort elects to enroll. Starting in 2011, the seminar narrowed to focus on specialized majors, including one section unique to STEM (science, technology, engineering and math) majors. In 2015, an additional section of the course just for engineering majors was added, due to the growing number of STEM majors in the program. In both the STEM and engineering only sections, all course lectures, assignments, and opportunities are geared towards those particular fields. Even the non-major specific section of the course, however, tends to have a high number of STEM majors.

The course description for the STEM section is as follows:

- Steering students towards majors and careers in the Science, Technology, Engineering, and Mathematics (STEM) disciplines continues to be a hot topic among educators, legislators, and the general public. In this course, we will work with first-year STEM majors to develop a unique action plan for deeper involvement in their discipline, whether that be through undergraduate research, internships, leadership, global experiences, or other opportunities. We will coordinate panels of current students involved in these opportunities, as well as invite representatives from various organizations in order to provide students with direct contacts in these areas.
- There will be intensive, guided work completed online outside of class, complemented by in-class activities, discussions, and panels facilitated by our two peer leaders. The combination of out-of-class independent work with in-class activities will allow students to develop a well-researched path to further involvement in the STEM fields. (Honors Program, 2015)

Course Highlights

While each instructor is able to put their own unique spin on their course, there are some common highlights among all sections. All sections spend at least a week each on seeking internships, undergraduate research, and global engagement (including study, research, and internships abroad). In addition, considerable time is spent helping students clarify their interests and career goals, learning how to network, and enhancing their resume and interviewing skills.

First-year honors students, especially in STEM fields, are anxious to dig deeper into their major and career interests. Knowing that they must take foundational courses during their first, and sometimes second year, they often look elsewhere to gain experience. Unfortunately, most students do not know where to look. As internships in STEM fields are typically reserved for rising juniors and seniors, most first-year students assume that they will have to wait at least two years to gain practical experience in industry. In this course, students are encouraged to attend the university's career day, where one entire day is devoted to technical careers. Instructors and peer leaders impress upon students the importance of bringing copies of their resumes, dressing appropriately in business attire, having a firm handshake, and preparing an elevator pitch.

While some students are initially discouraged that many companies are not recruiting first-year students, they are able to get pointers from employers about how to make themselves competitive in future years. And a few students do get called for interviews and even earn internships from their first-year career fair visit. Plus, many recruiters make return visits and remember the students they have talked to at previous fairs. Many students comment at the end of the course that they never would have known that the career fair was available to first-year students without learning about it in the course.

Undergraduate research is another mysterious area for first-year students. Conceptualizing the idea of research, finding available opportunities, and approaching faculty are all difficult for students who have never experienced work in a lab before college. Some students think they should take any position available, while others worry that no professor would take a first-year student. In class, instructors emphasize how to search for research positions, how to communicate with faculty, and how to best make decisions on accepting opportunities. They discuss how to review the prior research of faculty members to better understand the scope of current projects, as well as determine how research fits into their larger goals.

Due to the exploratory nature of the course, students are given flexibility to further investigate their areas of interest. Because the university has very high research activity, students feel pressured to be involved with research - even if that is not an area of interest or necessarily relevant to their career goals. Through this course, students can make better decisions about what they want to pursue and why. For those who truly want to get involved with research, however, they have been quite successful at finding opportunities as first-year students because of the background work they have completed through the course.

Global engagement is perhaps the greatest unknown for students, especially in STEM fields. Due to an already packed curriculum, as well as a dearth of study abroad opportunities related to STEM at the university, most first-year students in these fields worry that they must give up any idea of going abroad as an undergraduate. Through this course, student are challenged to attend the campus-wide study abroad fair to talk to faculty and advisors about opportunities that fit their interests. Most students return from the fair with new ideas about travel destinations.

In addition, instructors discuss options to research, intern, and volunteer abroad, widening the possibilities to combine academic interests with international travel. Finally, students are encouraged to meet with their instructors to look carefully at their program plans to find room for study abroad if interested.

Each of these opportunities - internships, research, and study abroad - tend to be nebulous ideas to first-year students. Through in-depth investigation, students learn the importance of adding these areas to their co-curricular plans, as well as the specifics for finding and obtaining them. Through hands-on work, students can gain a new understanding of their fields of interest, even though they are still only taking foundational coursework.

Students also learn how to enhance their skills in networking, resume development, and interviewing through the course. They are savvy enough to know that these opportunities are competitive, although initially some students worry that they are not going to be qualified to apply. Class sessions become mini-workshops where students practice interviewing, elevator pitches, and small talk with strangers in a safe, non-competitive environment. Instructors and class guests provide critical feedback to students to help them strengthen their skills. Students also learn how to adapt their communication strategies based on their audience. For example, interviewing with a company for an internship is very different from interviewing with a faculty for a research position. These sessions help students build confidence as they apply for opportunities, and are often the top-rated course activities.

Technology is used to provide extra resources on opportunities and skill development, as well as guided reflection and discussion. Students provide peer feedback to each other through discussion forums about their resumes, as well as responses to interview questions submitted as videos. Using a learning management system, students have quick and easy access to these resources both during and after the conclusion of the course. Social media is also utilized so that students and instructors can communicate in a more informal manner. Relevant articles from mainstream media, both campus and national opportunities, and general course reminders are all posted via social media.

Finally, relationships that form through the class are another important component. The broader learning community that is developed creates a cycle of support for engineering majors. Course alumni look forward to joining the first-year seminar as future peer leaders and panelists who share how the opportunities have helped to solidify their academic and career interests. First-year students see themselves in the faces of the course alumni and aim to emulate them. Connections through email and social media deepen the ties among current first-year students and course alumni. Frequent contact with an academic advisor as instructor provides the final layer of support, as students know they have regular access to a professional who can guide them through their academic planning, finding ways to take advantage of the opportunities they have learned about in class.

RECOMMENDATIONS

The aforementioned course certainly has limitations in solving some of concerns about diversifying engineering, especially among women. Not every student can be accommodated in the course, as eligible instructors are limited. The size of the class itself (25 students) does not always lend itself to personalized, individual attention for each student's interests and plans on a weekly basis. Because there are so many topics to cover, students only receive a broad overview versus in-depth discovery in a few areas. It only meets once per week for an hour, so time in class is limited.

On the other hand, there have been tangible outcomes of this course. Students have been able to apply for internship, research, and global opportunities earlier and more often than many of their peers. These opportunities are able to supplement their major at a time when they are still taking foundational courses. They have thought critically and reflectively about their major and career interests. Students find a sounding board to discuss their interests, not only with their instructors, but also with each other.

Moving forward, and in light of the need to continue outreach to first-year females in the program, a new section of the course is in the planning stages just for women in STEM. While the focus on engaging in various opportunities and building important skills will still remain, additional topics will be added on negotiating, building confidence, executive presence, and balancing personal and professional interests. The hope is to build a support network of first-year women in STEM that can continue after the conclusion of the semester. Women who have participated in the regular STEM course are being recruited to serve as peer mentors who can guide first-year students, share tips for success, and help them build their networks.

FUTURE RESEARCH DIRECTIONS

To address the nationwide shortage of trained engineers in the workforce, institutions must work to recruit and retain students in engineering majors and provide them with the tools and knowledge required to be successful in the profession. While much research is being conducted currently on STEM education, the crisis is still visible across the nation. Jones et al. (2010) argue that future research into how curriculum can support student's beliefs about engineering enjoyment, importance, and usefulness is essential. From a retention perspective, these areas must be examined early in a student's academic pursuit. Many institutional initiatives have been created to try to address this problem. Guidelines and best practices are helpful but more research is needed on how these initiatives directly impact the number of engineering students graduating and entering the field.

The issue of self-efficacy is another avenue where future research must be conducted. Several studies have examined and found a link between self-efficacy and student's persistence with engineering majors. As such, more research is needed to advance the findings and present real-world suggestions for practitioners. Determining how to best promote self-efficacy among younger engineering majors can have a life-long impact.

CONCLUSION

The economic demand for skilled engineers is high and these positions remain at the forefront of the list of hard-to-fill jobs (Jones et al., 2013). This shortage in the workforce is a problem that must be addressed from many levels. Colleges and universities play a critical role in helping to solving this economic crisis. However, to do so, they must make a conscious effort to procure proven techniques that will attract and retain more students in the field of engineering. Classroom instruction, in its current state, has not been a proven method to address this shortcoming. One reason may be the way in which the curriculum is designed.

For many incoming engineering students, the first year of coursework revolves around foundational math and science courses. While these courses help to build a solid base for upper-division, major specific coursework, they do not provide students with real-world application or understanding of the profession. As such, there is a clear need to engage students in activities that allow for exploration into the field. Providing this support during the transitional first-year, when students often struggle with the decision-making process, has the potential to make a powerful impact.

One institution has fought to address this issue through a first-year seminar course, designed specifically for high achieving STEM and engineering students. The course, which has been taught for many semesters, seeks to build a meaningful learning community for students to learn socially from their peers and faculty both in and out of the classroom. The course introduces students to high-impact that present opportunities for meaningful career exploration and skill development. It is designed to improve self-efficacy with regard to major and career decision making and help set students up for success as engineering majors. Providing this kind of support and demystifying the opportunities that exist within the field of engineering through a first-year seminar course is one way to address this challenge and retain and graduate more engineering students.

REFERENCES

Bandura, A. (1986). *Social foundations of thought and action*. Englewood Cliffs, NJ: Prentice-Hall.

Byars-Winston, A. (2014). Toward a framework for multicultural STEM-focused career interventions. *The Career Development Quarterly*, *62*(4), 340–357. doi:10.1002/j.2161-0045.2014.00087.x PMID:25750480

Chubin, D. E., May, G. S., & Babco, E. L. (2005). Diversifying the engineering workforce. *The Journal of Engineering Education*, *94*(1), 73–86. doi:10.1002/j.2168-9830.2005.tb00830.x

Cross, J. (2007). *Informal learning rediscovering the natural pathways that inspire innovation and performance* (1st ed.). San Francisco: Pfeiffer/Wiley.

Denson, C. D., Stallworth, C. A., Hailey, C., & Householder, D. L. (2015). Benefits of informal learning environments: A focused examination of STEM-based program environments. *Journal of STEM Education: Innovations & Research*, *16*, 11–15.

Honors Program. (2015). *Fall 2015 course guide*. Retrieved from http://www.honors.ufl.edu/courses/CoursesFall2015.aspx

Johnson, M. L., Pasquini, L., & Rodems, M. (2013). Connecting first year students to forman and informal learning experiences. In H. Wang & S. Wang (Eds.), *Cases on formal and informal e-learning environments: Opportunities and practices* (pp. 51–67). Hershey, PA: Information Science Reference. doi:10.4018/978-1-4666-1930-2.ch003

Jones, B., Paretti, M. C., Hein, S., & Knott, T. (2013). An Analysis of Motivation Constructs with First-Year Engineering Students: Relationships Among Expectancies, Values, Achievement, and Career Plans. *The Journal of Engineering Education*, *99*(4), 319–336. doi:10.1002/j.2168-9830.2010.tb01066.x

Jones, B. D., Paretti, M. C., Hein, S. F., & Knott, T. W. (2010). An analysis of motivation constructs with first-year engineering students: Relationships among expectancies, values, achievement, and career plans. *The Journal of Engineering Education*, *99*(4), 319–336. doi:10.1002/j.2168-9830.2010.tb01066.x

Kug, G. (2008). *High-impact educational practices: What they are, who has access to them, and why they matter*. Washington, DC: Association of American Colleges and Universities.

Miller, M. J., Lent, R. W., Lim, R. H., Hui, K., Martin, H. M., Jezzi, M. M., & Williams, K. et al. (2015). Pursuing and adjusting to engineering majors: A qualitative analysis. *Journal of Career Assessment*, *23*(1), 48–63. doi:10.1177/1069072714523084

Nicol, D. (2009). Assessment for learner self-regulation: Enhancing achievement in the first year using learning technologies. *Assessment & Evaluation in Higher Education*, *34*(3), 335–352. doi:10.1080/02602930802255139

Olitsky, N. (2014). How do academic achievement and gender affect the earnings of STEM majors? A propensity score matching approach. *Research in Higher Education*, *55*(3), 245–271. doi:10.1007/s11162-013-9310-y

Singer, K., Foutz, T., Navarro, M., & Thompson, S. (2015). Investigating the Extent that an integrative learning module broadens the perception of first-year students about the engineering profession. *American Journal of Engineering Education*, *6*, 99–112.

Xu, Y. (2013). Career outcomes of STEM and non-STEM college graduates: Persistence in majored-field and influential factors in career choices. *Research in Higher Education*, *54*(3), 349–382. doi:10.1007/s11162-012-9275-2

KEY TERMS AND DEFINITIONS

Career Development: The active process of planning and implementing the necessary steps to find and maintain an individual's professional and workplace goals.

First-Year Experience: A program, course, or similar opportunity designed by colleges and universities to engage new students and help them successfully transition from high school to higher education.

High-Impact Practices: Teaching techniques that have been found to make a significant contribution to student success and engagement.

Informal Learning Community: Experiential form of organic learning that is less structured and has no clear set objectives.

Self-Efficacy: An individual's belief in his or her ability to take action towards accomplishing specific tasks and reaching goals.

Social Cognitive Theory: The idea that individual's gain knowledge observing others, that learning is a social process, with exchanges occurring between people, environment, and behaviors.

Chapter 6
Engineering Teams:
Supporting Diversity in Engineering Education

Jennifer Loy
Griffith University, Australia

Simon Howell
Griffith University, Australia

Rae Cooper
Griffith University, Australia

ABSTRACT

Engineering education increasingly involves working in groups. This is partly because of a growing value placed on graduate attributes relating to effective team working, and partly a response to the practicalities of working with large groups in an educational environment and the emphasis on peer learning. This chapter argues that a superficial approach to understanding the drivers for establishing and managing groups during first year activities can have negative outcomes, including re-enforcing majority dominance. This will potentially contribute to attrition amongst minority students and undermine the outcomes for the engineering cohort as a whole. This chapter provides strategies for building groups in the first year focussing on team building, valuing diversity and cultural awareness. It emphasises the importance of transferable skills for students and of understanding themselves, their heritage, attitudes and values and their contribution to a team, building an approach to support diversity in teams throughout the engineering degree program.

DOI: 10.4018/978-1-5225-2212-6.ch006

INTRODUCTION

Contemporary ideas about graduate attributes across disciplines feature the ability to work in teams as an essential transferable skill. There has been considerable research into creating and maintaining these teams, both in the workplace and education. Building teams is a long established element of postgraduate studies for learning and teaching in higher education. Yet, there have been social and workplace changes that have been building up over recent years that will impact traditional views of teaching teams in education. For engineering, these revolve around the changing practices associated with an increase in internationalization and the extension of teams to include members with more diverse backgrounds. Essentially, teams are changing as the problems being addressed become more complex, and require input from a wider range of disciplines. In addition, there are changes to the way that users participate in problem solving that should impact how team-working strategies are developed for the engineering discipline, taking into account discipline-specific majority cohort profiles and attitudes.

Traditionally, there has been limited differentiation in the teaching of team working for individual disciplines in higher education. This has been the case for the education of academics as well as students, who study the same postgraduate qualification for learning and teaching in higher education regardless of discipline. This chapter considers a different approach, with the design of learning experiences and teaching strategies that foster team working specifically to recognise and respond to the challenges and aspirations for the engineering discipline.

Educational literature highlights the importance of a lecturer understanding the background, learning motivations and educational outcomes of the student cohort, and then adapting the learning experience to meet the needs of the cohort. Therefore, the education of faculty in learning and teaching needs to be based on an informed understanding of the particular disciplines they are working within. This extends to teaching—and working in—teams, based on the ideas and understandings that inform that discipline.

For engineering academics, there are challenges specific to educating students that should differentiate their education from that of other faculty and provide them with discipline-specific strategies to help students to work effectively in a team-working environment. Yet what that actually means post digital revolution (Gore, 2013) is still being explored, particularly in the light of changes to the way engineering is being taught by leading institutions such as Olin College, and the Singapore University of Technology and Design (SUTD), with their overt emphasis on experiential learning. This chapter aims to critically evaluate the role of team working specific to engineering education and presents practice-based recommendations on teaching informed by, and in response to, current theory.

RETHINKING THE TEAM-WORKING PARADIGM FOR ENGINEERING

Engineering education increasingly involves working in groups. This is partly because of a growing value placed on graduate attributes around the ability to work effectively in teams, and partly a response to the practicalities of working with large numbers in an educational environment, and a corresponding recent emphasis on peer learning. The growing imperative for effective team work skills in engineering is based on the changing nature of problem solving in the profession, with multi-disciplinary teams engaged in problem framing and solving in response to complex ideas. Project needs can include social and psychological imperatives, as well as the conventional practical, economic and environmental ones. The expectations of the profession are that qualified engineers will now work in international and interdisciplinary teams, and engage in non-traditional activities such as creative brainstorming. They can also be required to work in teams where empathy informs problem framing and design development. An example of the development of such teams can be seen in the study of design for disability provided by Pullin (2011). Once the National Health Service was established in the UK in 1946, teams of engineers and clinicians worked together to produce purely functional orthotics and prosthetics. However, Pullin tracks the developments of these teams and the impact of changes in attitude to orthotics and prosthetics, and how that has altered the design brief for assistive medical devices. He argues that the inclusion of user-centered design into the design brief for the development of functional orthotics and prosthetics is becoming more prevalent, and that this will increase as advances in digital manufacturing allow for more bespoke products. He gives the example of the experiences of the runner Aimee Mullins, who has functioning prosthetic lower legs that meet particular practical and emotional needs. Mullins' legs were created in multi-disciplinary teams that included the fashion designer Alexander McQueen, and filmmaker Matthew Barney. She is now able to alter her height as well as her appearance with a range of lower limb prosthetics.

This approach to design for disability recognizes that empathetic design improves the experience of the user, and can contribute to compliance (for example, in wearing hearing aids and spectacles), and therefore improve the medical effectiveness of the product. This approach has expanded the product development team to include designers and psychologists, and on occasion creative practitioners from other fields such as fashion, art and film. Another example of creating innovative interdisciplinary teams is demonstrated in the approach taken by Benyus (2002), a natural resource management specialist. Benyus brings together scientists, designers and engineers to more fully exploit the potential of biomimicry in the development of a systems approach to problem solving. This approach reflects current sustainability theory and the importance of understanding the embodied energy in products, and their social, economic, environmental and most recently political impact. Lifecycle assessment and product service systems thinking, and particularly those with an overtly biomimicry based approach, have significantly increased the complexity of problems for teams of which engineers are likely to be part of. These groups will be drawn from different backgrounds, with different working practices, values and ideas, and engineers need to be educated to ensure they can work effectively within these increasingly diverse teams.

The digital revolution is another factor that is contributing to breaking down the silos in education and practice in engineering, as it provides new opportunities for international collaboration. Students and graduates will be increasingly involved in working with culturally diverse teams as production systems continue to evolve, projects span countries, and the body of knowledge informing projects will be tempered by cultural attitudes. Engineers need to be aware of not only the cultural bias they bring from their own country, and possible ethnocentricity, but also the discipline bias they may bring to team work, and how that could influence the groups they work in.

The digital revolution is causing a reimagining of interaction that is impacting students and changing educational systems. What works for students today will not necessarily work for students tomorrow. For faculty, understanding each succeeding generation communicates is crucial (Dewart & Rowan, 2007). Working to adapt learning experience in response to students' technological preferences has a direct relevance to teaching communication during project work, and therefore impacts on the teaching of team working. Systems in engineering education are in transition, as the profession is being rethought for global communication, data generation (and mining, organization and application), and—significantly for engineering—distributed manufacturing. Changes to production systems and supply chains through the exchange of digital files for distant digital fabrication change what is possible to manufacture and the relationships within the supply chain, impacting engineering education (Kuniavsky, 2010). Consequently, the boundaries between engineering disciplines such as civil and electrical or mechanical and design are blurring, with teams involved in closer relationships throughout the entire development process, rather than as previously where engineers would have their place within a more serial approach to problem solving.

The educational response to the blurring of boundaries between engineering programs has been to bring together students for interdisciplinary learning experiences, though there is an argument that this approach is partly a response to the drive for economies of scale. Larger cohorts could in part be becoming necessary because of changes to university funding, and because of the political and social imperative to widen access to higher education, with alternative pathways for students to improve the educational opportunities of a broader demographic. The impact on teaching, with a focus on larger groups of students rather than small groups and individuals, is complicated by engineering accreditation requirements, but even there, recent directives from accrediting bodies have emphasized the importance of transferable skills, including team work and working with other disciplines. However, while research into the shift towards experiential learning in engineering education is evident, there is a dearth of professional development being offered specifically for engineering academics working with these changes, and more specifically the impact on teaching team work in engineering for these altered classroom environments. Institutions need to support engineering academics with adequate peer support and guidance, rather than assuming that the lecturers will be able to make this shift, or see the benefits, or even just the rationale. Staff development for academic staff was argued by Dewart and Rowan (2007) to be focused on teaching tools and promoting ways of sharing knowledge unique to the course or the teacher. Yet in reality, for the engineering academic discipline, working with these changes creates particular challenges that go beyond those addressed by the mainstream body of knowledge on teaching teams in a university environment. These challenges stem from the values and character that have become fundamentals in a long-standing discipline. They affect everything about the discipline itself—from teaching conventions to set work to learning experiences involving groups—and impact the profile of the academics likely to be considered by the faculty as a whole as successfully working within it and reinforced by selection criteria and performance reviews.

ENGINEERING TEAMS

As a consequence of the popularity of teams as early as 1990, 80 percent of companies with more than 100 employees (Gordon, 2002) have implemented teams, often without understanding the complexity and dynamics, causing constructive and/or destructive conflict that has to be managed to ensure efficient and effective cross-cultural teamwork, especially in multicultural environments. (Du Plessis, 2011, page.50)

A number of teaching publications (Race, 2007; Weimer, 2002) point to an underlying assumption in teaching group work, which is that issues, approaches, and outcomes are similar across disciplines. They also demonstrate a similar emphasis on creating equality in groups, treating everyone the same. However, just as learning styles within a cohort can differ, so there is a case for arguing that learning needs are shaped by the attitudes and aspirations of a discipline and require different ways of teaching team working. For engineering students, teamwork considerations should include:

1. Understanding the team-working needs of engineers graduating into the profession in the 21st century;
2. Understanding the discipline-specific attributes impacting team working and the teaching of team working; and
3. Positive support for minority groups within a dominant cohort appropriate to engineering.

Teams for the 21st Century

...mediated communication is used throughout the day by a great majority of young people, at school, in bedrooms, in shopping malls and in transit, in order to stay connected. This in turn has two main social implications: sheer connectivity of individuals is greater than ever before, leading to greater social awareness and responsibility; and control of those mediums is challenging—young people need to master the social networks and the technology to stay connected. (Mohan et al., 2013, p. 142)

In 2017, the generation born around the turn of the century, known as Gen Z, will begin enrolling in university education around the world. This raises the question as to whether educational teams should be built on the same ideas and underlying assumptions as for the preceding generation. For Gen Z engineering students graduating into the profession in around four years, team working will have very different characteristics from those experienced by previous generations. The digital revolution is breaking down traditional borders, and changing the interaction of team members working on a project. Supply chains and production systems are also changing (Loy & Tatham, 2016). Manufacturing can be global, based on communication across countries, and a value chain where products can be transported across regions during manufacturing as well as for delivery, with distributed manufacturing providing a more agile, responsive option to conventional, centralized manufacturing. Project teams working across countries, across disciplines, across time zones, and across cultural norms can address more complex problems than isolated teams, and have more understanding of the implications for systems (including natural systems) of product development. While conventional practices in engineering continue, increasingly innovative projects draw on digital technologies such as ubiquitous computing, scanning technologies, and interactivity. Teams are constituted very differently, and there are cross-cultural referencing issues (between professional groups and disciplines) that were not evident 10 years ago. To maximize the potential of working in newly formed teams, with colleagues with different working practices, values, and skills, it is increasingly important for engineers to understand how to work with others from very different disciplines. They may be working on projects with no precedent, be involved in brainstorming and problem framing, as well as problem solving. While interpersonal skills are still integral to success for the team, the ones they learn within a limited engineering educational environment, with peers and lecturers from the same domain in face-to-face learning, may not fully provide them with the exposure to alternative values, behaviors and practices they will require in the future.

Discipline Cultural Referencing

Discipline areas, by their very nature, support specific behaviors and attitudes as integral to their practice. For example, divergent thinking is identified as an essential component of practice, and to support this, designers are encouraged to engage in structured "play" (Brown, 2009), actively giving themselves over to "conscious unselfconsciousness". In contrast, scientifically based disciplines value documented repeatability as an essential basis for their study. As innovative engineering educational institutions, such as Olin, break the conventions of engineering education by integrating more design-specific educational practices into their curriculum, they are facing more than the practicalities of adding design into their teaching. They are also challenging the very core of discipline practice and attitudes that have been embedded in the discipline as it has developed over centuries of education, and fostered by definitions of professional practice and approach.

To support the development of project teams in engineering, group working is integrated into student learning. However, before it can be effectively part of engineering education, there are realities that have to be addressed for engineering specifically that will impact the feasibility of working in groups for the cohort and the academics involved, and then strategies to respond to those realities developed. The most fundamental is that the engineering discipline has a strong heritage spanning generations, providing a significant cultural infrastructure that has stood the test of time, and the reputation of outcomes determined by performance criteria and tested to agreed standards. To be an accredited engineer has social as well as economic value. Where leading designers advocate an ability to be playful during divergent thinking (as well as serious during the convergent phase of development), there is currently no such corresponding behavior advocating playfulness described within the traditions of the engineering discipline. Even where engineers discuss creativity as part of practice, engaging with divergent thinking and risk taking is a recent concept for the discipline.

Although there are exceptions, such as educational programs at Olin and Stanford, the reality is that engineering education has predominantly advocated inducting students with the practice of learning the fundamentals of mathematics and physics prior to engaging in the study of applications. Currently, group working tends to be restricted to labs that conventionally consist of students following preset procedures. Interaction is limited, and the scope of that interaction constrained by the nature of the activities. The important point here is that this restrictive, structured approach to learning has been highly valued by those in the profession and academia as preserving the integrity of the discipline. Much as training in the army creates confidence in the chain of command, so the argument in engineering is that a systemized approach to learning in the discipline ensures the survival of students who can follow procedures, maintain the integrity of the testing methods used in the profession, and demonstrate the engineering science knowledge essential for practice. Therefore, there is more at stake in changing teaching practice. Thinking that team-working applications for other disciplines are equally applicable to engineering is naïve. In fact, even more fundamentally, assuming that the majority maintaining the tenets of the discipline are a normal cross-section of society is misguided. Just as there are approaches to creativity and empathy in professional design practice that will be attractive to individuals who favor those ways of working, so in engineering are there systemizing approaches that have been developed over time, that will naturally attract students who are context learners, rather than active learners.

If the ability to systemize has been a preferred skill for engineering professionals, and the educational pathway has supported students who have the ability to respond to systemizing for generations, then it is likely that the dominant culture in engineering faculty favors systemizing. In any established discipline such as engineering, a culture of practice and learning evolves. This is not particular to engineering—for example, a culture of learning in languages has developed around immersive learning. For students with different learning preferences, this may be off-putting, and therefore the cohort learning style becomes a self-fulfilling prophecy with graduates being those that respond best to the dominant learning approach of the discipline.

The value of a systemizing approach for engineering is not the subject of this chapter, but rather the impact of that dominance in relation to the development of teams in education and for graduates. According to studies on developing effective teams, successful team membership involves developing a level of empathy with those in the team. Wheelwright et al. (2006) argue that empathy and systemizing are opposite ends of a spectrum, and developed a measure for testing empathy and systemizing as part of a study on Asperger's and autism. Leading researcher in this field, Baron-Cohen (2007) describes a systemizing approach as having an "engineer's brain", and argues that it is an inherent, and inherited trait. He also argues the more extreme forms of systemizing, characteristic of Asperger's syndrome, are positive—even vital—for scientific disciplines, and need to be acknowledged and understood:

Nobody wants to have Asperger's syndrome, a form of autism that limits a person's ability to communicate, often leaving them socially isolated and mired in weird-seeming behavior. But the disorder can sometimes be a strange sort of double-edged sword, coming coupled with powerful, if narrowly focused, intellectual gifts. Put those two together, and it's clear why many people believe any number of scientists and related geeks, from Bill Gates to Albert Einstein, may have or had the disorder. (Baron-Cohen, 2007)

While not necessarily purely on the systemizing end of the quotient spectrum, just as with language students who prefer immersive learning could be advantaged in a discipline where that is the dominant teaching approach, students who have a systemizing profile are arguably more likely to be attracted to engineering, and more likely to be retained in the study program. Successful students are arguably likely perpetuate that learning style if they become academics in the future.

The average number in the general population who have ASC (Autism) is approximately one in 200, but fathers and grandfathers of children with ASC are twice as likely to be engineers, compared with the general population. An early study found that of almost 1,000 families who were members of the National Autistic Society in the UK, 21.2 per cent of the grandfathers of children with ASC worked in the field of engineering, compared with only 10 per cent of grandfathers of children with different neurological conditions. (Acker, 2005, p. 40)

While not all engineers have autism, the poor social skills characteristic of Asperger's reduce career path options for those with the condition. Temple Grandin, a well-known spokesperson for autism and professor from Colorado State University, argues that there is a clear link between engineering and autism (Acker, 2013, p. 40). She notes that Asperger's sufferers tend to develop skills highly valued by the engineering discipline and that the discipline does not characteristically value social skills. However, if group work capabilities are based on relationship-building abilities, and are improved with a higher empathy quotient, then there will be particular challenges in teaching team working to engineering students, which should not be ignored. Rather, strategies for responding to this particular challenge should be central to first year university engineering education.

Many of the difficulties experienced as people working together in groups to solve problems and make decisions relate to how the members communicate and how the group or team conducts discussions on the problem to arrive at workable decisions. (Mohan et al., 2013, p. 216)

If the engineering discipline fosters systemizing approaches, and attracts students with a higher systemization quotient, and possibly a lower empathetic quotient, then creating learning experiences that support the development of team-working abilities directed at the evolving needs of the profession becomes particularly critical. The learning experiences need to be designed with the majority of the cohort in mind, and yet result in a learning environment that supports the minority students too. In engineering, the minority are likely to include women (perhaps based on links between systemizing and the male brain identified by Baron-Cohen), students from lower socioeconomic backgrounds (engineering is a profession that has been tracked as running in middle-class families), students from varied cultural backgrounds (these include any identifying with a sub-culture outside the dominant majority; e.g., LGBTI), and students who have entered the program through alternative pathways.

A number of observational studies (Guzzetti and Williams, 1996; Tonso, 2006) have shown how female students and others with low social or academic standing are prevented from fully participating in science and engineering teams while others (McHwee and Robinson, 1992; Tonso, 2006) have documented cases of outright sexism. Such marginalization not only limits female students from participating in active learning that leads to success, but also sends the message that they do not belong. Even when female engineering students do not report outright discrimination, they report feeling that they must work harder than their male peers to get teams to acknowledge the work they have done (Natishan, Schmidt, and Mead, 2000). Given this stress, it should not be surprising that team projects may do more to push women out of the profession than to keep them in. (as cited in Wolfe and Powell, 2009)

These issues have long been a factor for the discipline, but the situation is exacerbated by the shift in thinking around the discipline in response to the global changes in attitude and approach over the last 20 years, for example in relation to gender equality and diversity. Assuming that there are sufficient drivers for change, there arguably needs to be a comprehensive approach to developing strategies to meet these changes specific to the engineering discipline, rather than merely incorporating the strategies generated for other disciplines. This includes in relation to in relation to customized professional development for academics.

Supporting Diversity

It is particularly dangerous to bring to your role of learning facilitator any preconceptions about the different members of your groups, such as those based on gender, age, ethic group, perceived social status and any other area where assumptions may be unwise and unfounded. Treating people with equal respect is an important part of acknowledging and responding to individual difference. (Race, 2007, p. 145)

The most significant challenge in addressing diversity issues within an educational department is in ensuring that the faculty itself embraces diversity, and has experience and expertise in supporting the learning needs of a diverse cohort. Universities must continue to recognise the challenge involved in cultivating the cultural capital all students bring. This fundamentally involves needing to know what "treating us like adults" means to students and acting on that knowledge (Dewart & Rowan, 2007) for a diverse first year cohort in order to retain those students through the program. For engineering, one of the most high profile challenges at this time is around supporting women in engineering, but the reality is that issues around supporting diversity in engineering education and the profession go far beyond this to encompass broader diversity issues, learning preferences, attitudes and approach. In addition, team projects are places where personalities, interaction styles, and communication preferences all come to the fore as students can feel their differences more acutely in a team setting. In addition to the discipline characteristics, there is also arguably a research imperative within engineering academia that recognizes traditional research outputs over non-traditional outputs, or the scholarship of learning and teaching in higher education. This could potentially result in a faculty profile that does not favor academics for whom the focus is the scholarship of learning, or sociology and engineering, over scientific research.

There is little evidence that engineering academics are currently being provided with targeted professional development education in meeting the new ideas of the profession and of addressing the intent to support greater diversity in engineering student cohorts. In particular, there is little evidence of targeted professional development that addresses these issues and at the same time those that could be arising from the traditional profile of engineering faculty members, and the additional issues they face in working with more diverse student cohorts. Tackling the education of engineering academics needs to be addressed comprehensively. This includes providing self-reflection and research-based education for academics on the impact of assumptions embedded into set project work on students. It also involves reviewing faculty/student interactions and the presentation of learning materials for different learning styles, and building a supportive diversity framework in the learning environment from first year through to graduation. These involve some reversals in standard engineering discipline teaching strategies. These are likely to include moving from didactic learning and traditional laboratories, rethinking the type of projects set to address diversity, and working with deliberately "engineered" teams to support diversity. A flexible approach to learning, rather than restricted delivery and assessment, will increase learning outcomes and support student retention (Dewart & Rowan, 2007). There is evidence of leadership from institutions such as Olin, Stanford, and SUTD that provides a basis for a more formalized structure of learning for faculty around the world. Mapping these issues throughout an engineering degree will help track the process and create a body of knowledge for the engineering discipline. This chapter provides a starting point for this, with a consideration of the teaching of team working specific to engineering in first year, and offers initial strategies for establishing a pattern of practice and values as the basis for a more comprehensive framework.

A Strategic Approach to Teaching Team Working

Unfortunately, many Faculty lack the necessary skills to lead, manage, evaluate or train students working in teams ... as a result or consequently, students do not receive sufficient training or feedback to learn or master teaming skills. In many cases faculty just stick 3-5 students together, call them a team, assign a task and expect results. If lucky, students will complete their task, but it is unlikely they will develop good teamwork skills. (Adams, 2003, p. 2)

Team working has become common practice in first year university education. The educational reasons for this are especially valid in the first few weeks of the academic term for students, as in addition to any discipline learning or transferable skills, it provides opportunities for new students to socialize. There are many suggestions for organizing groups based on randomizing principles, such as names pulled from a hat. Equally, there are suggestions for the deliberate "randomizing" of team constituents, for example by splitting a minority such as women in an engineering cohort, so that there is one in each group. However, such superficial approaches to establishing groups during first year activities can actually reinforce majority dominance, potentially contributing to attrition among minority students.

There is some reason to suspect that team projects might actually accelerate attrition: women in engineering and technical disciplines frequently report negative team experiences which may make them question their place in the discipline (Tonso, 2006; Wolfe, and Alexander, 2005; Natishan, Schmidt, and Mead, 2000; Woodfield, 2000). (As cited in Wolfe & Powell, 2009, page 6)

The challenge is to provide a cohort made up of a majority who are likely to be systemizers and therefore less empathetic than the average study group, with constructive strategies for building groups in the first year that focus on team building and valuing diversity. Mohan at al (2013) describes how social sensitivity, measured by how well people could interpret emotions from facial expressions, was strongly linked with the performance of individuals within a group. This will be difficult to address, but will include providing learning experiences to enhance cultural awareness and communication. Study programs and assessment strategies need to provide students with the motivation for self-reflection to ensure that they are aware of what they themselves bring to the learning experience, and what impact that has on their views and behaviors in a group. Ensuring opportunity within group working is not about a superficial approach to equity, as the group in question here is not an evenly mixed group. Where a dominant majority within the student cohort exists, and is reinforced by an equally dominant majority in the faculty profile, a form of positive discrimination needs to be developed to prevent the minority students from becoming alienated as they are unlikely to be able to rely on the empathy of the majority to ensure their inclusion during group work.

COMMUNICATION

According to Mohan el at, "The self arises in communication and is a multidimensional process of internalizing and acting from social perspectives" (2013, p.151). The current engineering curriculum is content heavy, and investing the time in preliminary activities to support project work is arguably a challenge for individual lecturers. However, it will help to improve faculty acceptance and support if the intent is embedded in a first year teaching strategy that crosses courses, and provides the basis for a structured approach to developing these transferable skills within teams over the length of the full degree program.

Assisting students to understand their own strengths and abilities, or what they each bring as individuals interacting with others during a shared study experience, will support the learning of the students. Preparing for this can be done in a variety of ways. One example is a short course or class on evaluating the work of engineers and designers in history by studying the context in which they were working. Students can also be asked to explore their own "context" and cultural "baggage" in order to recognise what biases and interests they bring to group work, even if they may not even be aware of them. This will also highlight the differences between different cultural—and sub cultural—groups. One example of identifying these differences is an in-class activity where students are asked to work in groups according to the culture they identify with, and to list the characteristics that are valued by that community. Differences identified are ones that would clearly impact the development of group practices. For instance, in one example of this activity conducted within the Architecture department at the University of Tasmania (UTAS), Australian students highlighted "mateship" and "leadership" as highly valued, as well as "humor", though without a clear definition of what those things meant, or the implications they brought to interactions. At the same time, a group of Malaysian students identified "kindness" as the most important characteristic valued by their culture. Chinese students identified creating interactions that were good for all concerned and allowed for all involved to avoid any social embarrassment. In these cases, the contrast between the values identified the different groups of students' highlighted issues that would be faced in team working.

Social perspectives are constructed in particular cultures at specific times....the values that are endorsed in any society at a specific time reflect the prevailing values and prejudices of that era and place. (Mohan et al., 2013, p. 163)

By mapping the social, historical, aesthetic, technological, political and ideological influences that engineering students are responding to or reacting against at the time, students can be introduced to mapping their own context, and the impact of corresponding influences on their own work. Another example from the same students at UTAS Architecture department of the University of Tasmaniawas from a short project where students were asked to create an outcome to help them to understand themselves, and then communicate their thinking to others. Students chose a place anywhere in the world that had meaning for them, then designed a piece of street furniture, public art sculpture or a small "folly", or similar, that communicated their sense of themselves. There is always a risk involved for students in explaining themselves and being vulnerable to others that needs to be treated particularly sensitively with engineering cohorts for the reasons mentioned previously about empathy. However, the value is in developing connections and practising empathy for the group. This is unlikely to work unless academics are educated in working with, and responding to, this activity to lead the response of the group; for example, around clarification questioning rather than critical questioning. This will be likely to require academics to engage with the activity as a team themselves prior to facilitating the group, which will in itself highlight the issues they will face with the students and provide a basis for developing coping strategies for all concerned: "Reflected appraisal is our perception of another's view of us. How we think others appraise us affects how we see ourselves" (Mohan et al., 2013, p. 153). These activities lead into activities to help first year students (and faculty academics) recognise what they bring culturally to group working. Culture in this context refers not only to the culture of the country with which the students identify, but also the sub-groups. The students need to be supported in identifying and reflecting on the impact these connections have on their outlook and behaviors, and explore the potential impact on interactions they might have. They also then need strategies for working with others from different cultures, as do academics. Cultures are described by Thomas and Inkson (as cited in du Plessis, 2011, page 61), below:

- *Culture is something that is shared by a group of people;*
- *It consists of enduring patterns of behavior that are built up over a long period of time;*
- *Culture has a powerful influence on behavior, leading to behavioral patterns that are often difficult to break;*
- *Culture consists of an organized system of values, attitudes, beliefs and meanings that are embedded in an environmental context;*

Learning what is different about their own perspective and underlying assumptions, and how those impact their world-view and communication with others, is an important basis for developing effective practice in student teams. Educating faculty on cultural differences extends to considering issues such as the way in which a student may have been taught in high school in a particular country or situation, and the impact of this on their learning. For example, home schooling, or distance education will result in a set of life experiences different from those of a student brought up in city, attending a large school. However, it is important that academics are educated to base their cultural understanding on research and not be drawn into misconceptions where there is little supporting evidence, for example the idea of a dissonance between Eastern and Western learning approaches could impact decision making on developing learning experiences (Mathias, Bruce, & Newton, 2013). More fundamentally, culture and cultural values can influence what a student will focus on, or even be aware of during communication. It will also then impact the way students communicate with each other. Additionally, the way students respond to the communication style of cultures other than their own will be significant, impacting where the attention of the group will focus during discussions:

We can monitor our tendencies to attend selectively by remembering that we are more likely to notice stimuli that are intense, loud, unusual or that otherwise stand out from the flow of communication. This implies that we may overlook communicators who speak quietly and don't call attention to themselves. Intan, an Asian American student, once told me that Caucasians often ignore what she says because she speaks softly and unassertively. Westerners accustomed to assertive speaking styles may not attend to speaking styles that are less bold. If we're aware of the tendency not to notice people who speak quietly, we can guard against it so that we don't miss out on people and messages that may be important. (Mohan et al., 2013, p. 179)

There is considerable research into communication across cultures that is as relevant for engineering as it is for other disciplines—though maybe less than for humanities-based programs because of the emphasis on mathematics and physics in the discipline. For engineering, however, arguably more of a divide exists around the integration of women into the cohort. Statistically, the representation of women in engineering as opposed to the overall population is very low. One of the reasons may be around gender prescriptions that predetermine the roles women and men are encouraged to take in society and the way degree programs are marketed. Once studying, women and men face communication stereotyping that can inhibit their acceptance. If women or men do not behave as per traditional stereotypes, they risk being seen as non-competitive and marginalized. Also, if women behave assertively or men as less so, they can face social criticism:

Western cultures have strong gender prescriptions....women are expected to be caring, supportive of others, and cooperative, whereas boys and men are supposed to be more independent, self-assertive, and competitive.....consequently, women who assert themselves or compete sometimes experience social disapproval for violating gender prescriptions. Men who depart from broadly held social views of masculinity and who are gentle and caring risk being labeled "wimps". Our gender makes a great deal of difference in how others view us and how we come to see ourselves. (Mohan et al., 2013, p. 159)

Faculty need to establish a framework for team working that is non-judgmental and not prescriptive, supporting different ways of communicating, irrespective of gender, personality, perspective and interpersonal skills. Interestingly, research suggests that while there is evidence of a bias against "female speech patterns" in engineering culture, this bias is not rooted in the speaker's physical gender, but in the language itself, irrespective of gender (Wolfe & Powell, 2009), raising questions relating to ideas of normalcy, as studied in LGBTI literature.

Leading higher education learning educator, Phil Race, argues that although the human species has evolved on the basis of group learning, human nature also works against the idea that organizing learning into teams "cold"—that is, with no preparatory socializing activities—will be successful:

We hear much of collaborative learning, as if it's the most natural activity in the world. But it often seems like the least natural, particularly amongst strangers. Sociological research tells us repeatedly that it is human nature not to be involved with people we don't know....One key to working and learning with other people is, therefore, the ability to lower barriers and become friends with would-be strangers, while acknowledging differences and respecting different viewpoints. (Race, 2007, p. 126)

One strategy is to pre-empt group working challenges by providing structured opportunities for students to meet and socialize alongside their study. An example is the introduction of "common time"—that is, a timetabled meeting with loosely structured activities—in order to support the development of relationships between disparate groups. Essentially, they are an extension of orientation activities into the study periods to support non-competitive engagement of students with each other, to support in-class learning. Considering the potential of challenges specifically facing a significant proportion of engineering students (though by no means all), first year engineering students would benefit from structure for relationship building. This is especially the case with the growing diversity imperative, both because of potential issues supporting positive inclusion, but also because of the changing nature of the project teams for the engineering profession post digital revolution, as discussed earlier in this chapter.

Based on the issues highlighted for faculty working with engineering students, strategies for the organization of team working in the first year become even more important. Not only are faculty facing challenges around a cohort likely to include students with a preference for systemizing over empathizing, they are also working with a cohort with a significant gender imbalance, and where other minorities are present, such as mature students, or culturally or ethnically diverse students, who could bring diverse attitudes, behaviors, values, approaches and responses to working in groups. Engaging with these issues could be stressful for any academics, but for those with experience predominantly based on didactic teaching, reducing the distance between faculty and students can be an issue:

Group learning has never been as important as it now is. Yet we are still in a world where most teachers, educators and trainers are groomed in instruction rather than facilitation...Some lecturers find working with small groups more anxiety-provoking than lecturing, because of the necessity to work with students as individuals rather than in the anonymity of large groups. (Race, 2007, p. 126)

Educational research discusses team organization based on different approaches (e.g. Race, 2007, p. 132) such as deliberately hybrid learning teams, skills-based groups, and performance-based groups. For a student cohort where there is majority dominance and the distinct valuing of particular skills by the discipline and faculty that are at odds with the future of the discipline, team building clearly needs to be rethought. The teaching of engineering first year should arguably foster teams that address the particular skills, performance and team working that align with the future of the profession and provide an authentic, empowering learning experience that will be of value to the student (Weimer, 2002). Superficial learning is described as knowing "just how to slot in components to get the answer" (Ramsdon, 2003, p. 48). Constructing a learning experience that attempts to address the challenges that the profession will be facing in the future is more difficult, but is necessary to provide a deep learning experience that will engage both learners and faculty. In addition, by providing an overt basis for this approach to building teams to the cohort, not only will more relevant team working be facilitated, but there is also a possibility that personalities and skills outside the conventional engineering student profile will gain status. This could empower students and increase participation, ideally reducing attrition. As Race (2007, p. 133) states: "The most significant single enemy of small-group work with students is their non-participation."

One of the fundamentals of this approach is a rethink of the dominant strategy towards providing equity. Rather than trying to pretend that all students are the same, and identifying areas of weakness to develop to create an even cohort, this approach identifies and celebrates differences, and creates an educational environment that fosters individual skills, interests and personalities. Creating teams on this basis emphasizes the different roles required for complex project work and building "T-shaped" innovators (Demirkan & Spohrer, 2015), where team members have in-depth specializations coupled with an understanding of the related skills and discipline areas of colleagues. The benefit of working with diversity, in all its forms, also needs to be made overt, and discussed with students.

Race argues that "Helping students make sense of things that they are learning together, particularly by explaining things to each other, and making decisions together" (2007, p. 136) is an essential approach in effective learning. However, an important element of this strategy is to ensure that faculty and students are all informed and engaged not only with the organizational aspects of this approach to team working, but also the reasoning behind it.

Building on First Year Practice

If a majority of engineers do tend to be systemizers, then learning transferable skills should be approached in the same way—that is, planned and presented to students as a structured experience that evolves over the length of their degree program. Strategies include mapping examples of transferable skills experience (such as public speaking) and reviewing the key points of the courses at the end of each semester and mapping the build-up of skills and experience.

With regards to team working, it is important to revisit team building and maintenance strategies throughout the degree. This could help with retention. If it is not properly managed, it can have the opposite effect, with poor group experiences being shown to add to attrition, especially among minority groups, which in the case of engineering includes women. Team projects potentially accelerate attrition: women in engineering and technical disciplines frequently report negative team experiences, which may make them question their place in the discipline (Wolfe & Powell, 2009). Building team-working experiences across year groups should help with creating that continuity of learning and discipline attitudes. It should help to support minorities (including personalities and interests, such as in automotive), and providing opportunities for this type of interaction could help to build critical mass for some minority groups across the year groups, and provide role models.

The most important principle for effective interpretation is to be person-centred. To engage in dual perspective with a particular individual to whom you are listening is to understand that person's perspective on her or his terms. Certainly, you won't always agree with other people's ideas or how they see themselves, others and situations. Engaging in dual perspective doesn't require you to agree with others' perspectives; however, it does require you to make an earnest effort to understand them....Too often, we impose our meanings on others, try to correct or argue with them about what they feel, or crowd out their words with our own. (Mohan et al., 2013, p. 180)

One of the key issues that needs to be highlighted for the approaches presented here (as opposed to conventional group working and organizational strategies, including anonymous, random groups) is that there will inevitably be conflict as personalities and perspectives are celebrated rather than homogenized. Faculty and students will need to be educated on working with diversity during project situations where differences of opinion and complaints are inevitably aired. These situations will again highlight differences, especially in cultural behaviors and responses. In particular, Mohan et al. highlight that "research indicates that race-ethnicity is related to conflict styles and to interpretations of them" (2013, p. 208).

This is also the case with gender, where communication styles and reactions are influenced by perceptions of gender stereotypes that sustain attitudes that impact the way team members interact. Results of a study by Wolfe & Powell (2009) suggest that engineering males were significantly harsher than other groups in judging complaints that are typically spoken by women. Engineering men were more likely to see the speaker with female-typical speech patterns as whining, incapable, insecure, wasting time, and were less likely to want to be on a team with the speaker.

RECOMMENDING A STRATEGIC APPROACH TO COLLABORATIVE WORKING

...it does indeed appear possible to cultivate intercultural competencies in undergraduate engineering students, at least in the context of a civil structural engineering integrated design course. A series of three one-hour cross-cultural intervention modules and critical reflections were delivered during a single semester's offering

of a senior-level civil structural engineering integrated design course that also had significant group design project work. A comparison of findings from initial versus exit self-assessment and peer assessment instruments indicated that students in the class showed statistically significant gains over the course of the semester with respect to intercultural competencies related to their own behavioral flexibility and interaction relaxation when interacting with group members from different cultures; it also showed significant gains in the intercultural competency of group members from different cultures in terms of their identity maintenance. These findings were further supported qualitatively through the students' initial and exit open-ended responses about cultural differences navigated among their group's members. (LaFave, Kang, & Kaiser, 2015, p. 8)

The main recommendation for future practice, alongside faculty education, is the introduction of "engineered" teams into first year education. For these, students are engaged with preparatory exercises to promote self-reflection, and self-awareness. They are introduced to a range of group roles, and then given an activity where they are asked to observe the behaviors of other students within a fairly large group to ensure different behaviors are modeled and to gain insight into different roles. These are for roles taken on by other members of the group, and also roles they found themselves favoring. They are then presented with a pie chart of group roles and discuss the importance of the synergy created by bringing together all the roles within a group. They are asked to reflect on the previous group activity, and identify two roles for themselves, which they have to write on a card. They are then required to form into teams as per the pie chart, and based on their cards. In this way, the groups are consciously constructed based on a shared understanding of the strategy to support best practice for group working. Students know their role in the group, and have actively chosen that role for themselves based on a reflective experience.

This strategy has been tested with large cohorts of students from a variety of disciplines, from Architecture to Engineering and Industrial Design. The results have been found to be effective based on participation rates, and the inclusion of non-majority students such as mature students and females in team activities. Research needs to be undertaken to formalize this approach, and compare it to more generic practices. However, the link to systemizing and context learning should again appeal to the majority of learners in the group and provide support for the minority, while supporting the individual, and celebrating individual skills within a group. This provides a strong basis for developing collaborative group working throughout the degree that values and supports diversity in skills and approach. This should therefore provide a base for engaging in complex problem solving in diverse teams once in the profession.

FUTURE ENGINEERS

This book chapter provides a strategic direction for future research into rethinking collaborative practice in engineering education for the development of future engineers. It also provides the basis for research in learning and teaching in higher education more directly targeted at specific disciplines. Rather than homogenizing the students and attempting to treat everyone in the same way, it is about researching how to respond more overtly to individual needs. While classroom numbers are increasing due to funding pressures, blended learning is providing ways to re-imagine and individualize learning. Further research into eLearning and the opportunities for engineering education would assist in developing guidelines to effectively harness the technology.

CONCLUSION

This book aims to provide strategies for increasing diversity in engineering study programs and the profession. In order to do this, it is important to acknowledge that there are issues specific to engineering that impact the experience of minority students and in the profession. While there is considerable literature on the development and organization of teams and the teaching of group working, the philosophy underpinning learning and teaching in higher education arguing for an educational experience informed by a knowledge of the individual needs of the students does not appear, as yet, to extend to the teaching of team working for the academics themselves. Engineering academics are part of a long established discipline, with traditions and values, and favored ways of working. There is also a dominant male culture, and a majority of students who are systemizers rather than empathizers. In recognizing the issues involved, this chapter has presented a case for discipline-specific learning on team teaching in the support of academics and student diversity.

Organizing and managing team working in engineering education should not be a superficial exercise. The stakes are too high for a discipline that needs to adapt to the changing demands on the profession. In a faculty, the academics themselves need to work as a team, to approach the development of transferable skills for the engineering profession specifically with those changing demands in mind. A degree-length framework needs to be agreed upon, and mapped. However, the first year needs particular focus, as it establishes the attitudes and values for the cohort that will extend throughout the degree—and beyond. Academics need to invest time in understanding the preconceptions and attitudes they themselves bring to the learning experiences. Just as this requires dedicated time outside the normal activities the faculty will engage with, so the strategy for supporting diversity in the first year student cohort also needs to extend beyond the classroom. However, fundamentally, there needs to be recognition that providing equality is not the same as supporting equity. In a discipline with dominant attitudes, behaviors, strategies, and values already in place, supporting diversity requires swinging the pendulum of classroom culture beyond the center. Learning experiences need to deliberately accommodate minority students. Engineering students also need to be given additional learning experiences to build their understanding of the situation, and just as importantly, they need to understand why the dominant approaches and cultures need to change. Students need to be helped to accept—and preferably embrace—the changes based on a clear understanding of the reasons behind them and how it will benefit them in the future, as well as the profession as whole. Attention to detail in the first year has the potential not only to retain minority students in engineering, but also to contribute to the professional development of the faculty, and improve the discipline as a whole.

The world is changing and the engineering profession is changing with it. For established disciplines with a long history such as engineering, evolving to keep pace with change is more of a challenge than it is for younger, more agile professions. Engineering is weighed down by the expectations around content, delivery, and outcomes. Going forward, engineering has to re-evaluate its priorities and practices in an era post digital revolution. In an age where aspirations for society include empathetic, engaged design and international practice, engineering education has to establish new attitudes and practices for the profession. Engineering educators need to work together to provide a blueprint for future practice that supports diversity, re-imagines the future, and redirects the present.

REFERENCES

Acker, F. (2005, December). Autism and engineers: Is there a connection? *Ingenia, 25*, 39-42. Retrieved on 25. 8.2016 from http://www.ingenia.org.uk/Ingenia/Articles/343

Adams, S. G. (2003). Building successful student teams in the engineering classroom. *Journal of STEM Education: Innovations and Research*, *4*(3/4), 1–6.

Baron-Cohen, S. (2007). *Asperger's: The engineer's disease.* Retrieved August 22 2016, from http://wrongplanet.net/forums/viewtopic.php?t=45209

Benyus, J. (2002). *Biomimicry: Innovation inspired by nature.* New York, NY: Harper Perennial.

Brown, T. (2009). *Change by design: How design thinking transforms organizations and inspires innovation.* New York, NY: Harper Business.

Demirkan, H., & Spohrer, J. (2015). T-shaped innovators. *Research Technology Management, 58*(5), 12–15. doi:10.5437/08956308X5805007

Dewart, B., & Rowan, L. (2007). Improving tertiary student outcomes in the first year of study: What your students may not tell you. *Journal of College Student Retention, 9*(3), 259–271. doi:10.2190/CS.9.3.a

Du Plessis, Y. D. (2011). Exploring teamwork paradoxes challenging 21st-century cross-cultural conflict management in a multicultural organizational context. *International Journal of Cross Cultural Management, 12*(1), 49–71. doi:10.1177/1470595811413108

Gore, A. (2013). *The future*. London: WH Allen.

Kuniavsky, M. (2010). *Smart things: Ubiquitous computing user experience design*. Burlington, MA: Morgan Kaufmann.

LaFave, J. M., Kang, H. S. & Kaiser, J. D. (2015). Cultivating intercultural competencies for civil engineering students in the era of globalization: Case study. *Journal of Professional Issues in Engineering Education and Practice, 141*(3).

Loy, J., & Tatham, P. (2016). Redesigning production systems. In S. S. Muthu & M. M. Savalani (Eds.), *Handbook of sustainability in additive manufacturing* (Vol. 1, pp. 145–168). New York, NY: Springer. doi:10.1007/978-981-10-0549-7_7

Mathias, J., Bruce, M., & Newton, D. P. (2013). Challenging the western stereotype: Do Chinese international foundation students learn by rote? *Research in Post-Compulsory Education, 18*(3), 221–238. doi:10.1080/13596748.2013.819257

Mohan, T., Archee, R., Gurney, M., Wood, J. T., Tamparo, C., & Lindh, W. Q. (2013). COMU1030 Communication skills: Spoken language and interpersonal. Albany, NY: Cengage Learning Australia Pty Ltd.

Pullin, G. (2011). *Design meets disability*. Cambridge, MA: MIT.

Race, P. (2007). *The lecturer's toolkit: A practical guide to learning, teaching and assessment* (3rd ed.). Abingdon, UK: Routledge.

Weimer, M. (2002). *Learner-centered teaching: Five key changes to practice*. San Francisco, CA: Jossey Bass.

Wheelwright, S., Baron-Cohen, S., Goldenfeld, N., Delaney, J., Fine, D., Smith, R., & Wakabayashi, A. et al. (2006). Predicting autism spectrum quotient (AQ) from the systemizing quotient-revised (SQ-R) and empathy quotient (EQ). *Brain Research, 1079*(1), 47–56. doi:10.1016/j.brainres.2006.01.012 PMID:16473340

Wolfe, J., & Powell, E. (2009). Biases in interpersonal communication: How engineering students perceive gender typical speech acts in teamwork. *The Journal of Engineering Education*, *98*(1), 5–16. doi:10.1002/j.2168-9830.2009.tb01001.x

KEY TERMS AND DEFINITIONS

Culture: The shared understanding of distinctive values, ideas and attitudes of a social group.

Digital Revolution: The collective rise of digital technologies that has created new ways of communicating, building and organizing data to support new practices enabled by digital technologies.

Distributed Manufacturing: Decentralized manufacturing enabled by digital technologies.

Empathy: The ability to understand the emotional and practical experiences of another person.

Ethnocentric: The belief that a person's culture is superior to other cultures.

Systemizing: The drive to organize based on systems.

Transferable Skills: Non-discipline-specific interpersonal and professional skills.

Chapter 7
Addressing Cultural and Gender Project Bias:
Engaged Learning for Diverse Student Cohorts

Jennifer Loy
Griffith University, Australia

Rae Cooper
Griffith University, Australia

ABSTRACT

Engaged student learning is based on creating significant learning experiences for every student. Attracting a more diverse student body into Engineering requires a re-evaluation of the conventional project topics that dominate the discipline. Recognising and addressing cultural and gender bias in the development of project work allows for the education of Engineering faculty on the development of a range of project work opportunities that support the learning for a more diverse cohort. The selection of set project work has the potential to negatively impact the learning experience of minority students. This chapter considers the elements influencing set project work and provides strategies for understanding cultural and gender bias, and for redesigning project work that provides for a more diverse cohort.

DOI: 10.4018/978-1-5225-2212-6.ch007

DRIVING DIVERSITY

It is most beneficial to consider all students as individuals and provide an education that caters to a full range of personalities, interests, backgrounds and social groups. In supporting diversity, there can be an underlying assumption that the driver is for an equality defined as "anonymity" within the cohort. Many of the strategies suggested in learning and teaching publications, for example on group work, advocate randomization as a way of ensuring that all students are treated equally (Race, 2006) and there is an emphasis on language that does not differentiate between students. However, this homogenization both assumes a balanced cohort profile as a starting point and that the minority should be always striving to become the "same" as the majority. In practice, isolating minority students within a dominant majority will not provide them with equality, as their voice will count even less as individuals spread throughout a dominant group, than if they were grouped together. Similarly, providing all students with the same project work and assessment mechanisms does not provide equity if those tools bias a particular students" learning preferences over another and not support the development of diversity that the future of the profession needs to provide the balanced workforce. The argument that the outcome will be a spread of marks, with all having an equal chance, is superficial in that it does not take into account whether the dominant majority is repeatedly succeeding over minority groups. More fundamentally, it supports the notion that minorities with diverse learning preferences should adapt to suit the preferences of the dominant majority. This further sustains the existing paradigm.

According to scholarship on learning and teaching, Bloom's definition of deep learning is required in order for students to genuinely gain an understanding of any subject. This applies to engineering education as much as for any other discipline—arguably perhaps more so because of the need for engineers to understand the broader implications of their work for the development of viable, rigorous, systems outcomes for specific tasks. Leaders in educational research argue that for this to be achieved, students need to be actively engaged in their learning. Dee Fink (2013) describes this as the need to provide significant learning experiences for individuals; that is, activities that they can each personally relate to.

This is without doubt a challenge, and particularly for a discipline with an inherited body of knowledge and practice that has evolved very specifically for a dominant majority. Yet the drivers here are not only about improving learning and teaching, or even about supporting diversity for its own sake, they also relate to changing practice in the profession as a whole and the need to evolve the discipline to encompass more divergent thinking and practices as professional engineering challenges are become more complex and interdisciplinary. In addition, as engineering outcomes grow with the perpetual development of new technologies, the range of potential "end users" is increasing in diversity, requiring an understanding of the complexities of society in the global community. By addressing the aim to support diversity in the classroom, engineering academics are also opening the door to a rethink of the assumptions and practices that the academic discipline is dominated by and that may be in need of revision in response to the changing requirements of an evolving industrial, social and economic landscape.

ENGAGED LEARNING

Educational theory contends that engaged students, those who feel empowered through their learning, are more likely to achieve deep learning outcomes than those who are not (Weimer, 2002). For engineering education, this can be more challenging than for some other disciplines because of the way it has traditionally been taught. Essentially, students tend to be taught the fundamentals in the early years of their degree, in preparation for the application work they engage in once those fundamentals are in place. Dee Fink (2013) argues that for authentic learning, students need to always be involved in learning experiences that have personal significance for them. This becomes a particular challenge in the context of engineering education. It is in direct opposition to the dominant learning pattern where students must understand the engineering science that provides the underpinning for the discipline prior to applying that knowledge in a particular scenario. However, engineering educators are increasingly working to provide practical learning experiences earlier in the degree where students can be involved in active learning and engage in Kolb's (2014) learning cycle. What there is less evidence of, are challenges to the idea of engaged, significant learning as being limited purely to an expression of experiential learning. This narrow view of engaged learning in engineering education, as referring to practical engagement only, fails to take into account the issues relating to diversity within current student cohorts, and the understanding of what engagement for all students means.

If engaged student learning involves providing the opportunity for all students to be involved in projects that have some element of personal significance, then attracting a more diverse student body into engineering will require an extension of conventional project topics to appeal to a broader range of interests. Identifying cultural and gender bias and re-evaluating existing projects from this point of view supports the education of engineering faculty on the development of project work opportunities that support significant learning for a more diverse cohort. This chapter considers project themes and how the projects are framed, and it provides strategies for counteracting cultural and gender bias. It considers existing educational practices and student experiences and discusses redesigning project work for a more diverse cohort. It also considers the marketing of engineering to target audiences beyond the male gender.

The differences in the development of professions and their associated academic disciplines are determined by multiple factors that complicate an evaluation as to why a particular profession or discipline is dominated by a majority social group. There may be a temptation to predetermine, for example, the reasons for the majority dominance of women in fields such as Nursing and Education, as linked to ideas of a traditionally "female disposition". The reality is more complex and driven by social and historical factors. This is illustrated by historical documents relating to the frustration of women during the Second World War, where they took over traditionally male jobs in areas such as manufacturing, then were forced to relinquish them once the male workforce returned home (Thompson, 2016). It is also illustrated by the 1960s necessity and success of Dame Stephanie Shirley in developing a work-sharing mechanism for women with dependents, to contribute to the information technology workforce at a higher level than previously possible (Kavanagh, 2002). Yet even in contemporary societies where women have achieved a level of parity in the workplace, and parity in academic achievement in schools, this has still failed to translate into females choosing to study science-based subjects at University in equal numbers to their male counterparts. Whatever the issues, the reality is that females are still very much in the minority in post-high school science and technology education and are particularly under-represented in engineering in higher education programs of study.

At the secondary level, female students continue to lag behind male students in rates of physics course-taking in high school (Riegle-Crumb, Farkas, and Muller, 2006; Ma, 2011b). Indeed, this gender disparity has remained fairly constant over the past 30 years even as the overall percentage of students taking physics has increased (Freeman, 2004; Nord et al., 2011). This gap stands in contrast to trends in advanced math course-taking, as females have reached parity with male students in rates of completing high school calculus (Hyde et al., 2008), as well as trends in test scores, such that while small gender gaps in math and science achievement remain, they have shrunk considerably over the last several decades (Xie and Shauman, 2003). Despite progress on these other indicators....women have made little inroad into physical science and engineering college majors over the last 30 to 50 years (England and Li, 2006; Riegle-Crumb and King, 2010). (Riegle-Crumb & Moore, 2013, p. 253)

In addition, once enrolled in an engineering degree, those same females who achieved success in high school may be disadvantaged once enrolled in tertiary education. This is because whilst there is considerable research on the learning preferences of pupils in high schools relating to gender and the impact of different teaching approaches and assessment mechanisms on success (e.g., Baldiga, 2013), there is little evidence of this research impacting engineering education. Similarly, there is little evidence of culturally aware teaching approaches and assessment mechanisms, or of the significant inclusion of cross-cultural examples in lectures or projects. Culture is a significant factor in the structure and values re-enforced in a discipline and bound up with ideas relating to identity. An example would be the impact on Scandinavian aesthetics on form, material choice, construction techniques, etc., for study within that country. Similarly, the schools of thought in Japanese architecture illustrate the impact of cultural referencing and identity on market acceptance and value for specific places. The tenets of engineering education in the US, UK, and Australia are bound up with a Western experience and perspective of the Industrial Revolution that inform the learning activities within engineering education in these countries. In both respects, there is little evidence of academic programs actively support for diversity—gender or cultural—in the way that engineering is taught and assessed.

This chapter will focus on providing starting points for improving support for diversity in engineering cohorts, by considering the implications of re-invigorating project work. It is not a comprehensive review in responding to learning preferences for all students studying engineering, but rather it provides a provocation for challenging existing practices in engineering education. Disciplinary traditions have a major influence on the shape and values within a discipline, for example because of the impact of a train of thought or philosophy that has dominated its development, for example for disciplines subject to the influence of the Bauhaus. This chapter calls for academics to objectively consider their practice from outside discipline norms, to inform their understanding of the implications of their teaching practices. Genuinely providing for diversity requires a rethink in relation to many factors impacting the development and delivery of a program. Faculty need to be helped to see the value in addressing these issues and challenging the existing paradigms for all stakeholders, and then provided with clear strategies to do so.

Gender Bias

This study is guided by theories of gender as a social structure or system, one that is constructed across multiple levels or dimensions, ranging from the macro level of large-scale institutions and broad cultural beliefs to the micro level of local environments and personal interactions (Risman, 2004). Past research has documented how inequality is created and maintained by reinforcing connections between beliefs, norms, and behaviors across these various domains (Ridgeway, 2011). Yet it is also important to focus on variation in the extent of inequality and search for those places gender appears to be less salient in shaping expectations and behaviors (Eisenhart and Finkel, 1998; Deutsch, 2007; Risman, 2004). When viewing gender through this lens, theorists argue that gender can be alternatively constructed or deconstructed in ways that resist or subvert traditional paradigms, revealing that gender differences are neither inevitable nor omnipresent. (Riegle-Crumb & Moore, 2013, p. 253)

When considering gender in education, it is fundamental to understand gender as a continuum and not as a binary condition. This is not about a biological constant but about whereabouts on a continuum, from one extreme to the other, the students self-identify. As Riegle-Crumb and Moore point out "...gender can be alternatively constructed or deconstructed in ways that resist or subvert traditional paradigms, revealing that gender differences are neither inevitable nor omnipresent" (2013, p. 253). However, even within this approach, the idea of a continuum would have layers of interpretation—for example, would the continuum be based on male to female social stereotypes or would it be based on definitions of extreme male brain as systemizers to extreme female brain as empathizers? Is it explained by behavioral and attitude preferences, and how much are these impacted by socialization and cultural attitudes in different settings? For the scope of this chapter, the important element here is that understanding gender as a factor is not seen as merely providing for two opposites as "male" and "female" but rather as providing for a range of individuals with complex learning needs, interests and preferences. In improving diversity in the classroom, academics need to ensure that project work is not defined by limited ideas of gender that further re-enforce stereotypes, but support learning opportunities that individual students from diverse backgrounds and with diverse learning preferences and interests can identify and engage with. Academics need to cultivate a lack of predetermination, and foster inclusivity and supportive language in the classroom. This aligns with the professional development that engineering academics are likely to be engaged with at this time for the integration of experiential learning into the curriculum. This relates to the need to graduate increasingly innovative engineers who can work across disciplines in large, diverse teams on problem framing as well as complex problem solving.

Strategies to provide for cultural diversity in the educational experience are equally based on providing a paradigm where the basis for the learning experience is not predetermined by assumptions and assessment and not constrained by conventions. To provide a student with learning opportunities that are embedded within their identity requires structures that enable the student to tailor the project, rather than be faced with a fixed framework for learning that does not allow for their unpredictable experiences and background. With the rapid rise of globalization (Gore, 2013), it is the right time for the engineering faculty to embrace this challenge and provide a major shift in discipline thinking to foster a new learning paradigm, supporting classroom diversity. This will support the abilities of all students to work effectively in global teams, and on more diverse projects than are currently standard in some models of engineering education.

Project Bias

If ideas of gender are removed as much as possible from an evaluation of set project work in engineering, and gender instead considered as a continuum (even if the model used is not a complex or layered one) then it provides a starting point for considering the provision of projects that support diversity. For this chapter, the continuum being considered is based on Baron-Cohen's ideas on the relationship between systemization and empathy (Billington, Baron-Cohen, & Wheelwright, 2007). The intent is to provide an approach to developing study project options that work along that continuum without being gender specific.

A recommended starting point for self-evaluation would be for academic teams to plot the projects that form the basis of their in-class activities and assessments along a systemization and empathy continuum. If there is a bias towards one end of the continuum or the other, then there is an argument for providing a broader range of options in order to engage a more diverse cohort. However, the reality is that most project work can be reframed to allow the student to work along that continuum at where they best self-identify. The important element is how it is framed, expressed, supported, informed and then, most vitally valued and assessed. For example, a project that is based on bridge building can be presented as purely about structures, independent of any other factors. This learning experience could then potentially fail to engage students whose learning preferences are not at the extreme systemizing end of the spectrum. However, if the project was provided within a context that links it to a social imperative, such as providing a bridge for a specific purpose, then it could extend the learning benefits to include students with learning preferences towards the empathetic end of the spectrum. The purpose could be as unusual and diverse as to allow elephants to cross a chasm in a game park, or to provide a way for joggers pushing prams to cross a motorway. Not only could this approach to framing the project work engage those who identify more on the empathetic end of the continuum, but also ideally it could form the basis for academics to support individual students in tailoring the project to their own personal experiences and interests. This could create the opportunity for genuine significant learning and stimulate innovative thinking for all students, including the systemizers dominating existing cohorts.

There are implications for assessment with this approach, as mechanisms would need to be adjusted to provide marks for context research and for showing how this informed the project as well as the fundamental mathematics and physics involved. As importantly, the individual bridge design would need to be able to fail without jeopardizing marks for the students because it would not be pre-tested by the academic. This could add to the subjectivity of the marking, and that would add to the importance of moderation and a trained—preferably diverse in terms of the continuum—academic team. However, the benefits could potentially be more engaged, deeper learning outcomes for the entire cohort. In the longer term, this could lead to better retention. Changing the educational paradigm is not without risk or effort and would need to be part of a long-term strategy for the department for which all academics were educated and to which they are committed.

Project Strategies

Rather than focusing on narrowly pre-defined topics, a change in paradigm for engineering education needs to be based on providing students with a broader range of options within a manageable, directed study area. For example, the growing sustainability imperative provides a myriad of project options that would cater for the interests of systemizers and empathizers all along the spectrum. For students focused on fundamentals, the projects could be objective rather than sympathetic, whereas for students self-identifying as more interested in empathetic projects, there would be options for community or environment-based topics. For curriculum reasons, these would still include constraints such as the use of a motor, or levers and pulleys, but the context of the learning would become less rigid.

There are other advantages to this themed approach. Firstly, students would be able to be involved in real-world learning, and that has been found to be an effective learning strategy for engaging students. Secondly, it creates learning situations where the outcomes are not predetermined by the academics, with the students required to contribute more to their own learning. This changes the balance of power in the classroom, moving the academic from the role of expert, to the role of facilitator. This has also been found to be a strategy that empowers learners and improves learning outcomes (Weimer, 2002).

Stepping back from the immediate classroom situation and considering transferable skills, graduate outcomes and the changing nature of problem framing, problem solving and team work characteristic of the emerging professional engineering environment, this approach has the added benefit of supporting the development of pro-active, lifelong learners. Rather than following instructions, learners engaging in a boarder project approach such as this, will learn research, decision making and self-reliance as well as problem framing, solving and communication.

Language

As engineering education moves towards experiential learning and project based learning, then there will be more of a requirement for students to communicate and collaborate. If that is the case, then part of the professional development for academics and learning for students in terms of supporting diversity, needs to be around understanding and supporting the choices students make, and the values and life experiences behind those choices. This includes creating a safe, yet critically engaged learning environment, and teaching students to provide peer review and feedback on projects that have different priorities and values from their own. Language provides pitfalls for both academics and students in a majority dominant situation. For example, the way that students, including those who identify as LGBTI, who have been socialized as females will communicate differently from the way students socialized as males will. Studies have shown that language patterns relate to socialization and these can cause communication problems and bias. For example, language patterns typically used by females to make complaints can be characterized as "whining" even when spoken by males. Groups socialized as female will communicate differently from those with members socialized as males. Patterns of communication in minority groups, such as LGBTI, have been found to differ again. Culturally diverse groups also bring language patterns that are inconsistent, and colloquialisms can further complicate communication.

For project work to effectively support diversity, there is considerable groundwork that needs to be conducted within a learning group. The academic needs to establish the working practice, communication and values of the group in collaboration with the group itself. The challenge is to do this in ways that will empower the individuals and not merely re-enforce the views and ideas of the dominant majority. Short lead-in activities and engaging the students in deliberately diverse projects will help to build a shared language and values for the group. This may involve pushing students—and academics—outside their normal practice to deliberately engage in projects from the opposite ends of the spectrum as a way of rebalancing the group and indicating to the majority the level of acceptance for alternative views and interests. However, this is a challenging strategy to adopt and academics need to have experience of managing the discussions this will provoke.

According to Martin and Barnard (2013), women who choose male-dominated careers are likely to switch into gender-balanced or female-dominated careers. One of the reasons the authors describe is that women feel pressured to change their natural behavior in order to achieve in a male dominated environment. For example, Wolfe and Powell (2009) argue that male engineers have particularly negative perceptions of what they consider to be female-typical speech patterns and language. Women were expected to "mirror" male behaviors and attitudes in order to gain acceptance, yet Wolfe and Powell's (2009) research also found that engineering males were actively more critical of women who made mistakes than they were of their male counterparts. If that is the case, then team working on projects in engineering education without sufficient foundation work will perpetuate gender issues and, based on the research (Martin & Barnard, 2013), contribute to female attrition.

According to Prescott and Bogg (2011), and re-enforced by the research of Martin and Barnard (2013), women are motivated by different drivers to those of men. They identified men as being motivated by competition through ideas of status, power and social comparison, while women desired to do a their job well, and be recognized for that, and also to positively contribute to the function of the organization. If this is the case, then it also impacts the nature of projects and assessments and how they are organized in engineering education. Projects that have a competitive focus, such as those involving racing cars, or moving freight, are therefore less likely to provide motivation to female students. Projects based on social and community issues, such as improving safety, are more likely to be of interest to females than males.

Martin and Barnard (2013) argue that "organizations need to legitimize women's characteristics, natural behaviours and values and give them a platform in order to level the playing field for both genders". Adding in diversity across the gender and cultural spectrums the same imperative applies—to legitimize individuals' characteristics, natural behaviors and values, and then to provide opportunities for all to excel without compromising, and without being pressured to "mirror" the behavior of a dominant, entrenched majority. Increasing the sense of responsibility for supporting diversity within the engineering community, developing support strategies and networks, investigating responsiveness and increasing awareness of acceptable group behavior is necessary for a genuinely level playing field to exist.

The process of developing a holistic understanding of the issues surrounding gender is complex and potentially endless. Within the scope of this chapter and the topic of diversity in engineering, there can be limited development of feasible solutions for the complex challenges surrounding gender issues. However, it is possible to highlight the values and attitudes that exist within traditional disciplines such as engineering and those who work in the field. Academics can reflect on current teaching practices, how they impact on the diversity of student cohorts and the retention of minority groups—and how those actions and attitudes are contributing to the wider society.

In Davey's (2008) study, typical masculine behaviour related to success in a male-dominated environment included political game playing, aggressiveness, backstabbing, point-scoring, overconfidence and "stitching people up". The mechanisms women use to cope in male-dominated environments include mentorship and adopting participatory leadership styles like being caring, fair and encouraging, which is more attuned to their natural feminine inclination (Chovwen, 2007). Contrary to this, female graduates adopted uncharacteristic masculine behaviour, like self-seeking and individualistic behaviour, to survive (Davey, 2008). Ironically, some opinions point to the detrimental effect of adopting characteristic male behaviour on women's feelings of authenticity and work identity (Du Plessis & Barkhuizen,

2012). Therefore, women seem caught between resisting and accommodating masculine politics (Davey, 2008). (Martin & Barnard, 2013, p. 3)

Cultural Diversity

Mā te kōrero, ka mōhio.
Mā te mōhio, ka mārama.
Mā te mārama, ka mātau.

Through discussion we become aware.
Through awareness we gain understanding.
Through understanding we gain proficiency/expertise.

Whakataukī
Culturally responsive learning contexts are those where the learner can bring their own experiences into the classroom context. (Te ManaKōrero: Relationships for Learning, 2007)

The impact of cultural diversity on experiential learning planning goes deeper than providing a choice of projects to suit the interests of different groups of students. It is about academics being willing and able to embrace everything diversity means. This includes being open to the social context influencing the attitudes and values for a particular cultural group, as well as factors such as historical influences, attitudes to technology, customs and ideologies. These factors will impact what the student values, and their approach to framing problems, working through research and experimentation, and their final evaluation of the outcomes, as well as how they present them and communicate the results. This can be challenging for academics.

Even successful teachers can find it difficult to teach students and incorporate content from a culture that is not their own. Teaching for diversity, supporting minority students, and using culturally-responsive teaching practices are well-known concepts (for example, Airini, O'shea, Tarawa, Sauni, Ulugia-Pua, Sua-Huirua, & Curtis, 2007; Bevan-Brown, 2005; Bishop, Berryman, Cavanagh & Teddy, 2009; Cummins, 2001; Gay, 2000; Ramsden, 2003; Zepke & Leach, 2005). However, the transition from understanding the theory to implementing it in practice is often complicated—and when such practice requires teachers to step outside their cultural comfort zones, it can be downright intimidating. (McDonald, 2008, p. 5)

Different cultural groups have beliefs and traditions that will bias what they consider to be important. This is recognized across sub-cultural disciplines within engineering itself, where the priorities and approaches for civil, mechanical, electrical, etc., are different. The Victoria University of Wellington, New Zealand, Strategic

Plan, 2009 – 2014, emphasizes the importance of recognizing the different learning needs of students by using a variety of teaching approaches and showing sensitivity to cultural issues. Introductory activities to help everyone involved—students and academics—to gain an understanding of the different attitudes in a cohort are necessary prior to engaging in major project work.

Sensitivity to cultural issues requires providing students with an opportunity to illustrate, demonstrate and discuss what these are and where the points of difference may be. For many students there will be a lack of understanding of what the characteristics, attitudes and values of the groups they themselves are part of, and those their peers identify with. Preparatory activities can help students to explore these values and explain them to other students. However, these activities may themselves be fraught with problems. An example from first year studies in the Digital Media program at Griffith University, Australia, illustrates these challenges. It involves a project designed for first year students on George Orwell's novel *1984* ([1949] 2011). Students were asked to read the text, select a major theme relating to critical production of mass media and discuss, defend and critique their outcome within a group.

Students within this project experienced a number of diversity-based issues, the most dominant of which did not become obvious until after the project was completed. The brief asked students to critique their role within a political system that produced mass media and the relational impact of this upon its consumers. The "dominant majority" of students could relate to this topic within the context of their use of dominant social media (such as Facebook) and critical literature surrounding data collection and an individual's right to privacy. However, what was not fully considered by the teaching team was the background and political experiences of the cultural minorities. Students with diverse international backgrounds were unable to engage in the topic the same way, believed it was inappropriate, or found it ethically challenging. Within their teams, the minority students found it difficult to engage in all discussions and sometimes became the "quiet non-participator". In this example, the majority group reflects individuals who have the Western cultural capital, who often had little exposure or knowledge of international differences regarding freedom of speech or laws preventing the criticism of government. They "Western majority" expected their team members to voice their opinions freely and reflect on society in a critical manner and demonstrated little compassion or support to their team members who failed to contribute equally. The marking criteria added another level of majority bias by disadvantaging students who did not participate in critique or discussions.

Classification of Project Work Strategies

Engineering is a practical discipline. It is a hands-on profession where doing is key. Consequently, prior to the creation of engineering schools, engineering was taught in an apprenticeship program modelled in part after the British apprenticeship system. These early engineers had to design, analyze, and build their own creations—learning by doing. (Feisel & Rosa, 2005, p. 122)

The foundations of engineering education began with the apprenticeship model. Practical problem solving and experimentation through engagement with real world problems therefore formed the basis of learning for future engineers. As engineering moved into academia, the practical aspects have become to an extent institutionalized. According to Fiesel & Rosa (2005), engineering laboratories have become less about the exploration of engineering science principles and more about the qualification of data.

When students, especially undergraduates, go to the laboratory, however, it is not generally to extract some data necessary for a design, to evaluate a new device, or to discover a new addition to our knowledge of the world. Each of these functions involves determining something that no one else knows or at least that is not generally available. Students, on the other hand, go to an instructional laboratory to learn something that practicing engineers are assumed to already know. That 'something" needs to be better defined through carefully designed learning objectives if the considerable effort devoted to laboratories is to produce a concomitant benefit. (Feisel & Rosa, 2005, p. 121)

The recent recommendations of accrediting organizations, such as Engineers Australia, expand the substance of practical work in engineering education from fundamentals testing to project work that includes "engineering ability" and "professional attributes":

On the basis of previous recommendations by the Engineers Australia as well as national and international research, it is very important and critical that a framework is defined and is unique to Engineering projects. Based on Engineers Australia Stage 1 competencies it is proposed here that fundamental knowledge base, engineering ability, and professional attributes are the key elements of competency and part of the integrative learning principle for a project approach. (Chandrasekaren, Stojcevski, Littlefair, & Joordens, 2012, p. 6)

However, this definition is limiting. If engineers are to have the "big ideas" of the future, then engagement in practical and project work needs to foster a more ambitious intent for engineering education, and that should drive learning opportunities. Fiesel and Rosa (2005, p. 121) argue "The function of the engineering profession is to manipulate materials, energy, and information, thereby creating benefit for humankind. To do this successfully, engineers must have a knowledge of nature that goes beyond mere theory". According to Chandrasekaren et al. (2012), project work in engineering education can be organized into three classifications: university-based projects, industry-engaged projects, and community project work. Feisel and Rosa (2005, p. 123) disagree, stating that the three classifications need to be brought together: "A common goal is to relate theory and practice or to bring the "real world" into an otherwise theoretical education". The differences are essentially based on how the brief is established, sources for research and criteria for evaluation. For university-based projects, students have more freedom to work on their own brief, but potentially less objective research material to draw on. Industry-based projects tend to be more constrained, but the real word context can provide a motivator for students.

Industry projects are project-based activities in partnership with an industry organization. Students complete a research project focused on a given leisure, event or industry issue, problem or opportunity. Students mostly work independently on their selected topics. A teaching and learning approach where individuals conduct work activities (paid or voluntary), including research combined with intentional educational activities. The project involves a substantial research component, which can range from market research such as a questionnaire survey or focus groups, to observation, in-depth interviews or analysis of existing data. (Chandrasekaren et al., 2012, p. 4)

Community projects can be more complex to manage and the outcomes often less clear. However, they can provide students with more understanding of problem framing and project management, as well as problem solving.

These projects provide an opportunity to connect with other students from different faculties and get actively involved in the community. By doing these projects, students will not only be contributing their knowledge and skills towards helping others in the community, but they also gain a range of career related transferable skills such as networking, teamwork, marketing, supervizing, coordinating, organizing, public speaking and more. (Chandrasekaren et al., 2012, p. 5)

As highlighted in the directive from Engineers Australia, the inclusion of project work, and the organization of that project work, in engineering education expands the brief beyond the technical into a collaborative and leadership role in technological and human development. Engineers need to be able to view problem framing, research and problem solving holistically in order to maintain relevance in an increasingly complex professional work environment.

After World War II many of the great inventions that occurred as a result of the war were developed by individuals educated as scientists rather than engineers. The ASEE chartered a committee to "...recommend patterns that engineering education should take in order to keep pace with the rapid developments in science and technology and to educate men who will be competent to serve the needs of and provide the leadership for the engineering profession over the next quarter century". (Feisel & Rosa, 2005, p. 122)

A key issue here is the ability of engineers to be able to look at problem solving without traditional discipline boundaries and limitations. They need to be able to contribute productively to teams working on increasingly complex, cross-disciplinary problems.

...engineers who can speak many languages—and I don't mean French or Mandarin; I mean the ability to talk to lawyers, users, and policy makers—those were the most effective engineers. (Holmes, 2015)

In addition to building multidisciplinary teams, greater diversity in conventional engineering teams will provide new perspectives. This is a key issue identified for other disciplines where minorities—in engineering-related disciplines, these often include women—are under-represented.

If you have women on the design team, you're more likely to get discussion about privacy side effects and human/computer interface. Having a diversity of views and experience always leads to better results. (Holmes, 2015, p. 26)

In providing alternative viewpoints, diversity can challenge the ethnocentricity that can occur in majority-dominated groups. The recognition that there are alternative attitudes and modes of practice support the development of positive graduate attributes for internationalized professional practice. However, it is not enough to simply improve diversity within members of a learning group, the academics need to actively provide learning opportunities that provide positive re-enforcement of ideas, attitudes and preferences different to those of the dominant majority. In order to swing the pendulum back to the middle, arguably academics need to deliberately push the pendulum of the characteristics of learning experiences to the other side of the arc.

Engineering Empathy

In reviewing the profile of project themes in engineering education, one of the key characteristics is the reliance on competition and speed as a basis for project work. Racing cars—model and full size—is a recurring theme in engineering education. While there will be female students who are motivated by the same elements that drive an interest in competitive sport, there will also be females and males for whom the motivations are more subtle. Yet, if considered objectively, this can be seen as a gender bias, rather than based on genuine constraints imposed by the discipline. Deliberately choosing projects that have their foundations in feminized interests and activities sends a clear message to students that such projects are acceptable and even—ideally to help reset opinion—encouraged. Project work in areas that are less overtly feminized can also provide respond to the drivers that research identifies as motivators for women more than men.

Projects with foundations towards the empathy end of the continuum can be less overtly embedded in feminized areas of interest, but still provide a response to alternative motivators than competition. A good example would be project work that is linked to working with life-improvement for animals. One approach would be to combine opportunities for students to choose to work with sports technology and animal husbandry within the same project. For example, with projects framed from a review of the horse racing industry. From transport, to the construction of jumps, to safety barriers, to racing helmets, to the development of racing shoes, innovations in the treatment of leg breaks and the movement of horses for vets—a single theme can provide for a broad range of interests without being overtly one type of project or another. Even if this approach caters for one additional student, and the majority still choose projects that are competition based, or purely mechanically driven, it will provide the opportunity for those with more diverse interests to study engineering and new ways of looking at engineering problems in the world.

Engineering is really just a way of thinking about things. And when you go out in the real world after class, you see things differently. It's really cool to know what goes on behind the scenes in your everyday life. Engineers shape the world that everybody else lives in. They don't just live other people's ideas. And to be a part of that creation, to me, is just awesome. (Lindsay Perry on Engineer Your Life, 2016)

This approach would not only provide for diversity, it would contribute to the broader shift in paradigm for engineering education advocated by researchers in the discipline:

> *A cultural change is warranted in engineering, as is reform of the established theory-based curricula; currently there is little emphasis on experiential learning through design or the development of creativity. By contrast, design pedagogy fosters creativity by developing and nurturing problem-solving skills and providing regular opportunities for students to refine these skills through experiential project-based learning. Creative activities such as "reflection in action", problem framing, divergent thinking and open-ended problem solving are integral to the designer's education, but are notably absent from engineering curricula. Yet creativity is central to innovative problem solving and as such should be integral to the education of engineering designers. To be creative, engineers must desire uniqueness, accept unusual ideas, tolerate the unconventional and seek unexpected implications. In this regard, the engineering community can benefit from close observation of design pedagogy. (De Vere, 2009, p. 1)*

Projects that are deliberately embedded in cross-cultural situations can provide students with learning's that challenge their conventional perspectives. The fundamentals can be taught within an extreme variation of contexts; for example, based on an extreme challenge, such as supporting the latest generation of Gurung tribes people in Nepal. For centuries, the Gurung people, who live near Lamjung in the Himalayas, have scaled the severe cliffs to harvest honey from hives from underneath overhangs. However, this tradition is dying out because of the dangers associated with the climb. The feats of engineering required to tackle the task, plus the social issues around the changing priorities of young people in the area combined with the threats—and opportunities—of tourism, provide technical and social challenges. There are issues involved with teaching a project based in a different country that can impact the experiential learning involved—and the latest research into the detrimental impact of volunteer tourism (Loy, Tatham, Healy, & Tapper, 2016) discourages direct involvement of students with projects in remote regions—but academics can take the essential basis for the project brief and adapt it to their students" learning environment, and still have the opportunity to bring in learning on alternative cultural perspectives, and human-centred engineering. The mechanics involved would remain the same, while the context would provide a fresh approach to learning the fundamental principles. Feisel and Rosa (2005, p. 121) assert "the overall goal of engineering education is to prepare students to practice engineering and, in particular, to deal with the forces and materials of nature."

Mixed Messages

But it has few women, few minorities and far fewer women than other areas of computer science....this creates a chilly climate for both women and minorities. (Holmes, 2015, p. 27)

Marketing engineering to females is currently full of curiously mixed messages, from "Girls Only" days ("Girls' Day" 2016), which suggest that girls will not engage in a normal open day and yet somehow then cope with a predominantly male cohort, to images of women in hard hats contrasting the bright colourful graphics chosen for certain websites targeting girls for engineering (*Engineer Your Life, 2016*).

This confusion may in part be caused by confusion within the profession itself as to how it markets itself. The reality is that, as for all diversity issues, rebalancing the profile of those in power—in this the academics and professional leaders in engineering—there will always be confusion. For engineering more than other disciplines, this seems to be a challenge, which could in part be due to the funding-related priorities for appointments.

As so many engineering programs have developed an increasing interest in research, the faculty reward system, in the opinion of many, has shifted away from recognizing contributions to undergraduate education and toward rewarding research productivity. (Feisel & Rosa, 2005, p. 123)

During a series of interviews conducted by the author with potential engineering academics the response to a question on what the candidate would do to address the gender imbalance in the undergraduate engineering program, responses included the repeated comment that it would be good to have more women as they "make everything so pretty", that women would not be interested in engineering because it is a "dirty" job, and that women are "incapable of the thinking required for engineering". On questioning, it was found that many applicants had never taught a female, nor worked with a female, nor published with a female. This appeared to be a particular issue in relation to candidates from cultures where the opportunities for women were restricted and there were gendered expectations in the workplace.

Furthermore it also provides empirical support for the idea that the salience of gender varies across contexts, so that not only countries but also communities differ from one another in the extent to which traditionally gendered status expectations shape beliefs and behaviors (Deutsch, 2007). In communities where a higher percentage of working women are employed in STEM occupations, larger gender stereotypes at the societal level may be subverted by a picture of what is possible that differs from that typically associated with more traditional gender roles. Such alternative definitions of gender likely permeate the interactions and experiences not only of high school girls, but those of parents, teachers, and men and women throughout the community. (Riegle-Crumb & Moore, 2013, p. 268)

Standardized tests recorded in the US confirm that in schools girls score just as well as boys in mathematics (Hyde, Lindberg, Linn, Ellis, & Williams, 2008). Yet while statistics based on research in the Netherlands re-enforce mathematical parity, with 48% of undergraduate degrees in mathematics being achieved by females, the same research shows that the perception that females lack mathematical ability still persists in parents and teachers. Men remain highly over represented in science-based subjects in higher education—in particular engineering—whereas women are over represented in medical studies and social sciences (Riegle-Crumb & Moore, 2013). One argument is that personality types associated with gender socialization could be a factor. Research conducted in Austria suggests that different personality characteristics on their own continuum, such as extravert to introvert, are identifiably related to different professional interests and preferences (Boone, van Olffen, & Roijakkers, 2004).

Several studies have shown that there are significant differences in personality characteristics between boys and girls (e.g., Costa et al.2001for an international comparison in 26 cultures across the world; Goldberg et al. 1998 for a U.S. sample; Hendriks et al. 2008 for a Dutch sample). Scholars often use the Big-Five factor structure of Goldberg (1993) to describe people's personality characteristics, using five dimensions: Extraversion, Agreeableness, Conscientiousness, Emotional Stability (or inverse Neuroticism), and Openness to Experience (NEO-PI-R; Costa and McCrae1992) or Autonomy (FFPI; Hendriks et al. 1999a,b). (Korpershoek, Kuyper, & van der Werf, 2012, p. 623)

How quantifiable this is, and how much it is predetermined by nature versus nurture is still the subject of debate, but nevertheless it highlights the importance of ensuring that the learning environment and the professional working environment are overtly driven by open-ended learning experiences that do not favour one individual over another.

RECOMMENDATIONS

Despite an increase in calls for reform of engineering education, engineering faculties continue to focus somewhat narrowly on the science of engineering, without sufficient curriculum opportunities for students to develop either design skills, or creative methodology. Design is fundamental to engineering practice, and therefore should be a motivating factor in engineering learning. Without focusing on design activities and creativity, we will continue to graduate engineers who are competent technically, but not prepared for the practice of engineering nor the challenges of the 21st century. (De Vere, 2009, p. 1)

The essential recommendations for meeting the needs of student diversity through project opportunities for students are based on challenging the stereotypes that currently dominate project selections. Rather than replacing the current selection of projects with a broader range of projects that could label the student who chose the ones less stereotypical for the group, the recommendation is that academics develop an approach that allows for gradations in project work along a continuum from those appealing to pure systemizers to those drawn to empathetic projects, and a mix of projects in between. This allows students to self-identify academic practice and theory that meets their interests and breaks down a binary division between project profiles. Ideally, students would be encouraged to work across overlapping and related projects, and where possible identify shared research foundations on which they could build collaborations.

We collaborate all the time. If we didn't, we'd be lost! And once you get out into real-world engineering, it's not just going to be one engineer working on one problem. Collaboration is important. And that means communication skills are really important. You have to be able to present your ideas effectively. (Reeves, 2016)

Ideally, projects should introduce new topics each time. These would allow students to genuinely engage with challenges that are new to the department and academics, and also reduce the ability of students to plagiarise the work of their peers. Ideally themes should provide opportunities for students to genuinely engage with the work from a particular point of view. They could explore and explain their perspective without being penalized for being different to that of the academic or a dominant majority within the cohort. One approach is to constantly engage with real-word developments for project themes. For example, based on emerging technologies and theory on the development of future cities. Legge (2012) discusses developments that provide good examples for rethinking practice as project starting points. These include themes such as "collaborative urbanism" and the "pop-up movement". Other themes suitable for working across the systemizer to empathizer spectrum are identified in the 2015 publication *31 Brilliant Ideas for a Better World* (Nolan & van Lier, 2015), including "Emergency Building for People in Need" (p. 28), "Building Block for Habitat" (p. 68), and "Saving Nature by Designing It" (p. 78). By focussing on project work that has no traditions attached, students will be free to develop their own learning without prejudice. Riegle-Crumb and Moore (2013, p. 253) suggest "Local environments have the capacity to construct their own micro-level gender systems where individuals are less constrained by traditional definitions of gender". By providing projects that are less predetermined, more complex and require students to practice their interpersonal skills as well as demonstrate technical knowledge and research, engineering education is responding to the needs of the profession in repositioning it as outward looking, and engaged for the emerging inter-disciplinary context most engineers will be working in for the future. This repositioning should help universities to attract students with a wider range of attitudes and personalities to re-invigorate the profession as at the forefront of innovation and real world problem, community-based engagement.

The study of Morgan et al. (2001), for example, found that college students in the U.S. perceived science-related careers as less likely to afford interpersonal goals such as helping people. A practical implication of the results presented is that, as a consequence of the overrepresentation of the more introverted students in the SCIENCE profile, stimulating introverted students to choose the SCIENCE profile in secondary education might be more effective than trying to convince extraverted students to pursue a math/science career (Korpershoek et al., 2012, p. 12)

CONCLUSION

It is difficult to discuss diversity and inclusion without straying into assumptions and stereotypes. As discussed in this chapter, it is more beneficial to consider all students as individuals and provide an education that caters to a full range of personalities, interests, backgrounds and social groups. What engineering as a profession and academic discipline must not do is ignore the lack of diversity in cohorts and the attrition of students. As with all disciplines where a majority has dominated for generations, the reasons are deeply embedded and complex. Changing attitudes requires professional development and open collaboration and engagement. Positive discrimination could help to redress the gender balance for academics, which would in turn help support a more diverse cohort, but ideally the rebalance would come from a repositioning of the discipline to respond to the changes required for the evolution of the profession worldwide. By changing the priorities for the profession, the discipline faculty would change, as would the basis for project work, and subsequently, with appropriate marketing, the profile of the students interested in the degree. It is a long-term challenge, but one that engineers and academics need to engage with in order to maintain the relevance of the subject, and keep engineering at the forefront of leadership in an evolving world.

REFERENCES

Baldiga, K. (2013). Gender differences in willingness to guess. *Management Science*, *60*(2), 434–448. doi:10.1287/mnsc.2013.1776

Billington, J., Baron-Cohen, S., & Wheelwright, S. (2007). Cognitive style predicts entry into physical sciences and humanities: Questionnaire and performance tests of empathy and systemizing. *Learning and Individual Differences*, *17*(3), 260–268. doi:10.1016/j.lindif.2007.02.004

Boone, C., van Olffen, W., & Roijakkers, N. (2004). Selection on the road to a career: Evidence of personality sorting in educational choice. *Journal of Career Development*, *31*(1), 61–78. doi:10.1023/B:JOCD.0000036706.17677.ee

Chandrasekaren, S., Littlefair, G., & Joordens, M. (2012). Learning through projects in engineering education. *Proceedings of the SEFI 40th Annual Conference 2012* (pp. 1-8). Brussels, Belgium: European Society for Engineering Education (SEFI). Retrieved from http://dro.deakin.edu.au/eserv/DU:30048241/chandrasekaran-learningthrough-2012.pdf

De Vere, I. (2009). *Developing creative engineers: A design approach to engineering education*. Paper presented at 11th International Conference on Engineering and Product Design Education, Design Society, Brighton, UK. Retrieved September 12, 2016, from https://www.researchgate.net/publication/274718463_Developing_creative_engineers_a_design_approach_to_engineering_education

Dee Fink, L. (2013). *Creating significant learning experiences: An integrated approach to designing college courses*. San Francisco, CA: Jossey Bass.

Dewart, B., & Rowan, L. (2008). Improving tertiary student outcomes in the first year of study: What your students may not tell you. *Journal of College Student Retention*, *9*(3), 259–271. doi:10.2190/CS.9.3.a

Engineer Your life. (2016). Retrieved July 12, 2016, from http://www.engineeryourlife.org/

Feisel, L., & Rosa, A. (2005). The role of the laboratory in undergraduate engineering education. *The Journal of Engineering Education*, *94*(1), 121–130. doi:10.1002/j.2168-9830.2005.tb00833.x

Focquaert, F., Steven, M., Walford, G., Colden, A., & Gazzaniga, M. (2007). Empathizing and systemizing cognitive traits in the sciences and humanities. *Personality and Individual Differences*, *43*(3), 619–625. doi:10.1016/j.paid.2007.01.004

Girls' Day. (2016). *Griffith University: Women in Engineering*. Retrieved August 4, 2016, from https://www.griffith.edu.au/engineering-information-technology/griffith-school-engineering/women-in-engineering

Gore, A. (2013). *The future: Six drivers of global change*. London: W. H. Allen.

Holmes, M. (2015). Wanted: Diverse cybersecurity workforce. *Magazine of the Society of Women Engineers,61*(1), 24-29. Retrieved August 10, 2016, from http://www.nxtbook.com/nxtbooks/swe/winter15/#/30

Hyde, J., Lindberg, S., Linn, M., Ellis, A., & Williams, C. (2008). Gender similarities characterize math performance. *Science, 321*(5888), 494–495. doi:10.1126/science.1160364 PMID:18653867

Kavanagh, J. (2002). Dame Stephanie Shirley. *The Computer Bulletin, 44*(3), 20–21. doi:10.1093/combul/44.3.20

Kolb, D. (2014). *Experiential learning: Experience as the source of learning and development* (2nd ed.). Upper Saddle River, NJ: Pearson FT Press.

Korpershoek, H., Kuyper, H., & van der Werf, M. P. C. (2012). The role of personality in relation to gender differences in school subject choices in pre-university education. *Sex Roles, 67*(11), 630–645. doi:10.1007/s11199-012-0222-7

Legge, K. (2012). *Doing it differently*. Paddington, NSW, Australia: Place Partners.

Loy, J., Tatham, P., Healy, R., & Tapper, C. L. (2016). 3D printing meets humanitarian design research: Creative technologies in remote regions. In A. Connor & S. Marks (Eds.), *Creative technologies for multidisciplinary applications* (pp. 54–75). Hershey, PA: IGI Global. doi:10.4018/978-1-5225-0016-2.ch003

Martin, P., & Barnard, A. (2013). The experience of women in male-dominated occupations: A constructivist grounded theory inquiry. *South African Journal of Industrial Psychology, 2*(39), 1–12. doi:10.4102/sajip.v39i2.1099

McDonald, M. (2008). Developing relationships in a learning and teaching environment within Te Wānanga o Aotearoa: A philosophy of teaching/learning, a Pakeha perspective. In R. Hunia (Ed.), Toroa Te Nukuroa (3): Reflections from Within. Te Awamutu, New Zealand: Te Wānanga o Aotearoa University.

Nolan, B., & van Lier, B. (2015). *31 Brilliant Ideas for a Better World*. Amsterdam, Netherlands: What Can Design Do.

Orwell, G. (2011). 1984. London: Penguin.

Prescott, J., & Bogg, J. (2011). Career attitudes of men and women working in the computer games industry. *Eludamos (Göttingen), 5*(1), 7–28.

Race, P. (2006). *The lecturer's toolkit: A practical guide to learning, teaching and assessment* (3rd ed.). Abingdon, UK: Routledge.

Ramsden, P. (2003). *Learning to teach in higher education* (2nd ed.). Abingdon, UK: Routledge Falmer.

Reeves, M. (n.d.). *Engineer your life*. Retrieved August 23 2016. http://www.engineeryourlife.org/cms/makingithappen/6274/7039.aspx

Riegle-Crumb, C., & Moore, C. (2013). The gender gap in high school physics: Considering the context of local communities. *Social Science Quarterly, 95*(1), 253–268. doi:10.1111/ssqu.12022 PMID:25605978

Te ManaKōrero: Relationships for learning. (2007). Retrieved August 7, 2016, from http://temangoroa.tki.org.nz/tags/content?tags=te_mana_korero

Thompson, J. (2016). The sometimes neglected study of the role of women in World War II. *Ethos (Berkeley, Calif.), 24*(1), 24–25.

Victoria University of Wellington Strategic Plan 2009–2014. (n.d.). Retrieved July 26, 2016, from http://www.victoria.ac.nz/about/publications/strategic-plan2/strategic-plan.pdf

Wakabayashi, A. (2006). Predicting autism spectrum quotient (AQ) from the systemizing quotient-revised (SQ-R) and empathy quotient (EQ). *Brain Research, 1079*(1), 47–56. doi:10.1016/j.brainres.2006.01.012 PMID:16473340

Weimer, M. (2002). *Learner-centered teaching: Five key changes to practice*. San Francisco, CA: Jossey Bass.

Wolfe, J., & Powell, E. (2009). Biases in interpersonal communication: How engineering students perceive gender typical speech acts in teamwork. *The Journal of Engineering Education, 98*(1), 5–16. doi:10.1002/j.2168-9830.2009.tb01001.x

KEY TERMS AND DEFINITIONS

Diversity: The state of being different, a point of difference.

Dominant Majority: The driving sub-group within a larger group, based on size or profile.

Empathy: An understanding of another person's feeling and the projection of oneself into the situation of another to help that understanding.

Ethnocentricity: A belief that one's own culture is superior to others and central to activities.

Experiential learning: An approach to learning based on activity and reflection.

Gender bias: Sociological prejudice that is based on gender.

LGBTI: Refers to individuals who identify themselves as lesbian, gay, bisexual, transsexual and / or intersex.

Paradigm: A pattern or model of typical or stereotypical examples.

Section 3
Critiques and Administrative Suggestions to Restructure Engineering Education

Chapter 8
Revisioning the Engineering Profession:
How to Make It Happen!

Judith Gill
University of South Australia, Australia

Mary Ayre
University of South Australia, Australia

Julie Mills
University of South Australia, Australia

ABSTRACT

Beginning with a brief account of the value of diversity and inclusivity in a globalizing world, this chapter presents an overview of the current situation of the engineering profession in some English-speaking countries. The starting point addresses the enduring difficulty encountered by attempts to increase and diversify professional engineering. Drawing on a series of studies of engineering education, engineering workplaces and people, both in Australia and beyond, this chapter outlines barriers to entering engineering for anybody other than white mainstream males. Access and retention have long been recognized as serious impediments to increasing numbers of women in engineering. The particular breakthrough in this chapter describes the ways in which some Australian women engineers are working to sustain and enrich their professional status within the workplace by developing strategies that enable them to continue as professionals without diminishing other important features of their life worlds. The implications for all sectors of education, and employers, to emerge from this study offer a basis for redesigning engineering as a more diverse and inclusive profession.

DOI: 10.4018/978-1-5225-2212-6.ch008

INTRODUCTION

By the early 21st century people have become more or less acclimatized to the processes of globalization, which have increased population movements around the world, leading to unheralded heterogeneity in many Western cities and societies. In fact, multiculturalism has transformed many monocultural social settlements into centers of thriving diversity. Concurrently, education has become increasingly important in the new knowledge economy and heightened competition for jobs. Many of the established professions have been affected by these developments. For example, in the English-speaking world, the professions of medicine and law, once the reserve of privileged groups of white men, have attracted increasing numbers of women and people from backgrounds other than English. Such developments reflect not only the richer mix of peoples in the student population entering university, but also the breakdown of the traditional gender division of labor and the increasing likelihood of girls and women entering paid work in proportions commensurate with boys and men. Furthermore, many people have gradually come to accept the idea of women as equally able as men in many areas, leading to the logical argument that professionals work more effectively when their composition reflects the society they serve. A similar line of thought derives from research showing that corporate organizations work more productively when they include a diversity of staff at all levels, with a particular focus on gender (Concannon & Barrow, 2009). Such changes have led to the increasing acceptance of principles of inclusivity and gender equality as key features of diversity in professional development (Sharp, Franzway, Mills, & Gill, 2012).

BACKGROUND

In the general move towards greater professional diversity, engineering stands out in terms of its continuance as a male-dominated profession. Firstly, engineering has been markedly slower than other professions to respond to the calls for greater diversity and inclusivity. Across the English speaking world, the tertiary enrolment statistics consistently show that women comprise less than 20% of students entering engineering (Mills, Ayre, & Gill, 2010). A recent summary of the situation in the U.S. reports:

As a society, we are barely getting started. Women earn less than 20 percent of engineering bachelor's degrees in the U.S. In 2014, that added up to around 19,500 out of just over 99,000 bachelor's degrees nationwide (Crow, 2016).

Secondly, while the need for engineering to attract and retain more women as students and professionals has been increasingly recognized in the last two decades, there are increasing calls for more diversity in terms of race and background across all levels of the profession. The general need to improve was summarized most clearly in American Society for Engineering Education's (ASEE) Engineering Deans Council's statement of commitment to an action plan endorsed by more than one hundred Deans of Engineering:

While gains have been made in the participation of women, African-Americans, Hispanics and Native Americans in recent decades, significant progress is still needed to reach a level where the engineering community fully embraces all segments of an increasingly diverse and vibrant society (Blaschke, 2015).

At senior levels in both university departments and professional organizations, there has been strong encouragement to diversify and to promote gender equity (King, 2012).

At this point it is important to note that the research literature on diversity in engineering contains many more reports about the situation of women engineers than of other minority groups. Consequently, in this chapter reportage reflects this preponderance of studies of women engineers. In doing so the chapter seeks to develop an understanding of the way in which the study of women in engineering has changed and developed, not presuming that other minority groups will necessarily follow a similar pattern but in the hope that the understandings gained from the study of women's experience may provide a basis for investigating the situation of other minorities whose exclusion continues to haunt the profession. No doubt there will be some features which are unique to women's experience and thus do not apply to non-mainstream men. This project is to describe and detail the ways in which one minority group has challenged the culture of engineering classes and workplaces in the attempt to gain entry, in the hope that this work may assist others in a similar project. The chapter begins, therefore, by looking at the ways in which the profession has appeared to resist change. The ways in which resistance to diversification is explained in the research literature makes for interesting reading.

Research Attempts to Explain the Enduring Minority of Women Engineers

In the second half of the last century when the "problem" of women's under-representation in the engineering profession was first recognized, many commentators accounted for this in terms of women's lack of appropriate prior learning – or even lack of ability in the disciplines that underpin engineering: mathematics and science (Evetts, 1996; McIlwee & Robinson, 1992). In fact, most girls finished high school without studying math to senior levels. Not surprisingly then, sex differences appearing in the test scores repeatedly showed boys ahead in math. This situation attracted considerable attention, with explanations ranging from an innate female inability in mathematical thinking, to demonstrations of differences in early socialization practices of parents and caregivers. A formal review of sex differences in cognition raised doubts about the validity of these claims (Maccoby & Jacklin, 1976). In particular, the review revealed that the higher achievement of males was not a stable factor. Meanwhile, the range of hypotheses generated in the effort to retain the idea of innate male superiority in mathematical achievement ranged from efforts to establish a causal relationship between hormonal differences and differences in mathematical ability, to the possibly more productive idea of math anxiety causing many girls to avoid math classes (Hampson,1989; Wang, Korczykowski, Rao, Yong et al., 2007). In retrospect, the almost desperate hunt for research connecting math with innate sex differences revealed the depth of embedded cultural assumptions about the different values associated with different sorts of knowledge, a point returned to below.

Fairweather, a British expert in cognitive science, noted that sex differences in cognition were least likely to be found in studies of young children but appeared more strongly as the children became older – once again suggesting that socio-cultural climate had a distinct effect on the outcomes (Fairweather, 1976). Other research suggested the male math superiority was produced by the tendency of boys to enroll in more math courses than girls rather than any innate difference (Byrne, 1994; Phipps, 2008). In fact, since the 1980s the difference in favor of male higher achievement in math has diminished markedly, directly in proportion to more girls taking math courses to senior levels (Rogers, 2001). By 2006, Janet Hyde and her team could declare the difference in favor of males in math achievement had virtually disappeared (Hyde & Linn, 2006). Of course, the fact that considerable numbers of girls had shown themselves highly capable in math did not mean that they rushed into engineering courses, although a large study of women engineers in Australia reported that the majority of them had enjoyed math in high school and recalled a positive sense of achievement related to the subject (Gill, Mills, Franzway, & Sharp, 2008a).

Other explanations for girls not aspiring to engineering related to their early experiences which typically did not involve playing with wheeled toys and electrical circuits, consistent with the standard gender division found in 'educational' toys which dominated the market (McIlwee & Robinson, 1992; Evetts, 1996). Toys reflect general attitudes about gender-appropriate play and interests as well as parents' attitudes to sons and daughters in terms of life chances. Some research looked at the popular understanding of engineering as involving dirty work, with its operatives wearing hard hats and driving heavy machinery – an array of images assumed unlikely to appeal to girls (Myers, 2010). Other researchers suggested that parents and teachers were not inclined to encourage engineering as an appropriate line of work for girls (Harding, 1996; Kreinberg & Lewis, 1996). However, in a study conducted by the present authors, many women engineers reported that, in choosing a career, they were encouraged to do engineering by a father or other male relative in the engineering profession, while others were encouraged to do so by their teachers (Mills et al, 2010). Thus, support from a significant other could be instrumental in a young woman opting for a non-traditional career, a finding that confirms the degree to which engineering as a career was not widely seen as suitable for girls.

Some educational explanations for girls' avoidance of math included attention to the idea that elementary teachers were – and continue to be – mostly women, who are often not well prepared in math and perhaps did not display an interest in the subject or who, as role models, were more likely to be seen as warm and caring than as problem solvers who could do 'hard' subjects. Recent developments in neuroscience have confirmed the ways in which teacher attitudes can develop or preclude structural connections in the brain which relate to learning potentials (Doidge, 2010). Meanwhile the widespread attention to sexism in children's reading material and school texts (Gill, 2004) revealed another pervasive way in which schoolgirls were inducted into an understanding of a gender divided world of work in which men mowed meadows and male tilers tiled bathrooms, while women tended to the baby and worked in the home kitchen. Most of the more egregious examples of sexism in primary school texts have disappeared, but there are echoes of this conditioning in everyday examples of media images of work roles and of children's home experience. Certainly fewer girls enroll in senior math classes when compared with their male peers – and yet those girls who do so are usually highly successful. Nevertheless, only a minority of the very successful mathematically able young women proceed to enroll in engineering at university (Mills et al., 2010). Their choices demonstrate a preference for the more traditional professions of law and medicine, both of which show a 50% female enrollment in recent years, despite increasing concerns being raised about difficulties in finding jobs after graduating.

In recent times there has been a spate of publicity about women engineers, usually presented in the media as successful individuals who are able to command a salary much higher than their sisters in other professions. Why then are bright girls slow to capitalize on their best chance of doing work that is socially useful and well paid? The question brings into focus the ways in which engineering is seen by the general non-engineering public.

Thinking About a Future Career

As noted above, engineering is imagined by many young people as more typical of men than women because it is seen as hard and dirty and because its association with heavy machinery implies a need for physical strength and endurance. These misperceptions continue despite the changes in the working environment brought about by the many technological advances. These days, the professional engineer is more likely to be found at a desk or in a laboratory than in the field. The persistence of the outdated images demonstrates that the profession is powerfully gender-coded, more so than other comparable high-status professions. Girls at home and at school are much less likely than their brothers to be encouraged to think of engineering as a career unless, as noted above, there is a significant connection in their immediate social or family group. Above all, it seems true to say that the general public does not appear to have a clear idea of what engineers actually do. At one level, the English language usage does not help. In some places, certainly in Australia and in the UK, the person who does the mechanical maintenance on an aircraft is referred to as an engineer, as are many who work in the motor industry, IT and data management roles. While many of these people have high level technical skills, they rarely have tertiary qualification in engineering. For the purposes of this chapter, the term engineer refers to a person credentialed by relevant tertiary qualification and membership of a professional association of engineers.

In this connection it is important to acknowledge a widespread lack of awareness about the range and content of engineering work. Comparable professions appear to have benefitted from television soaps such as *Grey's Anatomy* and *CSI: Crime Scene Investigation* which transmit a fairly clear outline of the workings of professionals in medicine and law enforcement, but there have been very few popular media settings detailing the work of engineers. And yet, a cursory glance at any engineering association newsletter will detail a wide range of applications, all of which address particular problems, the majority with significant and immediate social applications. It is clearly an issue of some importance for the profession to sell itself more effectively to the general populace in the interests of attracting more newcomers to the field.

DEVELOPING EQUITY AND INCLUSIVITY IN ENGINEERING

There have been some significant advances in education addressing the need for schooling to develop high quality learning in the math and science areas. In the past decade, school curriculum developers have paid much more attention to the fields known as STEM (Science, Technology, Engineering and Mathematics) with a good deal of publicity and prizes for schools promoting these areas of study for all students, and especially girls. In a similar vein, toymakers have commissioned designers to encourage fewer gender-typed toys and to involve special versions of trains, planes and automobiles to appeal directly to girls. Probably the less said the better about *Barbie Does Science* but it is important to acknowledge that manufacturers are slowly but surely developing toys for girls that involve aspects of engineering. For example, in 2015, the top-selling toy in the US was set of circuitry labeled *Roominate: A Building Toy for Girls*, specifically designed for females. Another newcomer was even more specific: *GoldieBlox:* "A construction toy that gets girls into engineering early". It is yet to be discovered if these toys achieve their desired outcomes or if their very direct gender labeling manages to alienate the boys and thus has a negative impact on the market.

As noted above, children's books and textbooks are regularly monitored for inclusive illustrations and characters relating to science and engineering. (Finson, 2002). In Australia, much concern has been raised about the relatively poor performance of all students on the international comparisons on achievement in the sciences (PISA, 2015). Currently in the Australian press there have been increasing calls for mathematical achievement to be required for entry into primary school teaching, and for teacher education to include more attention to mathematics in its courses.

All these developments are definitely moving in the right direction to prepare a sound basis for increasing diversity and gender equality in the engineering profession. At the same time, it remains important to be aware of the barriers to this movement which continue to prevent the smooth transition to a more inclusive and diverse engineering workforce. The significant barriers identified here can be summarized as follows:

- Enduring image of engineering as a male occupation.
- Lack of knowledge about the applications of engineering to daily lives, especially in terms of social applications.
- Engineering positioned as dealing with dirty work and hard knowledge (i.e. mathematics and science at senior levels).
- Girls not encouraged to pursue engineering.
- Engineering portrayed as a pervasive masculine culture as symbolized in popular culture by tools, machinery, work overalls, hard hats.

Clearly there are implications for the profession itself and for the education of young people in the years leading up to career choice if these barriers are to be overcome. The profession needs to sell itself better by encouraging:

- Teachers to suggest engineering courses to students long before they are at the point of making choices.
- Making links between daily living and engineering applications.
- Broader recognition of engineers as significant in promoting social benefits and essential for humanitarian work in developing countries (Riley & Claris 2009; Hewlett et al., 2008).
- Public and social media to upgrade their offerings to include engineering applications in a range of work in drama, film and TV.

Building Diversity and Inclusion in Tertiary Level Education

Much research has been directed at the question of access in the effort to explain why so few women apply to do engineering. However, for those women who do enroll in engineering, the university experience is not always welcoming.

In research conducted in Australia, Gill and colleagues (2008b) found that women studying engineering tend to consider themselves as good students, especially with respect to math and science. Most describe themselves as having enjoyed their schooling and having relished the opportunities for competition and achievement. They account for their interests in terms of problem solving. They believe in their capacity to work out a solution and are prepared to spend time and energy on a problem. They persevere rather than give up.

And yet, across the English-speaking world, a higher proportion of females withdraw from engineering courses before completion when compared to their male peers (King, 2012). This situation is described by some researchers as a 'leaky pipeline' in which aspiring engineers are lost along a chosen career pathway (Blickenstaff, 2005; Mattis 2007; Kirkup, Zalevski, Maruyama, & Batool, 2010). It seems that women students do experience discrimination as minority figures in the classroom. They are generally the brighter students, as indicated by their entry scores, but many have difficulty being heard in class discussion and do not get a sense of being considered as important as their male colleagues (Gill et al., 2008a). Studies of retention in engineering degree courses demonstrate considerable differences in gender attrition between universities, raising the suggestion that the pedagogic style and ethos of some institutions might work better for some women than others, a situation that calls for more research (Ohland, Sheppard, Lichtenstein, Eris, Chachra & Layton, 2008).

Investigations into the process dynamics at university level reveal the need for more attention to the manner of teaching. For example, minority students have benefited from deliberate and consciously inclusive processes in the laboratories and classrooms (Mills et al., 2010). Professors are advised to avoid examples geared to traditional male pursuits such as fast cars and military applications which appear to still dominate in engineering classrooms (McLoughlin, 2005; Riley, 2008). Traditionally, university education in engineering appears to be driven by ideas of what every engineer needs to know and must know; a notional stockpile of essential knowledge that is continuously growing. Unfortunately, this core education progresses at the expense of explanations that make connections between different sorts of knowledge deriving from the search for solutions with social applications (Galloway, 2008). Researchers have noted the value of a problem-driven approach, which presents issues at a social level rather than just as abstract formulae (Mills & Treagust, 2003). A more socially-oriented presentation and discussion of engineering applications would allow greater identification with the content and hence improved engagement for a more diverse range of students. A similar conclusion was reached in a recent study (Wächter, 2012) which described the narrow approach to engineering education in terms of the mostly male faculty who are culturally traditional, resistant to change and who value technical approaches over social ones.

One potentially useful approach to developing a more inclusive learning environment would be to openly discuss with students the social dynamics of the engineering profession. For instance, engineering education ideally would include discussion of the lack of diversity and inclusivity within the profession. Debates about how this situation has arisen and why it can be a problem would be a useful preparation for any profession, but particularly engineering. Moreover, given that working in teams is a key feature of engineering work, professional preparation should include education about teamwork and collaboration. Organizing students into groups and requiring them to work together on a topic does not simply happen but depends crucially on the social mix and the specific skills of group members – or the lack thereof. Minority students can face problems in team situations. For some, the idea of speaking up can be threatening, while others (often women) report experiencing difficulty in getting their voices heard (Hartman & Hartman 2007; Stäuble 2004; Mills et al., 2010). Students need to be educated to work successfully in groups; they need to know how to set up acceptable rules for group behavior and how to speak up when rules are broken. (It is interesting to note that these rules are routinely on the agenda in social science classes, but only rarely encountered in engineering.) These features are especially important when group work is part of the assessment and evaluation. With proper preparation, all team members would develop skills in teamwork based on democratic and inclusive principles leading to real productivity. These skills would doubtless be highly useful in subsequent working lives.

A feature of the career trajectory of many engineers is the move into management after a few years of establishing themselves in an organization. Given the propensity of engineers to work in management, their professional education should also include an introduction to management theory where the current trend is towards collaboration and inclusivity rather than the traditional 'heroic' leader, nearly always a white male (Sinclair, 2007; Mills, Franzway, Gill, & Sharp, 2014).

Based on earlier work guidelines that have been published for implementing a gender inclusive engineering curriculum in Mills et al., (2010, 2012); and Ayre, Mills, & Gill (2014a). Guided by the principles that the curriculum and teaching methods of any course should recognize, acknowledge and respect the diversity of the interests, values, perspectives and expectations of class members, and use these differences to enrich the learning experience of all students, an inclusive curriculum is a continuous process rather than a final attainment. It involves addressing all components of the curriculum: classroom and laboratory management, teaching and assessment methods, aims and objectives of the course, as well as its content. The necessary change can be initiated by a top-down (management saying it must be done), or a bottom-up process – the latter occurs and is often the most successful. It can start with an individual lecturer banning racist and sexist jokes, and choosing a range of teaching examples and applications which are of interest to a diverse group of students, but to change the syllabus and assessment methods the support of departmental colleagues and management is an important element in the process. These more senior individuals are often persuaded by the improved course evaluations and examination results emerging from courses taught by enthusiasts. All of this takes time. And to date evaluations are mostly anecdotal rather than produce by rigorous study.

While there appears to be little research to date into the attitudes and teaching practices of engineering faculty there are some indications that the area is beginning to attract research attention. For example, at a recent European conference the question was raised of how receptive or otherwise academic staff in engineering and related disciplines are to the concept and implementation of an inclusive curriculum (Koppi, Roberts & Naghdy, 2012; Beddoes, 2015). Here one group specifically urged more research into the development of a pedagogy that specifically aims at integrating gender and diversity in different teaching formats in engineering education (Steuer, Berg & Leicht-Scholten, 2015), while another reported on an investigation of the attitudes to gender in engineering faculty. While gender and diversity were identified as topics that have not been widely researched in engineering education, these new concerns provide evidence of the move towards a newer and more activist direction being taken.

In summary, in this section we have argued the need for more attention to teaching and to the development of engineering faculty awareness of the issues of gender and diversity in the interests of effective and inclusive learning environments. More attention to the educational processes at tertiary level would be likely to benefit all students, but especially minority students. The business of building group inclusivity would be a useful way of combatting attrition, which is a problem long recognized in Engineering Departments in Australian universities (King, 2012), and one suspects elsewhere. Inclusivity fosters understanding and belonging and is thus a key element whereby minority group members can be drawn into feeling recognized as part of the team with shared tasks and responsibilities (Meyers, Ohland, Pawley, Silliman & Smith, 2012). It is also a key part of readying the student for working life as an engineer, where group work will be a significant feature of work projects and ongoing professional development.

Maintaining Diversity and Inclusivity in the Engineering Workplace

Earlier, the idea of the leaky pipeline was introduced as one image of the way in which women engineers are lost to the profession. By now, it is clear that the girl who succeeds in math and science in school and then follows through to engineering at university is likely to be a highly able, intelligent and determined person. And yet it seems that, for many women engineers, the way forward is still far from smooth. The final stage of the leaky pipeline becomes evident after graduation when some women find life in the engineering workplace an unpleasant and unrewarding experience, so much so that they decide to leave the profession (Fouad & Singh, 2011, Mills et al., 2014).

Most of the women engineers in the studies of Gill et al. (2008a, b), Sharp et al. (2012) and Mills et al. (2010, 2014) speak very favorably about their enjoyment of the work. In describing themselves, these women display a healthy sense of self efficacy and open minds – they are inclined to see themselves as solvers and doers. They do report experiencing difficulty and even occasional failure, but they are not destroyed by negative experiences. They claim to relish the challenge posed by some of the engineering problems they encounter and are continually motivated to try to find workable solutions. And yet studies continually find that, as a group, women engineers are more likely to leave the profession when compared to men of similar age. For some time, this phenomenon was explained by different stages in women's lives compared to men's, the thirties being a time for professional establishment in men's lives while for women this decade is often associated with having children. Indeed, this difference was frequently offered as the explanation by several of the male engineers in the studies of Gill et al. (2008a, b), Sharp et al. (2012) and Mills et al. (2010, 2014) when male participants were asked about the reasons for women leaving the profession – they told us the women had "gone to have babies".

In conversations with the women engineers in these studies, many gave a broader understanding of the situation. The women acknowledged the difficulty inherent in dealing with the work and time pressures and they spoke of the need to plan carefully in the effort to combine work and motherhood, in terms of timing their pregnancies and managing associated household transitions. However, on returning to work after maternity leave, many women encountered an unexpectedly changed relationship with their mostly male managers and colleagues. They found they were continually reminded that they did not really fit in or belong in the workplace; they were not regarded with respect as an engineer. They described being given jobs at a level lower than their experience warranted, being treated as a less-than-serious engineering colleague, and being resented for taking part-time or leave options connected with childcare responsibilities. Even though engineering organizations had adopted new policies and leave entitlements for workers with care responsibilities, the change had not apparently been picked up in the day to day working life of engineering teams (Ayre, Mills & Gill, 2011). The negativity the women encountered in workmates' attitudes served to discourage some of them from continuing in engineering.

In ways like this, the masculine workplace culture appeared to re-establish itself as distant from the woman who had a child. The idea of woman as mother seemed to position her further out of the 'real' world of engineers. She was reduced to tinkering at the edges. Not surprisingly then, the women engineers were hesitant to take up opportunities for maternity leave and part-time work and, in many cases, the policy provisions were undersubscribed (Ayre et al., 2011).

Researchers in the US and the UK have also found that the culture of engineering workplaces is increasingly identified as the reason for women leaving (Corbett & Hill 2015; Fouad & Singh, 2011; Kirkup et al., 2010; Sappleton, Takruri-Rizk, Dhar-Bhattacharjee, & Bezer, 2009). And yet, in some cases, women do manage to stay on in engineering and maintain themselves as professionals, committed to the work for which they were trained, and achieve a good working relationship with colleagues in the spirit of group membership and belonging.

A Study of Women who Persist in Engineering

The following paragraphs report on a study of women who have stayed in the profession since graduation. The research focused on one particular cohort, all graduates in civil engineering from one Australian university, in the effort to discover their reasons for staying. For the purposes of the study, individual women engineers who had found their own solutions for unsatisfactory workplace situations were described as "*persisters*". It is to be hoped that their persistence and solutions will serve as examples to others and thus contribute to changed attitudes within the profession (Ayre, Mills, & Gill, 2013, 2014b).

This cohort of women engineers who stood out from the trend of women leaving the engineering profession was drawn from all women graduates of a civil engineering program at one state university between 1974 and 2008. This group comprised 65 women in all, of whom 56 (86%) completed a survey about their work experience. It was found that 53 women in the cohort were still currently working as engineers – a retention rate of at least 70% which compares very favorably with women's retention in engineering nationwide.

The survey results were analyzed by comparison with results from an earlier national survey and showed that this group of women engineers expressed higher levels of commitment to stay in engineering than the national average. However, it also showed that a significant number of the women had moved round in a range of job placements, a picture suggesting they had exercised choice about their work situation. Sixteen of the woman engineers volunteered to be interviewed to allow the researchers to gain a more complete picture of the current position of women engineers. The one-on-one interviews were recorded and subsequently transcribed for analysis of the key themes that dominated in the responses.

Perhaps the most striking feature of the women's responses was the genuine pleasure they reported regarding their engineering work. They seemed to derive a sense of achievement from their situations, even if the problems they faced were neither always easy nor quickly solved. This sense was neatly captured by one who declared simply "I really like a challenge" (Ayre et al., 2011). Except for the very new recruits who were conscious of their inexperience, they all saw themselves now as good engineers; the idea of growing into the sense of being a competent professional emerged from many conversations.

Along with their sense of being able to do the work, the women registered the importance of effective communication – with peers, clients and the team hierarchy – as an important part of the engineering work. This perception led into an examination of the sociocultural context of the workplace which emerged as very significant in its effect on the women's sense of themselves as professionals. Not only was it important to be able to do the 'hands-on' work, it was crucial to be able to explain to stakeholders what was done, and to communicate effectively to the range of different people involved in any engineering project. For these women who persisted as engineers, it was very important to be in a context in which they were recognized, where their skills were acknowledged and where they were encouraged to learn and grow in the workplace. In fact, several women reported being unhappy in former workplaces and having moved around until a better environment was found. They were inclined to apply the problem solving orientation they had consciously adopted as engineers to their own situations at work and home – if it's not working do something about it. So they were not "stayers" in the sense of stuck in one place out of an incapacity to move on, but rather choosers with a clear sense of their own usefulness, a confident and clear skill set and a determination to work out the best possible solution within current constraints.

The project management orientation of their professional lives was also apparent in their arrangements in their personal lives, particularly around issues of time management in organizing child care. Several women had developed creative solutions to the question of managing a dual-income professional household. There were some interesting and unusual arrangements of time sharing between both parents in order to manage caring for young children, for example the father taking a lengthy 'shift' off work in order to maintain the currency and employability of the mother. These women were adventurous in their personal lives and relatively free of the traditional constraints associated with the gender division of labor. They expected equal sharing in living arrangements and recognized the value of working out how to achieve this.

A key thread in these discussions concerned the quality of management at their workplaces. If leadership in the organization was committed to building a strong and coherent team, there had to be a mutuality of respect and personal acknowledgement. These women were alert to any sense of being patronized as female and being considered inherently limited.

The analysis gives rise to implications for management in engineering organizations as an essential contribution to ensuring the values of mutual respect and recognition are part of workplace culture. Along with policy development, these implications offer significant strategies for changing the workplace culture so that newcomers can more readily become accepted and valued. An approach to management training that specifically focused on the promotion and maintenance of greater diversity and inclusivity in the profession would be a significant first step. Policy development is important but without the support of managers at all levels it is not enough to sustain women and minorities in taking opportunities to further their careers and/or manage responsibilities of care.

CONCLUSION

Starting from an overview of the problem of generating diversity within the profession of engineering the position of women as students and then as professional engineers has been examined in order to draw out implications for reform in the workplace and professional reform. Despite years of research and interventions, policy developments and educational improvements, engineering continues to be a profession numerically dominated by white mainstream males. Diversity is seen as increasingly vital for the current globalized world, and rather than feeling dispirited by the lack of progress towards diversity in engineering, it seems that the present time offers the best chance to date to push for important reforms. Central among these is the acknowledgement of the need for leaders in the engineering profession and in education to take responsibility for the furtherance of diversity and gender equity, a requirement to be continuously acknowledged in publicity and practice. The argument is supported on many fronts, including:

- The increased expectation of women to be in paid work.
- Increased media representations of women and minorities in non-traditional roles.
- Technological reforms reliant on knowledge and expertise rather than physical strength.
- Engineering's increased visibility as socially relevant and important.
- The capacity for engineering to offer women secure positions in the new economies.

In the analysis of the ways in which the women attended to their professional work along with issues in their life worlds there are important implications for engineering management if the profession is to reform itself in the interests of gender equity and diversity.

REFERENCES

Ayre, M., Mills, J. E., & Gill, J. (2014a). A Gender-Inclusive Engineering Curriculum. In D. Beanland & R. Hadgraft (Eds.), Engineering Education: Transformation and Innovation (pp. 87-89). Melbourne, Australia: RMIT University by arrangement with UNESCO.

Ayre, M. E., Mills, J. E., & Gill, J. (2011). Two steps forward, one step back: Women engineers in Australia. *International Journal of Gender. Science and Technology*, *3*(2), 293–312.

Ayre, M. E., Mills, J. E., & Gill, J. (2013). 'Yes, I do belong': The women who stay in engineering. *Engineering Studies*, *5*(3), 216–232.

Ayre, M. E., Mills, J. E., & Gill, J. (2014b). Family issues for women engineers. In D. Bilimoria & L. Lord (Eds.), *Women in STEM careers: International perspectives on increasing workforce participation, advancement and leadership* (pp. 79–100). Cheltenham, UK: Elgar.

Beddoes, K. (2015). Professors' Perceptions of How Men and Women Students Experience Engineering Education Differently…Or Not. *Proceedings of European Society for Engineering Education (SEFI).* Available from http://www.sociologyofengineering.org/wp-content/uploads/2016/01/Beddoes-SEFI-2015.pdf

Blaschke, J. (2015). *Texas State joins with ASEE in effort to foster diversity in engineering*. San Marcos, TX: Texas State University News Service, Office of Media Relations.

Blickenstaff, J. C. (2005). Women and science careers: Leaky pipeline or gender filter? *Gender and Education*, *17*(4), 369–386. doi:10.1080/09540250500145072

Byrne, E. M. (1994). *Critical Filters, Hidden Helix: Policies for Advancing Women in Science and Technology*. Brisbane, Australia: University of Queensland.

Concannon, J. P., & Barrow, L. H. (2009). A cross-sectional study of engineering students self-efficacy by gender, ethnicity, year, and transfer status. *Journal of Science Education and Technology*, *18*(2), 163–172. doi:10.1007/s10956-008-9141-3

Corbett, C., & Hill, C. (2015). *Solving the equation: The variables for women's success in engineering and computing*. Washington, DC: American Association of University Women.

Crow, M. (2016). Why we need more women engineers. *LinkedIn Newsletter*. Available from https://www.linkedin.com/pulse/why-we-need-more-women-engineers-michael-crow

Doidge, N. (2010). *The brain that changes itself*. Melbourne: Scribe.

Evetts, J. (1996). *Gender and career in science and engineering (Feminist perspectives on the past and present)*. London: CRC Press.

Fairweather, H. (1976). Sex differences in cognition. *Cognition*, *4*(3), 231–280. doi:10.1016/0010-0277(76)90019-6

Finson, K. (2002). Drawing a scientist: What we do and do not know after fifty years of drawings. *School Science and Mathematics*, *102*(7), 335–345. doi:10.1111/j.1949-8594.2002.tb18217.x

Fouad, N. A., & Singh, R. (2011). *Stemming the tide: Why women leave engineering*. Milwaukee, WI: University of Wisconsin.

Galloway, P. D. (2008). *The 21st-century engineer: A proposal for engineering education reform*. Reston, VA: ASCE Press.

Gill, J. (2004). *Beyond the great divide: Single sex or coeducation?* Sydney, NSW: UNSW Press.

Gill, J., Mills, J. E., Franzway, S., & Sharp, R. (2008a). Oh you must be very clever! High achieving women, professional power and the ongoing negotiation of workplace identity. *Gender and Education*, *20*(3), 223–236. doi:10.1080/09540250801968990

Gill, J., Sharp, R., Mills, J. E., & Franzway, S. (2008b). I still wanna be an engineer! Women, education and the engineering profession. *European Journal of Engineering Education, 33*(4), 391–402. doi:10.1080/03043790802253459

Hampson, M. E. (1989). *Contributions of gonadal hormones to human cognitive and motor skills: Evidence from the menstrual cycle* (Unpublished doctoral dissertation). University of Western Ontario, Ontario, Canada.

Harding, J. (1996). Science in a Masculine Strait-Jacket. In L. Parker, L. Rennie, & B. Fraser (Eds.), *Gender, Science and Mathematics* (Vol. 2, pp. 3–15). Springer Netherlands. doi:10.1007/978-94-011-0143-1_1

Hartman, H., & Hartman, M. (2007). Is teamwork a female-friendly pedagogy? In I. Welpe, B. Reschka, & J. Larkin (Eds.), *Gender and engineering: Strategies and possibilities* (pp. 73–91). Frankfurt: Peter Lang.

Hewlett, S. A., Luce, C., Servon, C. B., Sherbin, L., Shiller, P., Sosnovich, E., & Sumberg, K. (2008). *The Athena factor: Reversing the brain drain in science, engineering, and technology*. Boston, MA: Harvard Business Publishing.

Hyde, J. S., & Linn, M. C. (2006). Gender similarities in mathematics and science. *Science, 314*(5799), 599–600. doi:10.1126/science.1132154 PMID:17068246

King, R. (2012). *Engineers for the future*. Sydney, NSW: Engineers Australia, Australian Council of Deans of Engineering.

Kirkup, G., Zalevski, A., Maruyama, T., & Batool, I. (2010). *The UKRC statistics guide on women in science, engineering, technology and the built environment (SET)*. Leeds, UK: WISE Campaign.

Koppi, T., Roberts, M., & Naghdy, G. (2012). *Perceptions of a gender-inclusive curriculum amongst Australian Information and Communications Technology Academics.* Paper presented at the Fourteenth Australasian Computing Education Conference, ACE 2012, Melbourne, Australia.

Kreinberg, N., & Lewis, S. (1996). The Politics and Practice of Equity: Experiences from Both Sides of the Pacific. In L. Parker, L. Rennie, & B. Fraser (Eds.), *Gender, Science and Mathematics* (Vol. 2, pp. 177–202). Springer Netherlands. doi:10.1007/978-94-011-0143-1_15

Maccoby, E., & Jacklin, C. (1976). *The psychology of sex differences*. Stanford, CA: Stanford University Press.

Mattis, M. (2007). Upstream and downstream in the engineering pipeline: What's blocking US women from pursuing engineering careers? In R. Burke & M. Mattis (Eds.), *Women and minorities in science, technology, engineering and mathematics* (pp. 334–362). Cheltenham, UK: Edward Elgar. doi:10.4337/9781847206879.00025

McIlwee, J., & Robinson, G. (1992). *Women in engineering: Gender, power and workplace culture*. Albany, NY: State University of New York Press.

McLoughlin, L. (2005). Spotlighting: Emergent gender bias in undergraduate engineering education. *The Journal of Engineering Education*, *94*(4), 373–381. doi:10.1002/j.2168-9830.2005.tb00865.x

Meyers, K., Ohland, M., Pawley, A., Silliman, S., & Smith, K. (2012). Factors relating to Engineering Identity *Global. The Journal of Engineering Education*, *14*(1), 119–131.

Mills, J., Ayre, M., & Gill, J. (2010). *Gender inclusive engineering education*. New York, NY: Routledge.

Mills, J., Ayre, M., & Gill, J. (2012). *Gender Inclusive Curriculum in Engineering and Construction Management*. Canberra, Australia: Australian Learning and Teaching Council.

Mills, J., Franzway, S., Gill, J., & Sharp, R. (2014). *Challenging knowledge, sex and power: Gender, work and engineering*. New York, NY: Routledge.

Mills, J. E., & Treagust, D. F. (2003). Engineering education: Is problem-based or project-based learning the answer? *Australasian Journal of Engineering Education*, *3*(2), 2–16.

Myers, J. (2010, November 9). Why more women aren't becoming engineers. *The Globe and Mail*.

Ohland, M. W., Sheppard, S. D., Lichtenstein, G., Eris, O., Chachra, D., & Layton, R. A. (2008). Persistence, engagement, and migration in engineering programs. *The Journal of Engineering Education*, *97*(3), 259–278. doi:10.1002/j.2168-9830.2008.tb00978.x

Phipps, A. (2008). *Women in Science, Engineering and Technology: three decades of UK initiatives*. Stoke on Trent, UK: Trentham Books.

PISA. (2015). *A closer look at gender gaps in education and beyond.* Program for International Student Assessment. Retrieved July 15, 2016, from http://oecdinsights.org/2015/03/05/a-closer-look-at-gender-gaps-in-education-and-beyond/

Riley, D. (2008). *Engineering and social justice*. San Rafael, CA: Morgan and Claypool.

Riley, D., & Claris, L. (2009). From persistence to resistance: Pedagogies of liberation for inclusive science and engineering. *International Journal of Gender Science and Technology*, *1*(1), 35–60.

Rogers, L. (2001). *Sexing the Brain*. London: Weidenfeld and Nicolson.

Sappleton, N., Takruri-Rizk, H., Dhar-Bhattacharjee, S., & Bezer, R. (2009). *The organisational culture of NW engineering workplaces: The influence on women engineers*. Paper presented at the 6th Annual IAS-STS Conference: Critical Issues in Science and Technology, Graz, Austria.

Sharp, R., Franzway, S., Mills, J. E., & Gill, J. (2012). Flawed policy, failed politics? Challenging the sexual politics of managing diversity in engineering organisations. *Gender, Work and Organization*, *19*(6), 555–572. doi:10.1111/j.1468-0432.2010.00545.x

Sinclair, A. (2007). *Leadership for the disillusioned: Moving beyond myths and heroes to leading that liberates*. Crows Nest, NSW: Allen & Unwin.

Stäuble, B. (2004). *Towards gender inclusiveness in the teaching of science and engineering subjects: A case study*. Paper presented at the 27th HERDSA Annual Conference: Transforming Knowledge into Wisdom, Miri, Sarawak.

Steuer, L., Berg, T., & Leicht-Scholten, C. (2015). Breaking the habit–new approaches in engineering education. *Proceedings of Annual Conference of the European Society for Engineering Education*. Available from: http://www.sefi.be/conference-2015/CHAP%2016.%20Education%20concepts%20specific%20for%20Engineering%20Education/60823-%20L.%20STEUER.pdf

Wächter, C. (2012). Interdisciplinary Teaching and Learning for Diverse and Sustainable Engineering Education. In A. Béraud, A.-S. Godfroy, & J. Michel (Eds.), *GIEE 2011: Gender and Interdisciplinary Education for Engineers: Formation Interdisciplinaire des Ingénieurs et Problème du Genre* (pp. 47–63). Rotterdam: SensePublishers. doi:10.1007/978-94-6091-982-4_5

Wang, J., Korczykowski, M., Rao, H., Yong, F., Pluta, J., Ruben, C., & Detre, J. et al. (2007). Gender difference in neural response to psychological stress. *Social Cognitive and Affective Neuroscience*, *2*(3), 227–239. doi:10.1093/scan/nsm018 PMID:17873968

KEY TERMS AND DEFINITIONS

Career Trajectory: Path that each job and/or position of responsibility that a person undertakes as they move backwards, forwards or laterally during his/her working years.

Gender Equity: Parity in access, rewards, resources and opportunities regardless of gender.

Gender Inclusive Engineering Curriculum: A continuous improvement process applied to curriculum and teaching methods of any course that recognizes, acknowledges and respects the diversity of the interests, values, perspectives and expectations of class members, and use these differences to enrich the learning experience of all students.

Inclusivity: A key element of team building where each group member is recognized as part of the whole with shared tasks, responsibilities and benefits.

Masculine Workplace Culture: Emergence of a negative organizational culture where women engineers are assigned jobs below their abilities after returning from maternity leave, are resented for working part-time in connection with childcare responsibilities and are fearful of utilizing caregiver benefits of their company.

Social Dynamics: Study of the beliefs and behaviors that emerge from the interactions of and relationships among group members.

Teamwork: Action of group of individuals, who as a collective work effectively and efficiently to accomplish a vision or goal.

Workplace Culture: The unique attributes of an organization that emerges from the collective behaviors, values, attitudes, beliefs and backgrounds of employees and employers.

Chapter 9

A Globally Focused, Experiential Educational System for STEM Fields:

Measures for Intentionally Promoting Diversity

Aaron Sakulich
Worcester Polytechnic Institute, USA

Amy Peterson
Worcester Polytechnic Institute, USA

ABSTRACT

The diversification of the STEM labor force is essential to the health of the American economy, both because diverse teams are more adept at achieving project goals and because of shifting national demographics. Although the professional STEM workforce is currently more diverse than ever, sustained increases in diversity over the last 20 years has been relatively limited and may be decelerating. The focus of this chapter is specific programmatic components that can be implemented to intentionally supporting STEM students of diverse backgrounds. The hub of such an educational system would be Individual Development Plans (IDPs), which would empower students to follow an intentional path during their studies, combined with a strong focus on reflection and connection through ePortfolios and experiential learning both in academic projects and extracurricular professional activities such as internships.

DOI: 10.4018/978-1-5225-2212-6.ch009

INTRODUCTION

The diversification of the STEM labor force is essential to the health of the American economy. Diversification provides a competitive advantage, as it is well established that diversity on project teams leads to superior attainment of project goals (Tryggvason & Apelian, 2011). Further, industry and academic structures that are unable to train and support members of underrepresented populations in the STEM fields pose a threat to the American economy, as those populations are projected to grow most quickly over the next few decades – the U.S. Census estimates that by 2050, members of historically underrepresented ethnic and racial groups will make up more than half of the national population. Although the professional STEM workforce is currently more diverse than ever, sustained increases in diversity over the last 20 years have been relatively limited and may be decelerating; in some disciplines, the participation of conventionally underrepresented populations has been in decline since the early 1990s (Landivar, 2013).

The struggle to diversify the STEM workforce coincides with a period of gradual change in the fundamental focus of higher education. Conventional academic programs have focused almost entirely on professional preparation, providing students with the professional skills that were needed to succeed in the pre-internet business environment. With the rise of global connectivity, however, many in industry and academia alike have concluded that these skills alone will not be sufficient for student success, and are turning their attention to encouraging students to develop global competencies, the set of skills that will enable personal and professional success away from a student's home culture. Despite the widely recognized importance of producing globally competent STEM professionals, however, it is estimated that STEM majors make up less than 5% of the students that participate in off-campus programs, which are broadly regarded as the best mechanism by which to build global competencies (Parsons, 2010).

Although internships, co-op experiences, study abroad opportunities, learning communities, guest lecturers, project-based work, innovative classroom methodologies, student clubs, summer bridge programs, and similar activities can each contribute to supporting members of underrepresented populations as they build professional skills and global competencies in the STEM fields, such activities are often introduced as discrete components of an educational program and are almost always implemented and administered with goals other than diversifying an institution's student population. The complex and interconnected social, economic, and political processes that have resulted in the current demographic imbalance in the STEM fields is beyond both the scope of this chapter and the expertise of the authors, as are pre-college efforts that show merit in attracting members of underrepresented populations to the STEM fields. This chapter instead focuses on low-cost means, with which the authors have personal experience, of implementing specific programmatic components to build professional skills and global competencies in students while intentionally supporting STEM students regardless of gender, race, ethnic background, disability, sexual orientation, gender expression, age, socioeconomic status, nationality, or individual differences such as learning style and career aspirations. An educational system intentionally striving for greater diversity may use Individual Development Plans (IDPs) in order to empower students to follow an intentional path during their studies and would have a strong focus on experiential learning both in academic projects and extracurricular professional activities such as internships.

BACKGROUND

Greater diversity in the STEM workforce would provide a competitive advantage to the American labor force; it is well established that diversity on project teams leads to superior performance. Doug Wilde (2011) writes that teams exhibit superior performance when they are made up of individuals exhibiting

... the widest possible range of personalities, even though it takes longer for such psychologically diverse teams to achieve good cooperation. They must first cultivate an openness to opposing opinions and recognize the value of exploring a problem from various angles (p. 20).

Antonio et al. (2004) found that the inclusion of an African-American student in otherwise all-white small group discussions led to increases in integrative complexity, suggesting that diverse groups are more likely to produce creative, more broadly acceptable solutions to a given challenge. A study by Strayhorn et al. (2014) came to broader conclusions, finding that students that had been exposed to peers from different economic, social, gender, or racial/ethnic backgrounds were more likely to achieve greater learning gains. The authors noted that such exposure would optimally occur both in the classroom and in extracurricular activities. These results are consistent with those available in the broader literature, and clearly indicate that more members of underrepresented groups entering the STEM fields will contribute not only their personal technical skills to their field, but will enable those with whom they work to produce superior results.

Despite the well-established benefits of diversity in the STEM fields, and the importance of the STEM professions to the national economy, little progress is being made. In a 2013 report, the American Institutes for Research (AIR –a nonprofit, nonpartisan behavioral and social science research and assessment group) carried out a comprehensive analysis of data collected over the past 20 years by the U.S. Census. Non-U.S. residents have steadily received roughly 10% of STEM degrees, and are not included in the discussion below. At first glance, the numbers are encouraging: in 2010, more than three times as many American Indian/Alaskan Native individuals received STEM degrees than in 1990; the number of Latino/Latina degree recipients tripled; the number of Black degree recipients doubled (although the number of degree recipients levelled off for reasons unknown around 2004); and the number of women achieving bachelor's degrees in STEM fields also roughly doubled. It should be noted, however, that fewer than 2,000 American Indians/Alaskan Natives received STEM degrees in 2010, making numerically modest increases in numbers appear as significant gains, percentage-wise, and that more than twice as many white women received STEM degrees than women of all other ethnic/racial backgrounds combined.

Deeper analyses of the data reveal more complicated trends. Whites, who make up some 65% of the population, account for roughly 65% of the STEM degrees awarded, a value that has remained steady over the past two decades. Although overall numbers of STEM degrees conferred to members of underrepresented populations increased, these numbers did not keep pace with population growth; did not keep pace with the increasing number of individuals attaining a bachelor's degree in general; and did not keep pace with the increasing number of individuals attaining a bachelor's degree specifically in the STEM fields, with two exceptions. The American Indian/Alaskan Native population receives about 1% of all STEM degrees, and makes up about 1% of the national population; however, as noted above, this population size is too small to draw concrete conclusions regarding enrollment trends or to add significant diversity to the labor force at the national scale. Asian Americans and Pacific Islanders continue to be significantly over-represented; although only 5% of the national population, in 2010 they accounted for some 12% of STEM degree recipients. When viewed in the context of population growth, however, the Asian American and Pacific Islander population is growing at a slightly higher rate than that of attainment of STEM degrees; the increased number of degrees awarded in the past two decades therefore more accurately represents a decline. It should further be noted that "Asian and Pacific Islander" is an indefensibly broad and heterogeneous category, representing some 60% of the global population; it is all but certain that rates of attainment of STEM degrees are not identical between individuals of, for example, Chinese, Korean, Laotian, and Pakistani heritage.

Regarding measures of diversity beyond ethnic/racial and gender identity, veterans are slightly over-represented in STEM field majors (25.4% pursue STEM degrees vs. 20.8% of the overall population). It should be noted that the Post-GI Bill, which went into effect in 2009, substantially increased the number of veterans attending college; however, this is a relatively small population, so the number of veterans pursuing

STEM degrees only increased from 4.9% of the total STEM student population in 2008 to 5.2% in 2012. There is a positive trend between student budget as percent of income and majoring in a STEM field, with students for whom college is relatively more expensive being more likely to choose STEM fields of study. Between 2004 and 2012, veterans and students for whom college is relatively more expensive were just as likely to complete their STEM degree programs as the average student (National Center for Education Studies, 2016). Students with disabilities are somewhat underrepresented in STEM field majors (10.1% in STEM vs 11.1% overall), but delving into specific disabilities yields no trends in under- or over-representation, presumably because of the relatively small number of students reporting each type of disability. Students with a disability are less likely to complete a STEM degree program in a given year than students without a disability (17% vs 20.7% in 2012), but this difference is comparable to the difference in non-STEM fields (18.7% vs. 21.4%), meaning that this problem is not unique to STEM.

A focus on the workforce provides additional concerning trends. Although women make up half of the U.S. workforce and earn half of all STEM degrees, they account for only 25% of workers in STEM fields. Blacks and Hispanics, who together account for approximately 25% of the national labor force, are estimated to make up only 13% of workers in STEM fields (Landivar, 2013).

In summary, more members of underrepresented populations are receiving STEM degrees than ever before, but only in comparison to the employment statistics of 50 or 100 years ago can it be said that the professional STEM population in the United States has become more diverse. Further, not all of these degree holders are entering the STEM workforce. With changing national demographics, increased global competition, and a rapidly-transforming world facing challenges with the potential for global catastrophe unless innovative solutions are developed, educational professionals cannot afford to take a centuries-long view of STEM education.

COMPONENTS OF AN INTENTIONALLY INCLUSIVE EDUCATIONAL SYSTEM

The necessary factors for nurturing professional success in members of underrepresented populations in the STEM fields are well known and include the acquisition of knowledge, skills, and habits of mind; opportunities to put these into practice; an understanding of the societal benefits of STEM work; the development of a sense of competence and progress; a sense of belonging to, or self-identification with, the field; and access to information about academic progress, requirements, and opportunities. This section discusses various programmatic components that can promote diversity by focusing on one or more of these factors.

Individual Development Plans (IDPs)

IDPs are designed as a tool for professional and personal development through reflection and goal setting. In an IDP, an individual considers a long-term plan and sets relevant goals, assessing available development activities that can be used to complete objectives. It is important to note that an IDP is not a static document; rather, IDPs can and should be regularly referenced and evolve with the individual. IDPs have been used primarily for people who have already begun careers, or those with substantially more training than incoming undergraduates; however, the core idea of aligning activities with short and long term goals has value at any point in one's education or professional career. IDPs are widely used throughout the U.S. government and in industry. The Federation of American Societies for Experimental Biology (FASEB) proposed an IDP concept for postdoctoral researchers in 2003. Since 2014, the National Institutes of Health (NIH) have required all NIH-funded graduate students and postdoctoral researchers to use IDPs to "*identify and promote [their] career goals*". Research suggests that such structured oversight results in greater satisfaction and productivity.

One commonly used framework for Ph.D.-level IDPs is the American Association for the Advancement of Science's myIDP (https://myidp.sciencecareers.org/). This system begins with the assessment of, and reflection on, an individual's skills, interests, and values. Based on the results, myIDP suggests careers that might be a good fit and provides resources and tips for career exploration. After determining a possible career path, an individual sets career advancement, skill development, and project completion goals. myIDP also highlights the importance of a mentoring network for the successful completion of goals.

Undergraduate students differ from Ph.D. students in many ways; however, a similar framework can be useful for undergraduates, since both groups are interested in identifying career paths and completing goals related to the topics of career advancement, skills development, and project completion. An undergraduate student's IDP acts as a nexus for their curricular and co-curricular activities. Within the IDP, students identify specific, measurable, achievable, realistic, and time-bound (SMART) goals related to skills development, career selection, and project completion. As an example, a student pursuing their degree at Worcester Polytechnic Institute (WPI) may decide that they want to consider a career in water remediation. The student could plan to pursue a junior-year off-campus research experience at a location with water-related projects (e.g. Venice or Bangkok) or on-campus through the Massachusetts Water Resources Outreach Center; to pursue a capstone design project related to water remediation technology (e.g. at the Eilat or Panama City Project Centers or through an appropriate engineering department); to enroll in classes such as CE3060 (Water Treatment); to pursue undergraduate research with relevant faculty members; or to pursue an internship or co-op with one of many local companies or the Massachusetts Department of Environmental Protection. It is important to note that students are encouraged to identify goals (e.g. pursue a water-related off-campus project) rather than specific activities (e.g. carry out an internship with the Panama Canal Authority in July 2018) in order provide flexibility as well as meaningful direction.

IDPs are living documents. As a student's path changes, so do the goals. Therefore, a successful IDP will not necessarily consist of completed goals. The success of an IDP should be assessed through longer term markers. IDPs can be used as a reflection tool ("How does this career path align with my personal and professional values? What is the next step I need to achieve in order to attain my long term objective?) or can be used synergistically with other reflection tools, such as ePortfolios, discussed below, to assign meaning to activities.

While IDPs can be important tools for everyone, the future orientation, focus on SMART goals, and mentoring components of an IDP may be particularly powerful for underrepresented groups. Self-regulated learning, in which learning results from students' goal-driven thoughts and behaviors, results in greater student learning and achievement (Moore, 2013). Self-regulated learning and use of available college resources were the strongest predictors of college persistence in a study of first-generation college students. Both of these qualities are enhanced through the use of IDPs (Moore, 2013). While the study did not observe differences between minority and non-minority first-generation students, according to a 2016 analysis of 2012/14 Beginning Postsecondary Students Longitudinal Study data by Ifill et al., larger percentages of the parents of non-white college students have not received a four-year college degree (68.5% of Blacks or African Americans, 72.3% of Hispanics or Latinos, and 73.9% of American Indians or Alaska Natives vs. 51.1% of Whites). Therefore, incorporation of IDPs in a curriculum should benefit non-white college students proportionally more.

Professional Portfolios

ePortfolios are collections of significant deliverables, assembled and managed by a single user, used in some form by over 50% of U.S. colleges and universities (Dahlstrom, Walker, & Dziuban, 2013). A successful ePortfolio can serve many roles. At its basis, ePortfolios are a website where students compile significant deliverables that can be used to highlight accomplishments to better market themselves to the specific career field in which they are seeking employment. Student ePortfolio usage is also correlated with higher levels of student success (Enyon, Gambino, & Török, 2014).

ePortfolios can be also used within courses, often in combination with capstone experiences, learning communities, and other experiential learning activities. ePortfolios allow a student to curate a collection of their own learning deliverables, which is valuable within defined learning experiences, and may be even more valuable over the course of a student's career. Different versions of a student's ePortfolio can exist simultaneously, so there can be a public version, a university version, a version specific to a course, and the master version visible only to the student. This flexibility increases the potential advantages of ePortfolios.

ePortfolios can be used as a reflection tool to assign meaning to and make connections between activities, both compulsory and voluntary (Gerbic, Lewis, & Amin, 2013; Singer-Freeman & Bastone, 2016; Miller & Morgaine, 2009). Such skills enhance students' ability to learn independently, are important to professional success (in terms of "reflection-in-action") and are essential in ensuring competitiveness for prestigious fellowships (Miller & Morgaine, 2009; Schön, 1983). A Connect 2 Learning (C2L) survey found that students felt that IDPs helped them think more deeply about the content of courses, allowed them to be more aware of their growth and development as learners, and helped them to make connections between ideas. Furthermore, this effect was significantly heightened (from ~30% agreement to ~80% agreement) when students knew that someone was looking at their ePortfolio (Enyon et al., 2014).

While ePortfolios are beneficial to all students, they may be especially valuable to members of underrepresented groups. ePortfolios as a reflection tool enhance students' ability to make connections and achieve higher level understanding of content; to develop a sense of competence and acknowledge personal progress; to develop a sense of belonging to a community of scholars; and become more future oriented (Singer-Freeman & Bastone, 2016; Singer-Freeman, Bastone, & Krivanek, 2014; Shada et al. 2014).

Academic Advising

Although it is well established that contact with faculty members outside of the classroom is associated with higher overall retention and graduation rates, only a small body of literature exists explicitly investigating the best practices for advisors, particularly advisors from populations conventionally well-represented in the STEM fields, to apply when advising members of underrepresented groups. Museus and Ravello, in a 2010 qualitative study of 45 individuals from a community college, a state college, and a large research university, focused specifically on identifying trends from the perspectives of advisors (predominantly white males) and students from underrepresented populations (excluding Native Americans, due to the low number of individuals enrolled at the study sites, but including Asian Americans, who "*share common struggles with their Black and [Latina/Latino] peers*" p. 50). Although the study was limited in a number of ways, the authors identified three practices that students perceived as helpful: Humanizing the advising experience by faculty members taking steps to being seen by students as individuals and displaying concern or interest in the student's success; providing holistic advice that focuses not just on coursework and academic issues, but any difficulties that a student may be experiencing, even in their personal lives; and proactive advising, where the faculty member will take the lead in securing resources for the student, rather than providing the student with a starting point or contact with whom to follow up on their own.

Experiential Learning

Students must not only think intentionally about the paths that they want to pursue, but must build the skills necessary for professional success on those paths. Hoegel and Parboteeah (2007) divide professional skills into two categories: *domain-relevant skills*, generally referring to specific technical knowledge to be applied to specific technical challenges, and *creative thinking skills*, or the ability to identify unconventional approaches leading to innovative solutions. The authors of this chapter would include a third category, *workplace skills*, that encompasses such abilities as written and oral communication skills, personal and professional ethics, a well-developed management or mentorship style, and time- and stress management skills, among others. Generally speaking, professional skills were the focus of the pre-internet educational system because they were, on their own, sufficient for success in the relatively homogenous STEM professions. Global competencies, which have become a focus of both educators and employers over the last few decades, include the skills that enable a student to value, in the previously-quoted words of Doug Wilde (2011), "… [an] *openness to opposing opinions and recogniz[ing] the value of exploring a problem from various angles*" (p. 20). Global competencies include the ability to work in diverse teams with individuals that hold differing perspectives; communication skills including the mastery of secondary or tertiary languages and understanding of etiquette necessary to transfer knowledge through international networks; the ability to recognize and adapt to the social factors that may determine how project partners approach research problems and tasks; and an understanding of the effects of different technologies or policies on disparate stakeholders in a specific cultural, economic, and social context. In practice, these skills take the form of a spectrum ranging from basic survival skills, such as being able to drive a car or seek medical attention in a foreign country, to the ability to work for a foreign-based corporation or an overseas field office, to more nebulous concepts such as understanding the complex network of social, political, economic, cultural, and environmental factors in which the individual works in the broadest terms – how specific policies, technical solutions, and decisions may affect the local population. Van 't Klooster et al. (2008) identify the benefits of global competencies as going beyond the individual; in a world where social media enable rapid communication and feedback, globally competent individuals are also less likely to cause misunderstandings that may cause consumers to "*target corporate reputation*" (p. 691).

Project-based learning, activities that eschew conventional, educator-led lectures in favor of focusing on student participation and reflection in order to meet specific learning objectives, has been shown to more effectively promote the development of both professional skills and global competencies in students (Tryggvason & Apelian, 2011). Although there are several philosophical frameworks that have been established to guide experiential learning, they all take the same basic form: students identify a topic or problem of interest; they map relationships between potential stakeholders or anticipate outcomes; they apply knowledge to develop a solution or more thorough understanding of their topic, guided by educators with expert knowledge; and they reflect on their experiences. Experiential learning can take the form of academic experiences, off-campus experiences, or experiences in industry.

Academic Experiential Learning

The term 'academic experiential learning' is used here to refer to team-based educational activities taking place in the classroom that have the goal of meeting well-structured learning outcomes, most often occurring in the form of laboratory classes (as opposed to more creative, and more variable, project-based learning, described below). While such activities do not necessarily need to be team-based, students working together encourages the development of a number of professional skills and global competencies. Academic experiential learning activities have been shown to improve student outcomes in various professional skills, such as written and oral communication; to prepare students for self-directed research projects in industry or academia; to demonstrate to students the value of interdisciplinary inquiry; to encourage students in seeking graduate degrees; and to provide students with a better understanding of core concepts (Tryggvason & Apelian, 2011).

At the undergraduate level, experiential learning activities are often incorporated in conventional courses through the use of team projects, particularly in laboratory courses. Hoegel and Parboteeah (2007) identified the ability of individuals to collaborate effectively as determining both team effectiveness (the ability to find solutions to particular challenges) and team efficiency (the ability to find solutions while meeting budgetary or time constraints). The results of their study of 145 German software development teams showed that a team's ability to collaborate directly affected their ability to integrate domain-relevant skills into appropriate solutions, with low levels of collaboration negatively affecting end products. It was theorized that these relationships were due to the facts that not every member of a team is likely to have complete domain-relevant skills necessary for success and that members of teams that collaborate poorly due to poor workplace skills will lose motivation or resent perceived slights by their teammates, and submit inferior work as a result. An opposite relationship was observed regarding creative thinking skills: Highly collaborative teams seemed to represent teams where every individual was 'on the same page,' stifling creativity, whereas poorly coordinated or managed teams allowed opportunities for individuals to bring forward unconventional ideas. Overall, it was found that workplace skills were a better predictor of eventual success than either domain-relevant or creative thinking skills, and the authors recommended that innovative products requiring the development of creative solutions formally incorporate individual brainstorming-type work modes into the overall collaborative process.

Although members of conventionally underrepresented populations may have diverse backgrounds and interests that contribute to creative thinking in addition to solid domain-relevant skills, they face significant challenges in engineering teams. First and foremost, the relative rarity of such individuals means that they are likely to have few, if any, teammates of similar backgrounds with whom to commiserate during negative experiences. Further, the dominant demographic tends to assign menial or managerial roles to underrepresented teammates, while pursuing the most challenging, and most interesting, work for themselves. Such alienation not only deprives members of underrepresented populations of the benefits of teamwork, but reinforces their internal narrative of not 'belonging' in engineering. In a 2016 study by Seron et al., engineering students at four universities were asked to write twice-monthly diary entries reflecting on their experiences. While men "*describe[d] overwhelmingly positive collaborative experiences*," (p.199) women experienced exclusion "*often based on implicitly held (and sometimes explicitly expressed) gender biases about the 'natural' talents of men and women*." (p. 195) The Society of Women Engineering (SWE) was highlighted by many women in the study for its role in "*their socialization into the culture of engineering*." (p. 197) Whether or not student bias is intentional is wholly irrelevant; negative experiences on teams are the primary reason why members of underrepresented populations leave the STEM fields.

The only clear way for faculty to support members of underrepresented groups in experiential teamwork seems to be increasing their efforts on intentionally structuring coursework. Oakley et al. (2004) provide several suggestions, beginning with faculty selection of teams. When students are allowed to self-select, they tend to choose teammates with whom they have something in common, leading to relatively homogenous demographics. Although students may vigorously object to not being allowed to work with their friends, the instructor will be able to vary both the demographic composition of each team as well as the skill level; academically strong students are likely to learn more if their team contains an academically weak student that they must 'mentor.' Oakley et al. do honor requests from students regarding with whom they specifically *do not* want to work, and also recommend announcing that all teams will be dissolved and reformed mid-semester unless all students on a team agree to remain together. This provides an opportunity not only to shuffle members of perpetually dysfunctional teams, but also provides individuals, especially members of underrepresented groups who are having negative experiences, with a 'second chance' to join an inclusive team.

Another best practice identified by both Oakley et al. (2004) and Wolfe (2010) is the use of formal team contracts. These documents, written by the team and approved by the faculty member early in the course, establish mutually agreed-upon weekly schedules, individual responsibilities, expectations, and processes for dealing with issues that may arise. The faculty member can use these contracts to support members of underrepresented populations by mandating that positions or responsibilities be shared, ensuring that all team members have a relatively equal hand in menial or supervisory tasks as well as in project-critical, challenging work. For courses that include formal reports on a regular basis, requiring each team to submit a statement confirming that the distribution of work conformed to that planned in the team contract, or explaining why it did not, can be a useful tool to ensure that students do not simply produce a team contract and then ignore it. The proper use of team contracts, of course, requires additional faculty time and energy.

Project-Based Experiential Learning

In contrast to academic experiential learning, project-based learning is more variable, and allows students more opportunity for creative thought by identifying an issue or problem to be addressed, and then allowing students significant leeway in choosing which techniques or processes to investigate. Capstone design courses are the most common example. In addition to building conventional professional skills, project-based experiences are considered the most effective way to build global competencies. Off-campus projects are generally considered more effective than those carried out at a student's home institution; however, Parsons (2010) has shown that students engaged in coursework with well-planned and well-implemented internationally-focused content showed better attainment of learning outcomes related to global competencies, even when no travel was involved. It should be noted that off-campus projects are distinct from conventional study abroad activities, in which students pursue coursework at an institution other than their own, often with one or two tourism-type activities outside of the classroom. However, there is no way to ensure that a student who is taking a conventional study abroad class in a university setting, living in a dormitory, eating in a cafeteria, etc., actually interacts with other individuals, much less with other individuals from the host country.

Off-campus learning, historically, was the provenance of relatively few, presumably affluent, students travelling for an extended period of time. Excursions such as the multi-year 'Grand Tour' of Europe were considered the essential capstone to academic studies among the wealthy, although learning outcomes seem to have focused more on interpersonal networking and developing the social polish necessary for success among the upper classes rather than building skills on which to establish a career. Opportunities became more widely available as improvements in transportation, particularly air and rail travel, lowered costs and finally became almost commonplace during the post-war years, when the pursuit of a college or university degree became more widely available to the masses.

The past three decades have seen perhaps the most important changes in off-campus education. In 2015, roughly 300,000 individuals studied abroad, more than twice the number in 1985; and only 3% of students travelled for an entire year, down from 17% 30 years earlier. Over 60% of programs are of fewer than eight weeks' duration Dwyer (2004). A well-implemented program of short duration may have just as much impact on a student as less rigorous program of long duration. Chieffo and Griffiths (2004) and Dwyer (2004) each reported that length of time overseas is not necessarily a determinant of program impact; individuals reported strong, positive effects even 50 years after an overseas program of as short as eight weeks. Even shorter programs of one or two weeks, and those with no obvious educational content, can serve an important role by fostering student enthusiasm and interest in other cultures, leading to a greater likelihood of later participation in more meaningful off-campus travel. For example, while advising students during a 7-week project-based experience based in Morocco, one of the authors coordinated a weekend trip to the Sahara desert and the rural town of Ouarzazate. Although the educational content was limited at best, student feedback was overwhelmingly positive.

In addition to taking steps to proactively support students from underrepresented backgrounds on project teams as described above, there are additional steps that can be taken when organizing project-based learning (either off-campus or on-campus) that will help build global competencies. As these competencies generally focus on introducing a human domain to technical studies, doing so can help express to students both the usefulness of the STEM fields and the impact that their potential future work will have on people.

Most capstone design courses focus on developing technical solutions to technical problems. While this can be an effective way to build core knowledge and professional skills, global competencies cannot be built without giving some thought to the human context in which these technical problems and potential solutions exist. As an example, in 1970 WPI implemented a project-based educational system that included a junior-year 'interactive qualifying project' (IQP). This project is required for graduation, and despite the fact that over 80% of WPI students are in the STEM fields, it focuses on using social science techniques to understand the impact of science and technology on concerned stakeholders. Most interdisciplinary student teams partner with non-academic project sponsors, who identify a topic and provide access to necessary resources, while faculty members provide guidance on how to identify stakeholders, gather data regarding the risks and benefits that a particular solution may pose to those stakeholders, and make decisions based on that data. Developing technological systems is only a minor focus of the project. The implementation of this social-science based requirement for STEM students at WPI had roots in the death of post-war utopianism. Despite incredible scientific and technological breakthroughs, most of the world population was seeing little improvement in their lives.

Over 60% of juniors carry out their IQP off-campus, although every project, regardless of where it is carried out, follows the same format of focusing of technology's impact on human stakeholders. There is evidence that off-campus projects result in greater achievement of learning outcomes. In 2014 Vaz and Quinn asked 2,500 WPI alumni questions regarding 39 areas of long-term impact. This was followed by 25 in-depth interviews with alumni and their employers. The results were unambiguous: In projects carried out off-campus, students consistently achieved learning outcomes at levels significantly higher than on-campus projects, reporting a statistically significant higher level of positive impact in 33 out of 39 areas. Although research is ongoing, it appears that this is due to two factors: off-campus study is generally seen as an enjoyable experience, thus leading to greater student motivation, and working among communities significantly different from one's own creates difficulties that, when overcome, leave students with a greater sense of accomplishment and leave a greater impact.

In addition to ensuring meaningful student contact with project partners, relatively small logistical considerations can ensure student contact with the broader host culture, not necessarily directly related to the project. For example, 'home stays' in which a student lives with a local family in their actual home are preferable to housing students in hotels or dormitories. Living, eating, and spending downtime directly with a family provides an immersive environment that challenges the student every day to "*cultivate an openness*" to the local culture. At the absolute least, students (even if they are housed in dorms) should be provided only with cooking facilities – haggling in the marketplace or navigating an unfamiliar grocery store will rapidly prove unavoidable and provide at least a modicum of outside contact. Students will be best able to capitalize on such a situation only if they have received some formal, on-campus preparation. A course or series of courses in which students develop language skills and at least an introductory understanding of local culture, history, and politics, will enable them to not only converse more easily, but converse more deeply on more advanced topics.

Experiential Learning in Industry

Experiential learning in industry generally takes the form of internships (opportunities of a few months in length, generally pursued during the summer months, and requiring significant self-motivation on the part of students) or 'co-op' programs (opportunities averaging a half-year in duration, generally supported by an institution's career development center). Spending time in industry before graduation can help to train students in specific professional skills that are of interest to a particular employer; help build professional networks that can later be leveraged when the student is seeking employment; confirm a student's interest in a specific field and lessen the 'culture shock' that may be experienced during the transition from academia to industry, leading to greater satisfaction; and provide a source of income that can enable a student to worry less about finances and focus on academic work during the school year, though it should be noted that this is heavily dependent on student background, institutional costs, and the field in which the internship is secured. In a survey of 242 business school programs, Coco (2000) highlighted this final point, concluding that the rising cost of education has put pressure on institutions to do more to ensure employability after graduation, thus leading to the rise of internship/co-op programs.

Narayan, Olk, and Fukami (2006) proposed a conceptual framework for internships based on personnel transfer theory. In it, three actors (the university, the sponsoring company, and the student) undergo an experience that entails inputs (generally how well each actor is prepared for their role, and how well they understand the goals and organization of the other actors); to processes (the mentoring of the student by the sponsor and the university, and the student's dedication or commitment to carrying out the tasks they have been assigned); and outputs (benefits to, and capacity built in, the company and university, and skill development in the student). Generally speaking, greater preparation of all actors, more formal oversight, and effective communication all contribute to greater attainment of positive outputs, while the factors that would generally make one content with a job (e.g. autonomy, variety of tasks, or intellectual challenge) lead to greater student commitment. Overall, a properly organized internship experience can greatly increase a student's attractiveness to employers.

The conceptual framework of industrial experience as a relationship between three actors highlights an important point. Students are likely to pursue internships and co-ops in order to become more attractive to future employers or because they need the money; Sponsoring companies generally use such experiences as sources of temporary or low-cost labor, or to evaluate potential future employees; and universities coordinate such activities in order to improve student employability, thus making the institution more attractive in the eyes of students and their parents alike. Presumably, universities are also interested in building and maintaining relationships with industrial bodies who may become donors or sponsors of specific academic programs.

In a study of the usefulness of industrial experience in meeting academic learning outcomes (as opposed to building professional skills), Binder et al. (2015) summarized the difficulties that have faced previous studies attempting to correlate the benefits of industrial experience with student background. Foremost, many industrial opportunities are optional, and thus self-selecting; relying on student initiative to spur participation means that students that are already comfortable in the field and have already achieved a degree of success are more likely to seek such opportunities. That many organizations use academic records and, perhaps, interviews as their criteria for selecting students to fill available positions may lead to a further reduction in the ability of a member of an underrepresented group to secure employment. In the analysis of eight years of records from a large British university, however, Binder et al. concluded that students exposed to industrial internships accrue greater academic success in post-internship studies, and that self-selection played a secondary role at best (although it could not be ruled out entirely). Although "academically weaker" students saw the same improvements as their peers, previous academic success was also a strong predictor of future academic gains; thus, students with poor academic records gained more from their experiences than their peers. It was not clear if employers judged the work output of "academically weaker" students as equal to that of their peers; however, if an institution's goal is to ensure the success of all students, it follows that weaker students should be given preference for placement in industrial experiences, as they benefit from them the most. For a variety of reasons, and due the amount of energy and attention spent overcoming obstacles not present for their white peers, members of underrepresented populations are more likely to fall into this group.

SOLUTIONS AND RECOMMENDATIONS

An educational system that intentionally aims to support underrepresented students in the STEM fields through the use of established mechanisms would pair students with appropriate, proactive advisors; would have formal structures in place for the implementation of IDPs and ePortfolios; would focus as heavily as possible on team-based experiential learning in the classroom; and would provide opportunities, preferably off-campus, for carefully structured project-based learning and industrial experiences. Although many institutions implement these particular components, doing so in order to build diversity in the student population faces many difficulties.

Logistically, students should be paired with academic advisors with whom they identify, and that could serve as positive role models. It should be noted that it is not clear how such a thing could be implemented with the current state of STEM education: in 2010, the most comprehensive demographic analysis to date of tenured and tenure track STEM faculty showed that white men occupied over half of all faculty positions (Kena et al., 2016). Black and Latino/Latina professors of both sexes occupied only 10% of faculty positions, and females occupied only 44% of faculty positions (vs. 32% and 57% of undergraduates, respectively). There is an additional paradox: although students may identify most with advisors with whom they share some background, white men have disproportionately more influence. Students from underrepresented populations may therefore derive more benefit from advisors with whom they share no background, but who have the reputation and connections to help position a student in an interpersonal network that will lead to professional success. Until members of underrepresented populations make up a larger proportion of academic faculty, it may therefore be helpful for students to be assigned two, or even more, advisors (a recommendation, of course, that would strain the resources of practically all institutions).

The question of who has access to an IDP, and to what extent, is an important area of discussion in IDP implementation. IDPs contain personal, potentially sensitive information, so privacy is important; however, a lack of review may lead to low participation. At universities implementing IDPs at the Ph.D. level, there are a range of approaches. At the University of Massachusetts Amherst, the IDP is a central component of annual reviews between a Ph.D. student and their advisor; however, the Ph.D. student chooses which portions of the IDP to present. At the University of Massachusetts Medical School, Ph.D. students submit their complete IDP to the Assistant Dean of Career & Professional Development, who confirms receipt and may provide minimal feedback. In a pilot of IDPs performed by one of the authors at WPI, the students confirm that they have completed an IDP through the myIDP website, which generates a Certificate of Completion, but the sharing of any IDP content is completely in the hands of students. In spring 2016, WPI piloted an undergraduate course that helped students explore their strengths, personal values, and missions. At the end of the term, the students presented a visualization of the connections between their major and their mission to each other, the instructors, and the broader WPI community at a reception. The wide range of presented solutions to the question of IDP privacy highlight that there is no single answer; rather, the issue should be considered within the context of desired outcomes and university/program educational philosophy. Finally, the effectiveness of IDPs is difficult to assess over short time periods; it is likely that true best practices will only be identified in reviewing the experiences of current students some decades from now.

The challenges of ePortfolio implementation include technical (e.g. how to set one up, which features are most important to a given community), financial, logistical (e.g. transportability, lifetime of the ePortfolio), community buy-in of both faculty and students, and communication (what is appropriate to include, how to present it) challenges. Many of the articles and reports cited in this chapter have previously addressed these challenges and potential solutions; adapting potential solutions to specific contexts requires significant thought on the parts of faculty and administrators. Since the focus of this chapter is on low-cost strategies, it should be mentioned that there are numerous free ePortfolio options.

A focus on team-based experiential learning in the classroom will lead not only to better experiences for students of underrepresented populations, but also to better experiences for their peers, who will learn how to collaborate with diverse partners through experience. There appears, however, to be no clear understanding of how many individuals from underrepresented groups should be on each team. Individuals are more likely to become isolated; however, their gender or ethnic identity seems to become more pronounced as their demographic makes up a larger portion of the team population, leading to increasingly negative behavior from teammates. It seems likely that individual student personality may play an important role in guiding team selection; it may also be of value to include several members of underrepresented populations representing distinctly different populations on a team, if possible. A survey requesting demographic information, study habits, goals and professional aspirations, etc. may be useful in gathering such data, if faculty are teaching students with whom they are not familiar.

Establishing a system of off-campus experiential learning in order to promote diversity can demand significant resources. Establishing project partners for off-campus projects relies on faculty committing time and energy to developing personal relationships. Student preparation requires further investment, initially in advisors working with students to ensure that they can travel while meeting program-specific requirements; in administrative oversight of logistical matters such as scheduling flights and arranging housing, particularly if unconventional, home stay-type experiences are to be incorporated in the program; in preparing students with the technical background to meet their project goals; in preparing them with the emotional skills necessary to cope with culture shock; and in post-travel programming that helps them reflect on, and promote, their experiences.

As an example, off-campus experiences at WPI are overseen by the Interdisciplinary and Global Studies Division, a department-sized organization with more than 15 full time faculty members. IGSD personnel oversee a competitive application process and place over 700 students in opportunities each year; carry out a term-long, three-credit preparatory course; facilitate communication between students, project partners, and the faculty that will actually advise students on-site; and require student attendance at workshops focusing on sexual violence, traveler health and hygiene, and site-specific overviews of the specific socioeconomic, political, and cultural aspects of a site. Despite the massive resources dedicated to the off-campus program at WPI, it is not perfect; Although a few courses with relatively low student participation offer students a change to reflect on their experiences, there has been no widespread effort to establish post-travel programming. Kartoshkina, in a 2014 review of the study abroad literature, quotes a variety of sources as referring to re-entry after an off-campus experience as "problematic," "shocking," "difficult," "grieving," "painful," and even "traumatic" (although Kartoshkina identifies "bittersweet" as the one that seems the most appropriate). Although dedicating class time to pre-travel preparation, actual travel, and post-travel reflection may strain a student's ability to meet established program requirements, each component works synergistically with the others to maximize benefit to the student.

More than just the completion of projects is at stake during student preparation. At WPI's Panama City project center, for example, student teams have worked with the national air carrier, Copa Airlines, on developing appropriate strategies for meeting development and logistical goals at Tocumen International Airport. While working with this organization has resulted in some degree of culture shock among students, the shock has been mostly superficial, resulting in students having to deal with only mild psychological stresses. On the other hand, students have also worked with Footprint Possibilities, an Engineers Without Borders-type NGO, to bring potable water supplies to the populations of informal settlements at a landfill on the outskirts of Panama City. Engaging with members of such an extremely marginalized population, living in a level of poverty which the students have not likely previously encountered, generally results in more severe culture shock, leading to more serious stresses. The mental and emotional health of an unprepared student, unable to cope with such extreme stresses, can deteriorate rapidly, particularly in the absence of established support groups such as family and close friends.

Comprehensive student preparation, the cultivation of reliable and appropriate project partners, and the placement of students onto teams based on their personal attributes are all of extreme importance to off-campus experiential learning. The most common reaction to facing alien conditions such as extreme poverty, particularly in projects that bill themselves as service learning, is for students (and some faculty) to place themselves in a Powerful/Powerless frame of reference. They no longer perceive their project as having the goal of *working together* with a community to address specific needs, but rather as having the goal of *saving* the local community, a goal which the community is somehow inherently unable to achieve alone. Although it alleviates much of the psychological stresses that students encounter in such situations by focusing on their powerfulness and dehumanizing their project partners, even subconsciously subscribing to such a frame of reference reduces students' ability to build global competencies by placing a mental and social barrier between them and the project partners and reduces the likelihood of appropriate solutions being reached by dismissing stakeholder concerns. In the long run, projects that are posed as 'saving' project partners risk engendering a "White Man's Burden" worldview in students. To avoid this situation, institutions must dedicate resources to building faculty expertise in student preparation or hiring faculty with track records of successful research in off-campus locations.

The difficulties of using experiential learning in industry can only be overcome by conversations between the three stakeholders involved. In order to promote both diversity and the better education of students, institutions should be giving preference to the students who will benefit the most from co-op or internship experiences, despite the fact that these are the exact students that will least appeal to sponsoring organizations. Striking such a balance is so context-specific that there are no clear answers.

Finally, there is one difficulty that applies to the implementation of any of the programmatic components discussed in this chapter: Cost. Only large institutions are likely to have academic advising offices with sufficient resources to oversee the implementation of an intentionally inclusive advising system; the cost of implementing IDPs and ePortfolios, although there are several free frameworks upon which faculty-led efforts can be built, may be prohibitive; and project-based learning, whether on- or off-campus, requires significant initial investments in restructuring course catalogs and hiring or training faculty. There are no simple answers regarding the cost issue; only a concern that if the largest, most-well funded institutions implement the programmatic components discussed in this chapter while smaller institutions do not, increasing diversity in the STEM fields will come at the cost of increasing inequality, as wealthy students have increasingly better and better secondary education experiences, while students at smaller, less wealthy institutions do not.

FUTURE RESEARCH DIRECTIONS

Each of the different mechanisms discussed in this chapter are viable options for intentionally supporting students of diverse backgrounds in the STEM fields. Each, however, requires investment and the allocation of resources in terms of money, faculty effort, and administrators working to adapt policies to their own particular institutions. Improving academia's ability to support conventionally underrepresented students should, however, be a national priority.

There are obvious areas of future research, such as the identification of best practices through the long-term monitoring of various activities, or those outside the scope of this paper, such as the identification of how to breach the social, political, and economic barriers between underrepresented populations and the STEM fields. These would be particularly valuable at the large scale, comparing programs in different institutional contexts and drawing larger lessons from where they do and do not overlap. Further, more verification is needed to show that, in the long run, unconventional educational strategies such as those proposed here do in fact help students from diverse backgrounds enter the workforce, and confirm that after doing so, they contribute not only their own technical skills, but lead to the superior performance of their teammates. Such areas of inquiry are long-term, and are likely to be relatively low-impact, as general understandings have already been established, and the nation cannot wait another one or two generations for results to be generated.

One area of interest that could rapidly generate new data at a relatively modest cost regards the actual definition of the term *diversity*. Data regarding gender, race, ethnic background, and disability are relatively widely available. Lumping 60% of the world population into a single census category, of course, makes it difficult to identify the trends within the Asian American/Pacific Island population that have led to their over-representation in the STEM fields. Further, no comprehensive data could be found regarding the participation in the STEM fields of individuals in terms of sexual orientation, gender expression, or more nebulous, personality-related indicators such as professional motivation, professional experience, and personality.

Promoting the importance of a diverse workforce, and the quantifiable benefits that the American economy would see from doing so, may be an area of attention that would have a higher impact over a short period than pure research. If the benefits of diversity, and some of the relatively simple things that can be done to promote it, were more broadly known by both individuals and employers, it is likely that significant gains would be made simply due to organizations diversifying their workforce in order to secure an economic edge over their competitors.

CONCLUSION

The diversification of the STEM labor force is essential to the health of the American economy. To achieve this, STEM undergraduate programs must become more diverse and must improve retention of diverse populations, while STEM graduates must be motivated through their educations to pursue careers in STEM. This charge for the diversification of the STEM fields coincides with a push for inclusion of global competencies development within undergraduate educations. A number of programmatic components have been discussed that can build professional skills and global competencies in students while intentionally supporting STEM students from diverse backgrounds. The nexus of the proposed strategies is the IDP, which will empower students to develop and follow intentional paths of study. The proposed system is ambitious, but feasible given sustained commitment by administration and faculty towards the goals of greater diversity and global competency development.

REFERENCES

American Institutes for Research. (2012). *Broadening participation in STEM a call to action*. Washington, DC: American Institutes for Research.

Antonio, A. L., Chang, M. J., Hakuta, K., Kenny, D. A., Levin, S., & Milem, J. F. (2004). Effects of racial diversity on complex thinking in college students. *Psychological Science, 15*(8), 507–510. doi:10.1111/j.0956-7976.2004.00710.x PMID:15270993

Binder, J. F., Baguley, T., Crook, C., & Miller, F. (2015). The academic value of internships: Benefits across disciplines and student backgrounds. *Contemporary Educational Psychology, 41*, 73–82. doi:10.1016/j.cedpsych.2014.12.001

Chieffo, L., & Griffiths, L. (2004). Large-Scale Assessment of Student Attitudes after a Short-Term Study Abroad Program. *Frontiers: The Interdisciplinary Journal of Study Abroad, 10*, 165–177.

Coco, M. (2000). Internships: A try before you buy arrangement. *SAM Advanced Management Journal, 65*, 41.

Dahlstrom, E., Walker, J. D., & Dziuban, C. (2013). *ECAR study of undergraduate students and information technology*. Louisville, CO: EDUCASE Center for Analysis and Research.

Dwyer, M. M. (2004). More Is Better: The Impact of Study Abroad Program Duration. *Frontiers: The Interdisciplinary Journal of Study Abroad, 10*, 151–163.

Eynon, B., Gambino, L. M., & Török, J. (2014). What difference can ePortfolio make? A field report from the connect to learning project. *International Journal of ePortfolio, 4*, 95-114.

Gerbic, P., Lewis, L., & Amin, N. M. (2011). Student perspectives of eportfolios: Change over four semesters. In G. Williams, P. Statham, N. Brown, & B. Cleland (Eds.), *Changing Demands, Changing Directions* (pp. 423–436). Hobart, Australia: Ascilite.

Hoegl, M., & Parboteeah, K. P. (2007). Creativity in innovative projects: How teamwork matters. *Journal of Engineering and Technology Management*, *24*(1-2), 148–166. doi:10.1016/j.jengtecman.2007.01.008

Ifill, N., Radford, A. W., Wu, J., Cataldi, E. F., Wilson, D., & Hill, J. (2016). *Persistence and Attainment of 2011–12 First Time Postsecondary Students After 3 Years*. Washington, DC: U.S. Department of Education, National Center for Education Statistics.

Kartoshkina, Y. (2015). Bitter-sweet reentry after studying abroad. *International Journal of Intercultural Relations*, *44*, 35–45. doi:10.1016/j.ijintrel.2014.11.001

Kena, G., Hussar, W., McFarland, J., de Brey, C., Musu-Gillette, L., Wang, X., & Dunlop Velez, E. et al. (2016). *The Condition of Education 2016 (NCES 2016-144)*. Washington, DC: U.S. Department of Education, National Center for Education Statistics.

Landivar, L. C. (2013). *Disparities in STEM Employment by Sex, Race, and Hispanic Origin*. Washington, DC: United States Census Bureau.

Miller, R., & Morgaine, W. (2009). The benefits of e-portfolios for students and faculty in their own words. *Peer Review*, *11*, 8.

Moore, J. S., III. (2013). *Self-Regulated Learning and Ethnic/Racial Variables: Predicting Minority First-Generation College Students' Persistence* (Ph.D. Thesis). University of Rhode Island, Kingston, RI.

Museus, S. D., & Ravello, J. N. (2010). Characteristics of academic advising that contribute to racial and ethnic minority student success at predominantly White institutions. *NACADA Journal*, *30*(1), 47–58. doi:10.12930/0271-9517-30.1.47

Narayanan, V. K., Olk, P., & Fukami, C. (2006). Determinants of internship effectiveness: an exploratory model. In Academy of Management Proceedings (Vol. 2006, pp. E1-E6). Academy of Management. doi:10.5465/AMBPP.2006.22898555

National Center for Education Statistics. (2016). *National postsecondary student aid study*. Retrieved from https://nces.ed.gov/surveys/npsas/

Oakley, B., Felder, R. M., Brent, R., & Elhajj, I. (2004). Turning student groups into effective teams. *Journal of Student Centered Learning, 2*, 9-34.

Parsons, R. L. (2009). The effects of an internationalized university experience on domestic students in the United States and Australia. *Journal of Studies in International Education*, *14*(4), 313–334. doi:10.1177/1028315309331390

Schön, D. A. (1983). *The reflective practitioner: How professionals think in action*. New York, NY: Basic books.

Seron, C., Silbey, S. S., Cech, E., & Rubineau, B. (2016). Persistence Is Cultural Professional Socialization and the Reproduction of Sex Segregation. *Work and Occupations*, *43*(2), 178–214. doi:10.1177/0730888415618728

Shada, A., Kelly, K., Cox, R., & Malik, S. (2011). Growing a new culture of assessment: Planting eportfolios in the metro academies program. *International Journal of ePortfolio, 1*, 71-83.

Singer-Freeman, K., & Bastone, L. (2016). *Pedagogical Choices Make Large Classes Feel Small* (NILOA Occasional Paper No. 27). Urbana, IL: University of Illinois and Indiana University, National Institute for Learning Outcomes Assessment.

Singer-Freeman, K., Bastone, L., & Skrivanek, J. (2014). ePortfolios Reveal an Emerging Community of Underrepresented Minority Scholars. *International Journal (Toronto, Ont.)*, *4*, 85–94.

Strayhorn, T. L., Long, L. L. III, Williams, M. S., Dorimé-Williams, M. L., & Tillman-Kelly, D. L. (2014). Measuring the educational benefits of diversity in engineering education: A multi-institutional survey analysis of women and underrepresented minorities. In *Proceedings of the 121st ASEE Annual Conference and Exposition*. Indianapolis, IN: American Society of Engineering Education.

Tryggvason, G., & Apelian, D. (2011). *Shaping our world: Engineering education for the 21st century*. Hoboken, NJ: John Wiley & Sons.

van 't Klooster, E., van Wijk, J., Go, F., & van Rekom, J. (2008). Educational travel: The overseas internship. *Annals of Tourism Research*, *35*, 690–711.

Vaz, R., & Quinn, P. (2014) Long term impacts of off-campus project work on student learning and development. In *Proceedings of the 44th Annual IEEE Frontiers in Education Conference.*Madrid, Spain: Institute of Electronics and Electrical Engineers. doi:10.1109/FIE.2014.7044128

Wilde, D. (2011). Personalities into teams. *IEEE Engineering Management Review*, *1*(1), 20–24. doi:10.1109/EMR.2011.5729969

Wolfe, J. (2010). *Team writing: A guide to working in groups*. Bedford: St. Martin's.

KEY TERMS AND DEFINITIONS

Academic Experiential Learning: Team-based educational activities taking place in the classroom that have the goal of meeting well-structured learning outcomes, generally in the form of laboratory or design courses.

Creative Thinking Skills: One of three categories of professional skills, composed of the skills necessary to identify unconventional approaches leading to innovative solutions, particularly important when working on non-routine projects.

Domain Relevant Skills: One of three categories of professional skills, composed of the ability to apply specific technical knowledge or fluency with information technology or technological processes to specific technical challenges.

Global Competencies: The set of skills necessary to work effectively in diverse teams with partners that hold differing perspectives, have different social norms or values, and approach research problems and tasks in different ways, a growing theme of post internet-era education.

Individual Development Plan: A tool that can be used by students to identify an intentional pathway through their academic and professional career.

Industrial Experiential Learning: Experiences in which a student hones professional skills by being temporarily employed in a relevant position, either through relatively brief internships or more formal 'co-op' programs.

Project-based Experiential Learning: Educational activities that allow students more opportunity for creative thought by identifying a single issue or problem to be addressed, and then allowing students significant leeway in choosing which techniques or processes to investigate.

Professional Skills: The ability to perform tasks leading to the efficient and effective meeting of project objectives, generally the focus of the pre-internet education paradigm.

Workplace Skills: One of three categories of professional skills, composed of the skills necessary to work well on a team, such as communication skills, ethics, and management.

Chapter 10
Diversifying Engineering Education:
A Transdisciplinary Approach From RWTH Aachen University

Linda Steuer
RWTH Aachen University, Germany

Sonja Gaedicke
RWTH Aachen University, Germany

Anna Bouffier
RWTH Aachen University, Germany

Carmen Leicht-Scholten
RWTH Aachen University, Germany

ABSTRACT

Engineers and therefore engineering education are challenged by the increasing complexity of questions to be answered globally. The education of future engineers therefore has to answer with curriculums that build up relevant skills. This chapter will give an example how to bring engineering and social responsibility successful together to build engineers of tomorrow. Through the integration of gender and diversity perspectives, engineering research and teaching is expanded with new perspectives and contents providing an important potential for innovation. Aiming on the enhancement of engineering education with distinctive competencies beyond technical expertise, the teaching approach introduced in the chapter represents key factors to ensure that coming generations of engineers will be able to meet the requirements and challenges a changing globalized world holds for them. The chapter will describe how this approach successfully has been implemented in the curriculum in engineering of a leading technical university in Germany.

DOI: 10.4018/978-1-5225-2212-6.ch010

INTRODUCTION

Global challenges, megatrends and political developments are influencing societies worldwide. Engineers create technical solutions for the resulting diverse challenges and therefore are shaping societies and trends with their work. As global developments change, so does the complexity of professional requirements set on engineers (Thursby, Fuller & Thursby, 2009). Consequently, internationally, engineering education is discussed beyond the mere imparting of technical knowledge, but also integrates skills that prepare future engineers to create socially responsible, sustainable solutions for global societies. Despite the commonality, there are variety of strategies and opinions on how to approach this challenge in order to realize this engineering education.

This chapter describes a new approach which aims at yielding socially responsible engineers, through the distinguishing means of gender and diversity integration into engineering education, research, and practice.

The aforementioned approach is based on a theoretical concept which is transferred into a practical approach in teaching and research. On the following pages, said practical concept will be elucidated by a best practice example of a course which has been very successfully applied at RWTH Aachen University, one of the leading technical universities in Germany and, one of the seven nation-wide Excellence Universities.

"MADE IN GERMANY" - ENGINEERING EDUCATION IN GERMANY

Historically German engineering had its origins in military crafts, evolving into the civic sector when a growing demand for technical applications in a broad scope of fields emerged due to the increasing use of technology during industrialization. To meet these new challenges engineering education found its way from military academies to universities. Since its commencement thru up until today engineering has always been a key factor for technological progress and economic prosperity (Knoll & Ratzer, 2010).

"Made in Germany" stands for the high reputation of German engineering worldwide. Hence, German engineering graduates are ranked among the international elite because of their profound professional expertise and leadership competencies. Thus, engineering presents the flagship discipline within the scientific profile of amongst leading German universities. The longstanding tradition of women absence within the military constitutes a strongly male-dominated and conservative disciplinary culture, shaping persistent institutional structures, stable curricula contents, and learning formats through many decades (Franzke, 2010; Leicht-Scholten 2008; Steuer, 2014).

Even today German engineering studies are significantly characterized by small numbers of women. The imbalance of sex in engineering programs is represented by only 22, 8% of female graduates in 2014 (Statistisches Bundesamt, 2015, p. 13; see Figure 1). However, the count of females has been rising minimally extends due to extensive efforts the relevant institutions started, especially since the turn of the millennium, all in all there is a great persistence concerning the level of minor female amounts in engineering education (Leicht-Scholten, Breuer, Tulodetzki & Wolffram 2011; Leicht-Scholten, 2011a).

Consequently, the bridging of the imbalance of sexes has been the main field of action concerning the attempts to increase diversity within the spectrum of engineering students at German universities; implemented by means of gender mainstreaming initiatives, quotas, and women promotion programs. A national initiative has been founded to empower young girls and females for STEM subjects[1]. Even though the scope of action has been extended towards further categories of social difference with the introduction of the term 'diversity' into the discussions and initiatives concerning social justice and equal opportunities within the German educational system, there still remains a strong focus on the category of sex, dominating and neglecting the consideration of other categories (Andresen, Koreuber & Lüdke, 2009; Jansen-Schulz & van Riesen, 2011; Krell, Riedmüller, Sieben & Vinz, 2007). The authors consider *gender* as one of many categories of social difference depending on specific contexts which are summarized under the umbrella term *diversity*. From this perspective, the term gender is inherent in the term diversity. Due to the

Figure 1. German engineering graduates 2014
(Own Figure, data based on Statistisches Bundesamt, 2015a, p. 13)

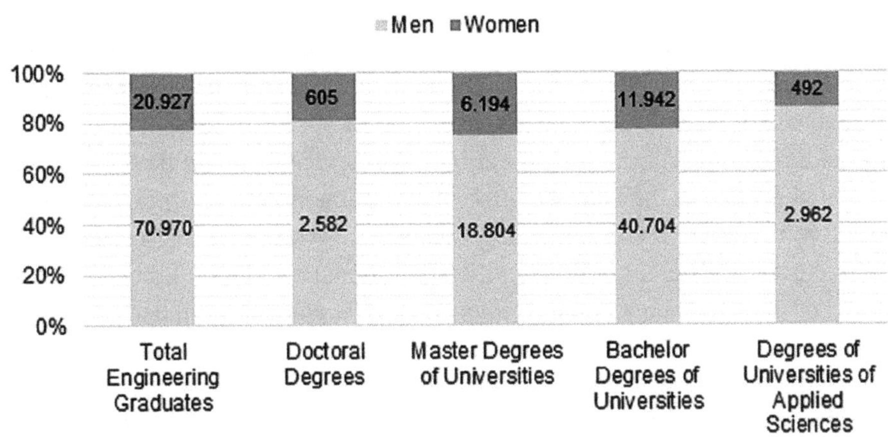

still existent underrepresentation of women in the engineering field, the category gender is especially stressed in this context.

The last ten years, in particular, have seen a number of efforts complying with a broad concept of student diversity addressing a wide scope of categories such as, social, economic, ethnic, culture, as well as disability and other physical impairments. Thus, the improper limitation of a comprehensive concept of (student) diversity seems to be increasingly overcome. Today a variety of diversity categories are taken into account with increasing tendency, but at many universities they are not standardized, yet, and still lay behind the attempts aiming on the disproportion of sexes.

The underlying lack of awareness for the necessity to consider further diversity categories beyond gender can also be identified in an insufficient statistical documentation of differentiated data about the quantitative representation of diversity groups among students. While the differentiation between women and men is standard in the quantitative analysis of statistics considering academics over the years, the documentation of further student diversity qualities and representation of other minorities besides women is just beginning in Germany (Bakshi-Hamm, 2007; Bakshi-Hamm & Lind, 2008).

In addition to gender, nationality is another common characteristic German statistics often include. They differentiate between German or foreign nationality; whereby, a further differentiation of different nationalities is not seldom missing. This is reinforced by the fact that the German nationality is strongly coupled to familiar ancestry and not linked to the place of birth like it is for example in the United States. Furthermore, the academic system is characterized by a high level of international exchange and mobility resulting in an un-usually large number of foreigners in academia. The current number of German and foreign engineering students at German universities, differentiated by gender, is pictured in Figure 2.

In order to document and analyze diversity, and inclusion or exclusion, respectively, it is crucial to be able to further differentiate between 'internationals', defined as people with a foreign nationality residing in Germany temporarily for studies or research, and people with a 'migrational background', a term used in Germany describing people who own a foreign nationality but continuously live in Germany, both educated or born in Germany (Bakshi-Hamm & Lind, 2008).

In studies aiming on the identification of disadvantaged and therefore under-represented groups in the German tertiary education system it is pointed out that especially children from families without academic background, and a critical socio economic status as well as children with 'migrational background' are still under-represented in reference to their amount of pupils achieving higher educational achievements, particularly the higher education entrance qualification (Autoren-gruppe Bildungsberichterstattung, 2016). Therefore, they have minor chances to participate in tertiary education. In practice this implicates that a majority of people

Figure 2. German and foreign engineering students at German Universities, Winter Terms 2010/11 until 2014/15
(Own Figure, data based on Statistisches Bundesamt, 2015b, p. 33)

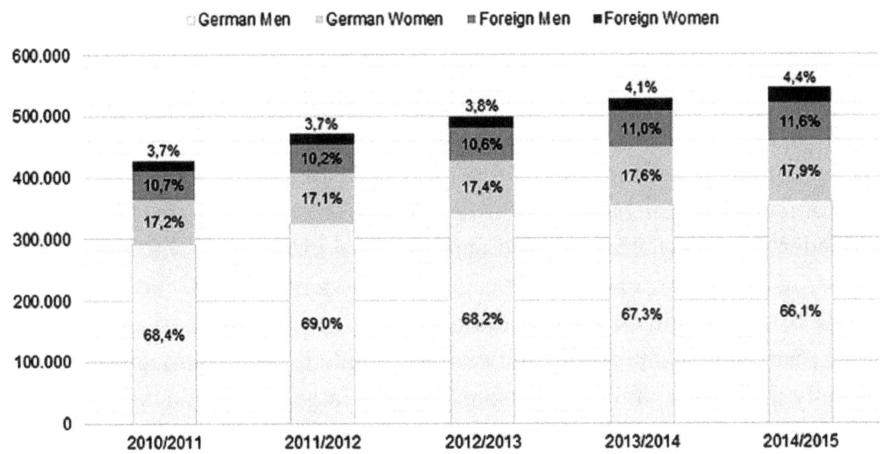

originally born and educated within the country are statistically documented as foreigners because they do not own German citizenship and cannot be differentiated from immigrants or 'internationals'. Consequently until today there is a great lack of valid and differentiated statistical data providing an incomplete picture of the representation of minorities within the German data collection in general but also within the academic system concerning student diversity.

Given the historical background and resulting perspective on engineering as field of study, the chapter offers a framework to look at diversity beyond successfully recruiting and retaining a more diverse cohort of students. It argues that diversity can drive innovation and produce better solutions for society's global issues, if engineers have a diverse view on society, as users of engineering products and as designers.

THE AACHEN WAY

Why Diversity Matters: Two Diversity Biases

Despite common commitment in regards to broadening engineering education, there is currently no common approach or action plan on what this enhancement should look like and how it should be implemented[2]. In particular, the integration of diverse perspectives is not yet part of a common agreement, despite this being taken into account consequently, since it is an inherent element of a comprehensive and coherent concept of a holistic engineering approach. The integration of concepts and theories from gender and diversity studies, as well as science and technology studies in engineering disciplines are approaches discussed for many years (Harding, 1986; Haraway, 1991; Faulkner, 2007), but have not 'reached' engineering culture and curricula.

The need for the consideration of diversity in academic contexts can be argued from two different starting points. The first perspective would be the one of equality. From a humanistic, Humboldtian perspective education represents a universal value from which a general right of every human emerges to participate in education (Andersen, 2004). This entitled and universal claim applies to every human being - or in the context of democratic societies - to any inhabitant according to the principle of equal rights, chances, and opportunities. This is stressed by the educational mandate educational institutions are awarded with in democratic societies (especially if they are publicly sponsored like the large proportion is in Germany). Moreover, institutions of tertiary education are required by law to ensure equal access for every citizen regardless of his or her individual backgrounds, to provide equal chances and opportunities. Thus, following the right for equality universities take diversity into account especially in the context of equal opportunities measures and initiatives to correspond to their educational mandate.

However, approaches behind the shared aim to strengthen diversity and bridge the proportion of underrepresented student groups is often addressed economically. The so called 'Diversity Management', addresses diversity from a human resources perspective. By means of this concept, differences between the backgrounds, needs, competencies, interests and perspectives of the personnel are regarded as an important potential for creativity and innovation (Hewlett, Marshall, & Sherbin, 2013). Even if diverse working teams may lead to new and ambivalent challenges in management and leadership, the human potential approach aims to open up resources towards this working force to the benefit output and success. Having first found access in enterprises the approach is also used in the academic context as an instrument to realize excellent research and teaching (Chun & Evans, 2016).

In Germany, the first discussed approach has a long historical and political tradition. Starting from attempts to open the tertiary education system towards broader parts of the population, the so-called '68 generations', a German student movement, initiated an extensive change of values referring to societal norms, civil, as well as in particular women's rights, and fought against a conservative and dominantly male supremacy within society, politics, businesses, and the academic system. In succession, university institutions received more democratic structures, oriented to democratic values and equality standards. Since the 1990s, equalization was conducted under the term of gender and the European gender mainstreaming, which focuses on the bridging of a persistent underrepresentation of females in studies, research, and academic institutional bodies and processes, especially in relation to STEM fields. Until the turn of the millennium equality discussions and attempts were more or less limited to women's interests, whereas other categories of social difference received much less consideration. This changed when the US concept of diversity found its way over the Atlantic, where it already had a long historical and political tradition of implementation. At the same time, a set of new liberal developments in higher education policies obtained entry into the German academic system converting the former governmentally imprinted institutional structures to a more entrepreneurial governance of universities, which caused a lot of skepticism. Influenced by this development also the concept of diversity, including its human potential bias, was associated with these developments as a disliked, entrepreneurially characterized process, which resulted in an overall rejection. In the center of its criticism stood the concern that the diversity concept might displace the gender movement and its human potential bias might dominate the objectives and achievements of equalization. Although the two concepts are often discussed in a contradictory way (Andresen et al., 2009), both concepts can also be brought together complementarily in a fruitful manner (Kehr & Leicht-Scholten, 2013). Thus, today diversity is on the advance and experiences a rising implementation within the German academic system (Leicht-Scholten, 2011b).

By doing so, the value of heterogeneity is closely connected with equality and leading to a broadening focus on further categories of social difference beyond the mono-dimensional concentration on sex and gender.

Changes Require Changes

Necessity for diversity consideration not only stems from aspects of social equality or the human potential approach, but also from the concept itself. Moreover, external reasons promoting a consideration of diversity, closely related to its potential bias, exists. This raises the question of 'why' especially engineering should take diversity into account and moreover take a look on other dimensions of diversity next to its relevance for social equality.

Fundamental societal changes and global megatrends lead to new problems and requirements for sustainable solutions and innovations addressing the creativity and power of the full range of academic disciplines. In this setting engineers obtain a special role. As inventors and entrepreneurs engineers are rated among the powerful designers of global developments in the past. Fostering technical improvements and innovations they thereby influence economic growth and societal prosperity and change the face of our world in many ways (OECD, 2014).

Given this decisive power, shape and change is affiliated with a great responsibility. Where opinions will be deeply divided when it comes to answering the question of how far engineers met their responsibility until now, it is indisputable that there is a rising awareness within the global society that there have still many fields of work remained which have to be improved and connected with concrete expectations to ensure the wellbeing of the planet in the future.

In order to meet these challenges future actors in charge require abilities which are up to now not part of the current regular education concepts and curricula. Conversely, future engineers must be prepared for future tasks. Therefore academic teaching needs to be reshaped and up-dated towards a holistic and transdisciplinary approach. It has to be expanded by essential competencies which go beyond a purely technical expertise. To ensure a future-oriented perspective, urging problems engineering has to take into account such as, the complex interrelations between technology, human, culture, society, and environment must be taken into account. Rethinking engineering towards a more holistic view is the precondition for the development of sustainable and socially responsible technical solutions. Furthermore, it provides an important potential for technical innovations with societal and economic added value. The integration of gender and diversity perspectives is an essential part of this approach and offers methods and concepts for its concrete implementation in research and teaching.

Changes Challenge

In engineering and technology, there emerges a need for a paradigm shift towards a holistic approach regarding technology in connection with its context (see Figure 3), and associating it with gender and diversity issues. The complex interrelations between technology, humans, culture, society, and environment are an essential relevance for the development of future-oriented, sustainable and socially responsible engineers producing technical solutions and innovations with societal and economic added value.

To comply with their interface position, leading universities consider these insights and respond to them with different approaches to advance engineering education such as, the CDIO initiative (Kamp, 2014). Even the necessity for change is clearly evident and recognized, advancements seem to require a long-term change process as there is still potential for improvement. The traditional origin of engineering sciences follows the conclusion, that it takes a long time to change the grown structures and policies even the need for adapting engineering education to the global challenges as well as the advantages are obviously.

Whereby, the overarching question is how to bring the German engineering education, which is known for the mediation of technical competencies and deep qualifications, together with core competencies such as the abilities of thinking transdisciplinary and intercultural (Hippler, 2011). The 'Verein Deutscher Ingenieure'

Figure 3. Range of interfaces and topics in engineering context
(Source: GDI)

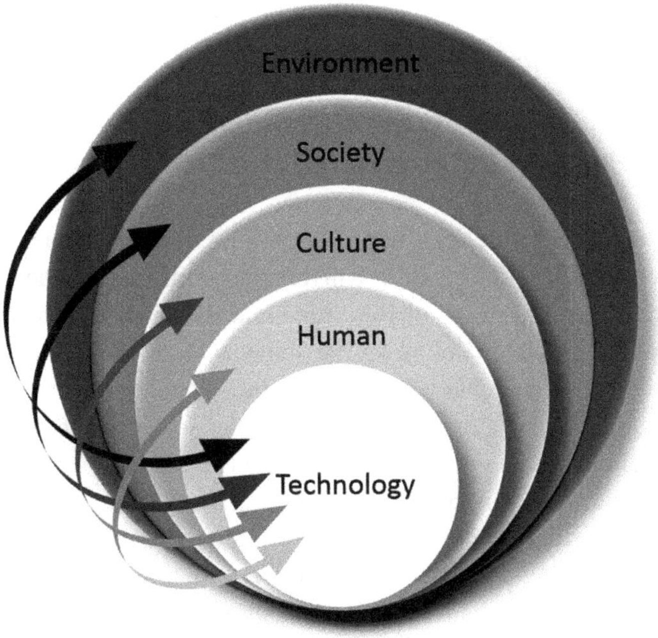

(VDI), a 160 years old technical and scientific association and the biggest engineering institution in Germany, identified trend statements in the frame of a large-scaled study (VDI, 2016). These conclusions are pointing the way for a demand-oriented alignment of engineering education and reveal a need for action. Moreover, different national and international institutions, such as, TU9; or the nine leading technical universities in Germany, Acatech; the German Academy of Technical Sciences and the IDEA League; a network of leading European universities of science and technology, created different approaches for adjusting engineering education to their individual perspective on external requirements. The "League of European Research Universities" (LERU) analyzes the role of gender and sex analysis in research and innovation (R&I); arguing that it needs to be better integrated into R&I funding, content and implementation process (LERU, 2015). Although, there seems to be a variety of perspectives on how to change engineering education towards a future-oriented education system, one would almost agree that there is a necessity.[3]

In Germany identified as driving factor for widening the engineering science perspectives, diversifying engineering education is seen as an element of excellence engineering education (Leicht-Scholten, 2013; Leicht-Scholten & Wolffram, 2010). However, this also points to a diversifying of engineering research as well as engineering contents. Although there seems to exist tension between the required engineering profile and the persistence in university structures, RWTH Aachen University decided to follow new ways and implemented an innovative format of chair, addressing this new approach of technology.

The GDI Approach: RWTH Aachen University Enters New Paths in Engineering Education

RWTH Aachen University belongs to one of the nine leading technical universities in Germany distinguished as an Elite University within the Excellence Initiative of the German Federal Ministry of Education and Research and the German Research Foundation[4]. With 57% of its 43,721 students in the winter term 2015/2016 the greatest amount of students were registered for engineering studies, followed by students signed up for natural sciences, computer sciences and mathematics at 23%, 13% studying humanities, social sciences and economics, and 7% of students signed up medicine and dentistry (RWTH, 2016a, p. 37). Not only is RWTH Aachen University distinguished for its cutting-edge research with the highest amount of external research funding in the national comparison, but it is also highly esteemed for its excellent education in national and international rankings, especially in engineering and natural sciences (RWTH, 2016b). Therefore, RWTH Aachen University continually aspires to maintain its leading position with innovative approaches. In 2012, it established the 'bridging professorship' "Gender and Diversity in Engineering" (GDI).

Structurally located within the faculty of Civil Engineering, the GDI team is working at the intersection of gender and science and technology studies. The chair presents an innovative format aimed at overcoming common disciplinary gaps in order to open up new transdisciplinary perspectives within engineering and technology through the consideration of gender and diversity issues. By unlocking this innovative potential the added value of the integration of diversity issues and transdisciplinary perspectives provides for scientific research; however teaching can be exploited.

With social and political science considerations taken into account, research at the GDI is always focused on issues touching upon interdisciplinary fields between society and technology. To open up the benefits of crossing perspectives of different disciplines gender and diversity aspects are identified and integrated in technological contexts. By linking these new perspectives approximations towards technology can be provided and mono-disciplinary engineering views can be enriched by social facets of technology, which as of thus, have not been taken into consideration so far. There are always social factors and dimensions embracing every technical issue, which should be taken into account to contribute to a comprehensive technical solution. In solving complex problems there is a need to understand them in-depth along with a regard to multidimensional contexts. For sustainable solutions the consideration of social components is crucial. For example the integration of heterogeneous interests of single persons, (user) groups, as well as public and environmental concerns provides a precious pool of information strongly relevant for suitable solutions, techniques, and applications to be developed. This represents an important potential for technical innovations in research and development.

The institutional structure of the chair "Gender and Diversity in Engineering" (GDI) reflects the transdisciplinary approach of diversity and science and technology studies (see Figure 4). Disciplinary transfer of concepts, theories, and methods of the social sciences, particularly the gender and diversity studies, build the scientific basis of the GDI approach. Whereby, it concentrates on the non-technological perspectives of technology, which are often underestimated in current research approaches in engineering. This might lead to the consideration of perspectives of target groups and the concrete technology users (e.g. Acar & Lopik, 2009). By analyzing and differentiating their specific needs and interests, the social acceptance and economic success of future technological inventions can be raised (Himmel & Ziefle, 2016). Furthermore, concerning the contents a focus set on the analysis of social, societal or environmental contexts of the respective technical research task and its interrelations. Thus, aspects of social justice, responsibility, and sustainability are discussed. As a result, technical solution approaches are ensured to be ethically reflected and in compliance with the diverse requirements of the interests and needs of their users, as well as that of society and the environment.

Looking at its portfolio, GDI, comprises the working fields of research and development, inclusive engineering education, and innovation processes (see Figure 4).

In the GDI approach research and education are closely linked to one another, as although the two fields traditionally are separated in Germany, the mutual enrichment is obvious. In contrast to the fact that universities are committed to their educational mandate the sector of academic teaching has been underappreciated within the German academic system for many decades. Higher education policies have recognized this imbalance. In the last decade, a range of initiatives have been initiated to improve the quality and to bridge the disproportion between the value and reputation of teaching in contrast to research (Bouffier et al., 2014).

Strongly committed to the advancement of engineering education, the GDI approach provides students of different subjects within engineering to gain competencies referring to gender and diversity. These competencies are crucial for meeting the multiple requirements future engineers have to face. Ultimately, an ability to capture diverse perspectives and holistic views is not only essential for leadership but also for responsible and sustainable technical solution approaches. Educating students and teaching them the necessary competencies obtain a key role to ensure

Figure 4.Structure and outline of the "Bridging Professorship" gender and diversity in engineering
© 2016, GDI RWTH Aachen University

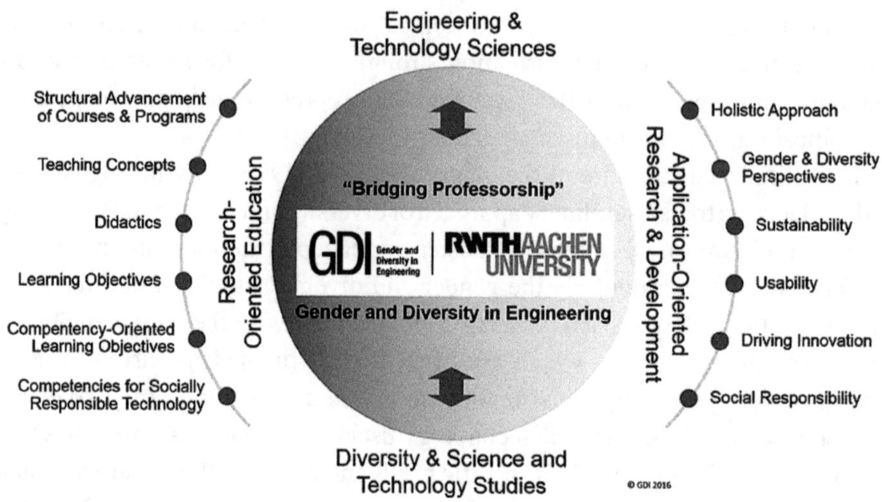

that they will be able to meet the future challenges our societies will have to face. With the implementation of a compulsory gender and diversity module within the curriculum of its bachelor study "Civil Engineering" RWTH Aachen University complies with its pioneering position among the German technical universities, and allows a social responsible engineering education. Given the high level of value placed on proper education forums, the GDI developed a teaching concept aimed towards student empowerment.

DIDACTIC APPROACH

In engineering education "Mismatches exist between common learning styles of engineering students and traditional teaching styles of engineering professors. Consequently, students become bored and inattentive in class, do poorly on tests, get discouraged about the courses, curriculum, themselves, and, in some cases, change to other curricula or drop out of school. (…) Most seriously, society loses potentially excellent engineers." (Felder & Silverman, 1988, p. 674). Although these findings by Felder and Silverman are dating back to the year 1988, and engineering professors and teaching staff have tried to change the teaching style in engineering education over the last few decades, the underlying problems described still hold true today given the high-level of drop-out rates in engineering study programs are counted among the highest of all study fields not only in the USA but also in Germany and other European countries (Heublein, Hutzsch, Schreiber, Sommer & Besuch, 2010; Marra., Rodgers, Shen & Bogue, 2012; Matthews, n.d.; Knight, Carlson & Sullivan, 2003). To match teaching styles of engineering teaching staff with the learning styles of engineering students an all-inclusive, holistic didactic approach must be established and applied.

Educating students to be able to cope with global challenges, follow different career paths as well as acting in a socially responsible manner, requires a close connection between teaching and research and is strived for and realized by a continuous adjustment of learning contents. Engineering students view interdisciplinary content in the curriculum as necessary preparation for employment, despite being taken out of their 'technical-comfort-zone' (Barnard, Hassan, Dainty & Bagilhole, 2013). The implementation of concrete research activities and findings, as well as current social and political developments, is necessary for a reflection of personal actions. The mediation of different personal or disciplinary perspectives and the creation of tasks, which are inspired by real concrete problems and accompanied by cooperation partners like NGOs and companies, enable an experience which gives students an insight into their later field of work. The GDI's teaching approach is characterized by its competence orientation instead of pure knowledge transfer.

With the help of the "Holistic competency-based approach", GDI broadens engineering education to fix the mismatch described above, which seems to be present between learning and teaching styles in engineering training. The approach provides students with competencies beyond, as traditional engineering skills considered capabilities, which are inquired by companies and for creating solutions for global issues alike. This holistic competency-based approach helps students to not only think critically outside the box, in interdisciplinary ways, but also to develop creative and innovative ideas, work both independently and in a team setting, and to keep morals, ethics and social responsibility in mind. Basic non-technical skills like communication skills and scientific writing are also implemented in the courses at GDI. Through the integration and coverage of diversity-concepts, all those different competencies are trained and the students learn to apply the competencies to face global challenges while keeping the human factor in mind.

The holistic competency-based approach of the GDI conveys diversity-aspects not only through course contents but also from a didactic perspective. The GDI approach leads to the integration of all students into the course. The active inclusion of diverse students allows the individual to feel involved, included and empowered. From a context-based perspective the whole class learns about diversity concepts and how to reflect upon individual diversity in this context. This individual experience is embedded in an application-oriented frame, where the course participants learn about different concepts and models of gender and diversity theory in a - for non-social sciences students - customized way. Basic knowledge referring to sex, gender, and intersectionality, as well as the social construction of stereotypes, social labelling, and discrimination are being conveyed[5]. They learn to think beyond their own borders and needs in order to reflect their personal habits. After these cognitive processes, students are able to transfer their experiences and knowledge to the range of responsibilities of their future field of work. To make individual firsthand experiences, an essential part of the GDI seminars consists of teamwork. This achieved knowledge is connected with their future work area; whereby, students get a problem-oriented task, which they should solve in a diverse team. In this context, diverse means an interdisciplinary, intercultural and gender-mixed team composition. Thus, especially the active and passive self-reflection, promoting a sustainable knowledge production.

From a didactic perspective GDI pursues a participatory, research and application-oriented approach. Empowering, application-oriented teaching concepts introduce students into concrete research practice and motivate them to develop their own solutions. In this frame, students have the possibility to develop their own thoughts and to discuss them under professional inspiration.

Maxim and Theoretical Teaching Concept

To emphasize the practical relevance of competencies beyond technology, integrating diversity perspectives into engineering education within the frame of practical applications in research and practice is a necessity. In order to ensure this, a close connection between research and teaching must be established, to help reveal research gaps in the field of diversity in engineering. Besides the close link between research and teaching, the authors' aim for engineering education focuses on keeping the human factor in mind, while meeting challenges and developing socially responsible and sustainable innovations with and for societies. Based on this maxim, GDI developed a theoretical teaching concept which functions as a framework for seminars and which helps to convey competencies beyond technology for future engineers. Said theoretical teaching concept should be treated as an ideal type as defined by Max Weber. Weber's ideal type is a tool which helps to analyze and capture certain parts of real life. Additionally, it categorizes sections of social realities and serves as an indicator for actual occurrences. The most important purposes of an ideal type with regard to the theoretical teaching concept are to gain new insights and to stimulate the creation of new hypotheses in research (Weber, 1968). In regards to Weber's ideas the theoretical teaching concept serves as an ideal type for creating courses and seminars for students because it helps the students and also the teaching staff to obtain new findings and hypotheses for their future work and research. The theoretical teaching concept consists of different components like a research-based but also practice-oriented input for the students, the definition of competence-oriented learning objectives, gender and diversity-integrated didactics, a student-focused support, independent and practical work experience, an interdisciplinary approach and empowering and self-reliant learning – all of those components lead to the inclusion of gender and diversity aspects into the course contents. The practical implementation and ideas behind previously mentioned components of the theoretical teaching concept will be discussed in the following part.

Besides, the definition of competence-oriented learning objectives is an essential part of every course at the GDI because they help the students and the teaching staff to focus on the central themes of each session and to strengthen the competence-oriented approach. The students know, what kind of knowledge they are supposed to gain during the course and the lecturers know, what they can expect from the students in terms of competence development. Competence-oriented learning is one of the most important factors for improving the educational system according to the Bologna reform (Gaethgens & Zervakis, 2007).

The didactic setting can vary only generally, the style used in the courses is not a traditional teacher-centered teaching style, but more of an interactive self-learning process, moderated by the lecturer. Gender- and diversity-aspects are integrated in the didactic style to help create an inclusive learning environment. This ties into the concept of close, individual support for the students, which is especially important in courses which deal with interdisciplinary topics. The (engineering) students get introduced to a way of analyzing problems which is mostly new to them - namely a social science based perspective. This requires a solid support and help system so that students do not feel left alone. Furthermore, the teaching staff creates a safe space for the students in which sensitive topics might be discussed. Given this setting, a code of conduct is introduced which urges the students and lecturers to adhere to certain rules to create a safe working environment. In spite of the close personal support, the teaching staff gives the students space to work independently, to create their own ideas and solutions for real life cases. Those cases enable the students to gain practical experiences while learning and developing new ways of problem solving. To help the students in creating innovative ideas, they work together in interdisciplinary teams. Moreover, analyzing transdisciplinary problems is another component of the theoretical teaching concept.

The compilation of the different components explained above lead to empowering and self-reliant learning, an important factor in generating confidence in one's own abilities.

The main objective of the theoretical teaching concept is the development of competencies beyond classical engineering skills which is done by including gender and diversity perspectives and connecting those with engineering topics.

One of the overall characteristics of the theoretical teaching concept is that it can be adopted from different faculties, or, other (interdisciplinary) teaching fields and subjects, to help diversify education in those fields as well.

Different Learning Types

The theoretical teaching concept is influenced by academic approaches to engineering education and by the assumption that there are not just one but many different ways in which people learn best (Vester, 1998). Although learning style theories have been criticized by various scholars (Willingham, Hughes, & Dobolyi, 2015; Rohrer & Pashler, 2012), the development of the theoretical teaching concept is based on various learning style theories. The result is that with the help of the theoretical teaching concept, the different needs, requirements and approaches of diverse students are taken into account.

In particular, seminars are designed according to the method of experiential learning and action learning by Kolb (1984) also called "The Kolb Learning Style Inventory". According to Kolb's (1984) method, successful learning requires skills of the following four areas:

1. Practical involvement and concrete experience,
2. Reflective observation,
3. Action orientation and active experimentation,
4. Analytical detachment.

Kolb's concept describes an experiential cycle of learning which embraces the four dimensions mentioned above. As these are based on different learning types (comparable to Vester, 1998), the approach considers diverse needs of students with regard to their learning behavior. For this reason, GDI developed a chain of knowledge generation, which implements different learning types and styles in an individualized way. The participatory model allows the active implementation and consideration of students' individual experiences, personal backgrounds, realities of life as well as their diverse needs. Figure 5 illustrates this correlation by showing the different steps of knowledge generation in the GDI lecture concepts.

The theoretical teaching concept also adapts to societal trends and developments.

The Blue Print - Putting Theory into Practice

After introducing and discussing the concepts and models of learning and how-to-design courses theoretically, in the following paragraph the transition between theory and practice will be described.

Figure 5. Chain of - knowledge generation
(Source: Leicht-Scholten, Steuer & Bouffier, 2016)

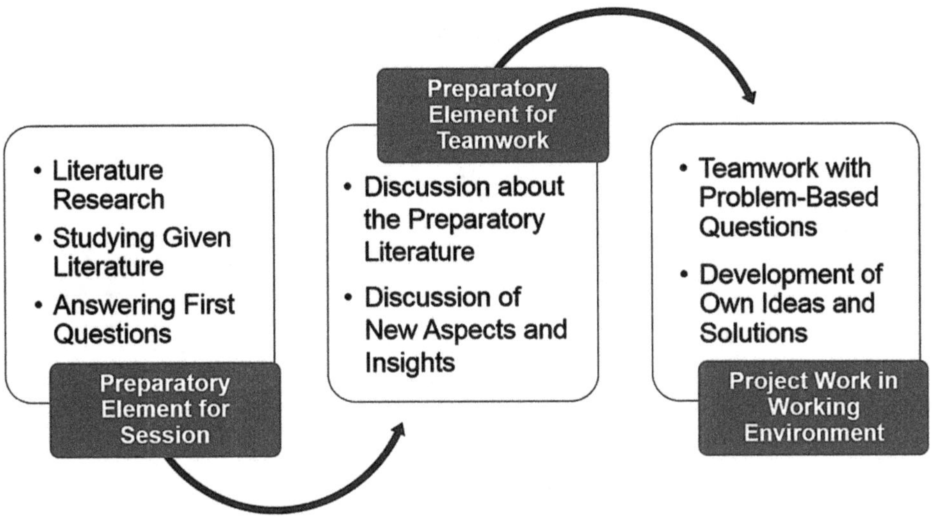

The graph (see Figure 6) below shows, how the theoretical approaches translate into practical measures. The overarching goal of this maxim is to create socially responsible engineers for the challenges of tomorrow.

The question about the relevance and reasons for implementing the practical measures as shown in Figure 6 should be discussed in the following section.

The first step in constructing Figure 6 is *Sensitization and Self-Experience*. In this step, the students should be sensitized for the specific topic of the course. In the seminar "Skills for a Social and Sustainable Technology Design" for example, a representative of a cooperation partner talks about the course topic which will be the object of the courses' analysis. During this first step, empathy is created because the students are being confronted with the topic directly by people who are actually working in that specific practical field which helps to sensitize the students for the issue at hand. Sensitization, self-experience and empathy are key factors for keeping diverse user groups in mind while creating technological solutions (Walther, Miller, & Kellam, 2012). As a result, these factors are essential in order to avoid a dislocation between the creation of infrastructures or products and the needs of the actual target group.

The second step is *Essentials of Gender and Diversity Theory*, which translates into the mediation of basic knowledge concerning concepts about gender and diversity. This knowledge helps students to gain a profound understanding for heterogeneous user groups and their diverse needs. In this step, the theoretical concepts of gender and diversity are laid out. Aforementioned concepts help to support the empathy

Figure 6. Holistic and transdisciplinary approach
© 2016, GDI RWTH Aachen University

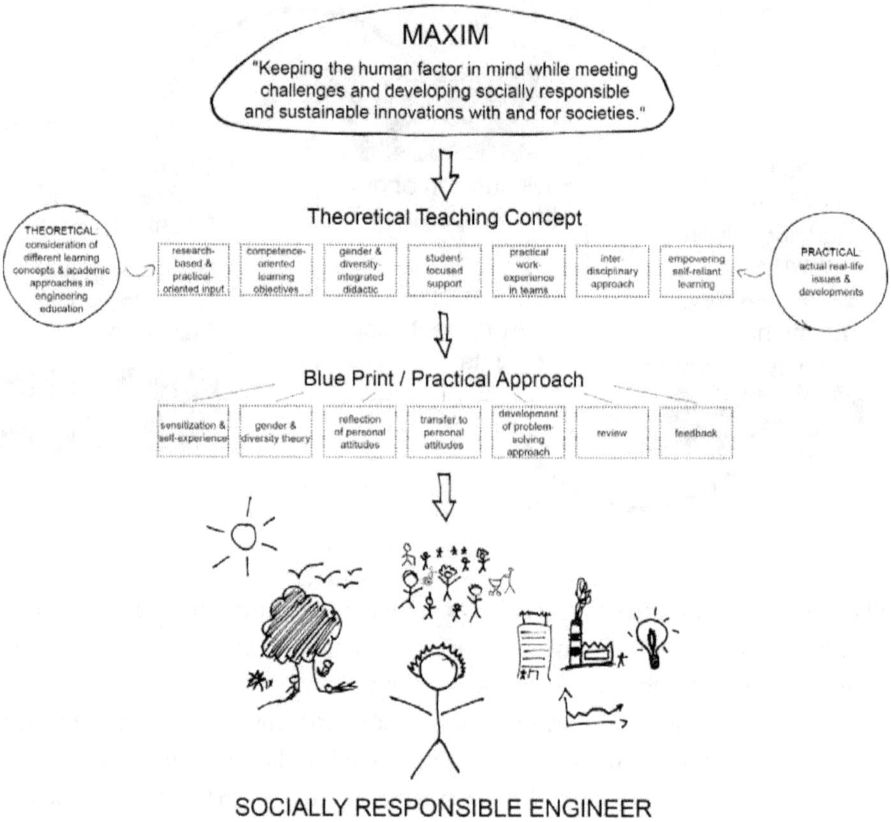

which has been already created in step one. The introduction of gender and diversity concepts is important because they help to broaden the students' perspectives, and to develop skills beyond technology which are demanded by companies and also by the global challenges societies are facing and for which engineers develop solutions (Kamp, 2014).

Reflection of Personal Attitudes and Subject-Specific Cultures constitute the third step of the blue print. The students get together in groups to work on their specific topics. During this process they should reflect critically their own personal attitudes to reveal potential stereotypes or biases which they might have in mind concerning their topic. Moreover, empathy plays an important role again when analyzing subject-specific cultures. This step is essential to help students develop a critical mind.

The forth step is about the Transfer to Personal Attitudes and Subject-Specific Contents. In this step students should take their findings of step three and transfer those to concrete subject-specific problems. Transferring knowledge is stressed during the tasks the students have to work on because it is a prerequisite for thinking outside the box in interdisciplinary, creative ways. Therefore it is one of the most important criteria for the grading of the students' work.

In step five the Development of Concrete Problem-Solving Approaches is paramount. The reflections and observations resulting from the previous steps should now be put into practice. A solution for the given task has to be developed.

The Review of the Outcome is the sixth step, which is set up as a peer review process. Here comes into play the ability for critical reflection because students have to present to each other, review their own ideas and solutions critically and integrate improvements if necessary.

The last step is about Feedback and Final Reflection. The students have to deal with the feedback given to them by the professional teaching staff. Those abilities are important for the future careers of the students because it is crucial to be able to deal with feedback and to transfer it into improving the solutions they develop further. The feedback is given in a very transparent way with the help of evaluation sheets.

To illustrate the "Holistic and transdisciplinary approach" the seminar "Skills for a Social and Sustainable Technology Design" will be described in detail below.

"ENGINEERS WITHOUT BORDERS CHALLENGE" - BEST PRACTICE

Based on the discussed teaching methods, course elements of GDIs teaching concept will be demonstrated in the following chapter section. In order to achieve this, a best practice example will be presented to illustrate the different teaching components in the frame of a coherent concept. First of all, structural aspects considering the integration of these teaching concepts will be described. Afterwards, the course contents will be specified. Building on the contents the course elements will be presented and explained with an established course.

Engineering education from a content based and didactic perspective in Germany is focusing on the mediation of engineering discipline-oriented knowledge exclusively (Steuer, Berg, & Leicht-Scholten, 2015). A key premise for defining the course contents is to select a topic that allows the consideration of scientific engineering issues with the inclusion of diversity aspects and the needs of the potential target group. To describe this approach of engineering education in Germany, the training of I-Shaped (Grasso & Burkins, 2010) professionals seems to be the pursued education aim. The research group "Gender and Diversity in Engineering" deviates deliberately from the usual procedure, as the claim is to educate engineers for the global challenges of tomorrow in an interdisciplinary setting (Leicht-Scholten, Steuer, & Bouffier, 2016).

From a structural perspective, the integration of the research institution GDI in the faculty of Civil Engineering, representative for all engineering faculties at the RWTH Aachen, as well as the faculty of Philosophy allows a teaching approach which enables a mediation of social science perspectives and methods from the engineering sector. As a result, the teaching format is adjusted to the needs of students from diverse engineering fields and makes a subject-organized in the social sciences. Rather, it is the aim to convey sociological knowledge which stands in a direct connection to the later field of action and the individual engineering discipline.

The course, "Skills for Social and Sustainable Technology Design", was developed and implemented in the winter term of 2014. The intention is to offer a course that allows students, from different fields of study, the linkage of engineering core topics with societal requirements, to achieve a multidisciplinary course setting, a structural frame had to be established.

Structural Frame

To create a working atmosphere which is comparable to the working environment and team situation of future engineers, an important structural aspect of the course is to include students of different disciplines. In order to achieve this, the course is structurally implemented into different curricula as an interdisciplinary module. This implementation makes it possible for students to enrich the individual professional profile with transfer competencies that go beyond the common engineering education.

To combine complementary disciplines and to ensure a holistic perspective in discussion processes, the in Table 1 presented faculties are taken into account.

The resulting interdisciplinary frame allows problem-based tasks to be discussed from different perspectives. This is aided by the fact, that every course participant has a pronounced professional habitus which is shaped and evolved in the frame of the studies (Paton, Wagner & MacIntosh, 2012; Bourdieu, 1977; Jamison, 2013). Given that students have to work in multidisciplinary teams, they have to learn to incorporate different disciplinary cultures in their teamwork. On the one hand, students learn how to deal with diverse perspectives and how to transfer these diverse approaches into a solution. On the other hand, they learn about the varying aspects of diversity which includes, handling differing needs and appreciating other perspectives. The participants should learn to take different attitudes and stances into account when thinking about engineering innovation in a social responsible way. The element of group work enables students to reflect team building processes and to find personalized ways on how to work in a team, dealing with discussions and disputes.

Table 1. Participating faculties and subjects. © 2016, GDI RWTH Aachen University.

Department	Faculty of Civil Engineering, RWTH Aachen			Faculty of Economics	Technical Faculty	Philosophy Faculty, RWTH Aachen
Course of study	**Civil Engineering**	**Mobility & Transport**	**Environmental Engineering**	**Industrial Engineering with Business Studies**	**Technology Communication**	**Political Sciences**
Specialisation	Traffic	Road & Vehicle	Urban Water			
	Construction Systems & Construction Mngmt.	Transport Logistics	Water Resource Management	Constructive Engineering		
	Constructive Civil Engineering	Rail System Engineering	Energy & Environment in Civil Engineering	Water Management		
	Structural Engineering	Airport & Aviation	Recycling	Building Operations & Geotechnics		
	Constructive Hydraulic Engineering	Mobility of Persons	Envrionmental Process Engineering	Traffic & Spatial Planning		

With regard to the structural implementation, the timing for offering such a highly demanded course in the curriculum is of vital importance. Given the set of personal requirements placed on the students such as, self-reflective capability, broadened interest beyond individualized specialist disciplines, content-related competencies in the field of study, and future field-related activity, the GDI decided to offer the course in the corresponding Master Programs. Furthermore, the high professional competence in this stage of studies can lead to fewer application-oriented problem-solving approaches.

Course Format and Contents

Conceived along the lines of the Design Thinking approach, the Context-Based course format, an established method that allows for the creation of new ideas and means of solving problems from a user-centered perspective, was developed. The seven steps described above correspond in a way with the process of Design Thinking, as the five pillars of the Design Thinking Approach are: Emphathize, Define, Ideate, Prototype and Test (Plattner, Meinel, Leifer, & L. Leifer, 2015; Grots & Pratschke, 2009; Brown, 2008).

The Design Thinking Approach focuses on designing creative and innovative ideas in an inter- and trans-disciplinary setting and has been shown to be very successful in achieving innovations (d.school, 2010; Plattner et al., 2015). As Design Thinking is not the primary topic of this chapter it should not be discussed any further at this point. Nevertheless, it serves as a reference for the Practical Implementation Approach developed by the GDI, which enriches the approach with a strong enhancement of the inclusion of gender and diversity perspectives.

Students are geared to work with an application-oriented approach in the frame of the course, in order to combine their disciplinary knowledge with diversity studies; however, the course contents have to meet different requirements. Firstly, the course contents have to be suitable for an application-oriented transfer between diversity studies and engineering. These criteria are best met by real projects in companies or organizations. In order to ensure this, the course should be conducted with a partner organization or enterprise that is actually confronted with real-world examples of application-orientation in regards to the engineering sciences. Furthermore, students should learn to deal with the diverse needs of real users. In this context the focus emphasizes users who have other cultural backgrounds or requirements differing from the daily life of a German student. As a result, students have to empathize and to put themselves into the position of potential users.

According to the requirements and learning objectives, a cooperation with the non-profit organization "Engineers Without Borders e.V. Aachen"[6] an organization carrying out engineering projects in development cooperation, was established. As part of the collaboration, a project actively dealing with diversity categories in gender, culture, age and disability was selected. Thus, students have the possibility to develop different concepts in an engineering context for a concrete project with "Engineers Without Borders e.V. Aachen" EWB.

The overarching project consists of planning and establishing wastewater-free, dry toilets in a village in Sierra Leone. The system separates urine and feces into separate storage cells. The advantage of this approach is that urine becomes fertilizer and feces is composted. Whereby, toilet use becomes part of a sustainable system. This main project is divided into different task groups; where concepts, plans, and measures are developed beforehand and real implementation occurs on the spot by the "Engineers Without Borders".

The in Table 2 presented projects have been developed with the aim to bring engineering in a social responsibility context. Students should gain an ability to reflect on their social responsibility and consider diversity in their later field of action.

In Figure 7, the GDI course design is described in each step, on the basis of this concrete course. Based on the chronological course expiry, each phase will be placed in a concrete connection with the resulting educational intention.

Table 2. Included Projects. © 2016, GDI RWTH Aachen University.

Project 1: Innovative Usage and Operating Instructions for Dry-Toilets - Creating a construction manual under the consideration of local accessible materials, building capabilities, language barriers, metric system conversions, etc. - Creating a user manual for promoting the correct use of the dry-toilet system, taking into account differing users
Project 2: Design of Dry-Toilets (from a user perspective) - Design of dry-toilets with regard to differing users
Project 3: Fundraising Strategy and Individual Measures - Implementing a financial strategy for how to fund the project
Project 4: Hygiene Promotion - Taking into account: Illiteracy, Target Age-Group, Language, Religion
Project 5: Innovation Design Plans and Blueprints for Dry-Toilets
Project 6: Structural Foundation for Schools

Figure 7. 7 Phases of social responsible engineering education
© 2016, GDI RWTH Aachen University

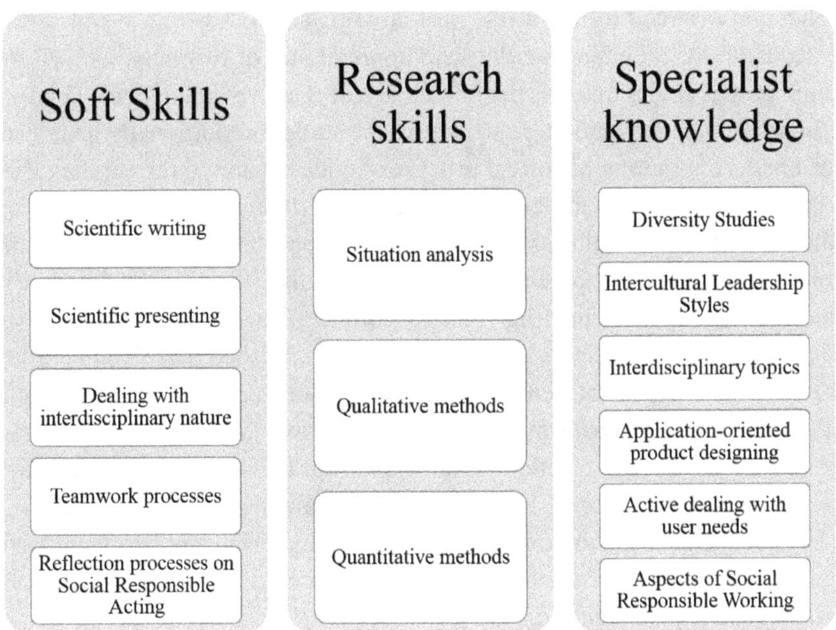

The first step of the content based GDI teaching concept is to allow students to make self-experiences concerning diversity. Thereby, under diversity gender aspects, culture as well as physical peculiarities, educational background and financial opportunities are concluded. This step is initiated primarily through interviews and personal contact with users. In this specific context, the target group is people from the village of Makali in Sierra Leone.[7] In the presented course this process is made possible by the provision of information, material, and experience from the Aachen regional group of "Engineers Without Borders". Course participants have the chance to discuss with members of "Engineering Without Borders", whom had previously volunteered in Makali Town, and report their personal experiences and impressions. Furthermore, photographs and material on the concept of dry-toilets is available in this frame. In this sensitization and reflection process students have the chance to see their task in an intercultural context. For example, the important role of local organization such as "Madame" as ambassador between the target group and the foreign German organization, is clearly evident after the exchange. In addition, students learn about the cultural, religious, and gender aspects related to common bathroom practice. The teams have the possibility to implement this important connection to the users in their approach development. Furthermore, students obtain background information on the "Engineers Without Borders" organization and learn the functionality behind sustainable development cooperation. This is crucial, as students learn to invent concepts that enable the target group to be independent and to create solutions from a non-European perspective, which are not applicable in an application-oriented frame of the local circumstances.

Prerequisite for an understanding and appreciating of diversity, as well as the reflection process build upon, is the knowledge of the own culture and social structures. In order to understand these structures the students additionally hear lectures and read papers about the historical process of equality and to reflect the different steps that lead to the contemporary legal and social situation. To consider social structures as well as the cultural aspect of the development of equality in Germany, it is compared to the development process in the United States. In this step, the triad of equal opportunity, sustainability, and economic efficiency is considered under a global perspective. This allows a reflection of as established perceived approaches and aspects, but cultural influenced perspectives and leads to a critical altercation with the own culture. Furthermore, experience shows that it is also important to create a basis for communication in an intercultural frame. To demonstrate the diverse requirements and needs with regard to organizational constructs, the Globe Study, is also a discussed topic in this context (House, Hanges, Javidan, Dorfman & Gupta, 2004). Thereby, students learn to develop customized solutions and to put down their individual perception.

In an actual paper the authors show that engineers who participate in activities as "Engineering Without Borders" seem more diverse and more broadly interested than the larger population of engineers in the US. Therefore the authors suggest including socially engaged engineering activities in engineering may increase the number and broaden the diversity of engineering students and practitioners (Litchfield & Javernick-Will, 2015; Litchfield & Javernick-Will, 2016).

In regards to future field-related activity, students begin by learning to empathize with potential users of a product and to recognize needs. It is particularly imperative to contemplate varying societal factors, since a narrowly defined target group could lead to the exclusion of potential users and needs. Especially in regards to the innovation context, the internalization of this approach constitutes a core competence of innovative engineers.

In the second phase of the GDI teaching approach, students learn to connect personal experiences with theoretical background. Topics related to diversity studies and management, gender mainstreaming, stereotyping, and label-processing are addressed. The ultimate goal of this learning unit is to enable students to reflect on individual roles in social processes, resulting in social responsibility and the already mentioned social responsible engineer. This also includes that students recognize social barriers and stereotyping constructions in societal structure.

Building on the first two phases students come to reflect in the third step personal attitudes and subject specific culture. This reflection process is enabled through the previously mentioned interplay between the interdisciplinary settings of the course. This procedure is in particular initiated by the addressing of the construction of stereotypes and barriers. Gaining of often new insights leads to a detailed inner perception of alleged realities, as well as, open discussion of the subjective world view. Subsequently, after the reflection process the fourth phase leads to the transfer of personal attitudes and subject specific contents. This phase is supported by discussions which are embedded within the concept of the course. Experts from "Engineers Without Borders", whom are representing the target group, are available to answer further queries during this stage. Consecutively, the detailed analysis of potential problem solving approaches begins, participants need specific information.

In the fifth phase students have to develop a concrete problem solving approach, where the strong application-oriented focus of the course comes to light. Students must summarize their ideas, inspired by the steps before and by the individual reflection processes, into one model. This aim is also influenced by the processes within the group and therefore a result of teamwork. As different characters in one group lead to a combination of diverse rolls within the team (Belbin, 1981; Belbin, 2010) the course participants learn to deal with this set of diversity and to participate in the resulting process of approach development. As the needs amongst the diverse users still represent the core of task, the problem solving approach should include all diversity aspects that were discussed in prior phases. At this stage, the teams work autonomously. This approach leads to an intensive altercation within the group and trains the ability to discuss, to work in a group and to grapple with a specific topic.

In the next step the review of the outcome is subject to discussion. Normally at this point, the solution approach is presented to the potential users. Since a direct contact with the users is not possible in this case, the solutions are presented to "Engineers Without Borders" who will implement the invented approaches in Makali and are, from a point of view, an indirect target group. This procedure allows the groups to get a direct feedback from the peer group that generates an expertise in the frame of the case. Thereby, the resulting critical reflection in this frame must be clear differentiated from the last phase of the concept.

In the sixth step, the course instructors evaluate the results by regarding the bigger frame and the context between diversity studies and the case content. In contrast to the feedback by the peer review in the fifth phase, the bigger frame is focused in this phase. This step is simultaneously part of the examination, as the approaches have to be presented in a customized way. The second element of the examination process is a written draft of the solution approach. This examination requirement introduces the course participants to the textualisation of ideas and findings, competencies which are not trained in most of the engineering seminars. Building on to this step, in the last phase students receive professional and structured feedback from the course instructors. The feedback is given by a structured guide. This approach has two advantages. On the one hand, grading becomes transparent as evaluation criteria becomes transparent. On the other hand, this process allows participants to again self-reflect through critical discussion of the problem solving approach from the users, as well as, the other students.

Insights and Experiences

Although the concept has a challenging character, the experiences with this concrete course concept have been very promising. Considering the various involved disciplines, it becomes apparent that the course concept is challenging particular students within the engineering field. Research projects of the institute have also shown that concepts which allow different accesses to knowledge and contents, according to the learning types after Vester (1998), not only include different students, but allow through this broad participation and discussion basis a knowledge production for the whole course (Steuer, Berg & Leicht-Scholten, 2015). In addition, engineering students, especially, seem to gain first-hand team experiences and have to cope with diversity in teamwork for the first time. For these students the course is the first possibility to apply their own knowledge in an interdisciplinary setting and to gain experience for later field action. The ability to participate in a real project and to develop an approach with strong research background seems to motivate students to take the course. Ergo, particularly motivated students registered for the course. Thus it is hardly surprising to find a very positive evaluation for this course concept, as teaching contents and structures as well as the strong application-orientation of these skills with the later field of action are clearly evident. Table 3 illustrates in summary the different skills students gain in this course.

CONCLUSION

Our universities, schools and faculties are committed to high-level research-based engineering education, research at the front edge of knowledge, close co-operation with business, industry and the public sector towards innovation, continuous improvement in and integration of education, research and innovation in the knowledge triangle. Contributing to sustainable economic, societal and environmental development is high on our agenda. (CESAER, 2016, p. 1)

This actual statement of CESAER, the conference of European Schools for Advanced Engineering Education and Research[8], demonstrates that the necessity to foster sustainable development still seems to be one of the great challenges for both research and education in engineering sciences.

The "League of European Research Universities" (LERU) has recognized that one of the main innovation roles of universities is in developing 'human capital'. It pointed out that as university-level education is mainly focused on incremental innovation; engineers do not have the opportunity to develop the skills and experience for the other types of innovation that society urgently requires. Students therefore need to be taught the principles of radical innovation, but it is important that they

Table 3. Acquisition of competencies © 2016, GDI RWTH Aachen University

<table>
<tr><th>Soft Skills</th><th>Research Skills</th><th>Specialist Knowledge</th></tr>
<tr><td>Scientific writing</td><td rowspan="2">Situation analysis</td><td>Diversity studies</td></tr>
<tr><td>Scientific presenting</td><td>Intercultural leadership styles</td></tr>
<tr><td>Dealing with interdisciplinary nature</td><td rowspan="2">Qualitative methods</td><td>Interdisciplinary topics</td></tr>
<tr><td>Teamwork processes</td><td>Application-oriented product designing</td></tr>
<tr><td rowspan="2">Reflection processes on social responsible acting</td><td rowspan="2">Quantitative methods</td><td>Active dealing with user needs</td></tr>
<tr><td>Aspects of social responsible working</td></tr>
</table>

also build up practical experience by working on real-life projects. It is stressed that dealing with complex challenges requires people with deep technology expertise and people with broad thinking capabilities to work together in multidisciplinary teams. Innovation education needs to be based on experiential learning techniques with teams of students addressing real challenges from business or wider society. It is crucial that students work in multidisciplinary teams across departments and schools with engineering (Boulton & Lucas, 2008). In 2015 LERU even launched an advice paper to integrate sex and gender analysis in the research process. As they pointed out the aim was "to significantly contribute to creating new knowledge and to finding solutions to global challenges, such as climate change, security and public health. LERU universities realize that in order to do so effectively, they need to take into account potential sex and gender differences with respect to the way research is designed, carried out and implemented. Without considering sex and gender analysis in research and innovation (GRI), the scope, impact and utility of research results may not be equally valid for both men and women." (LERU, 2015, p. 3)

Based on the requirement toward educating engineers for the future we have argued in our article that this only can be achieved by linking research and teaching. In order to be able to implement gender and diversity perspectives in research and innovation processes, future students have to be educated more broadly being able to understand the social embeddedness of technology they are developing. We are recommending a holistic approach with a teaching concept that not only delivers technical skills. Engineering students also have to reflect their specific epistemic culture and not only focus on the technical innovation process but also on the specific social political processes, the developed artefacts are embedded in and be aware of or even promote democratic processes the innovation process is linked with. We are arguing for socially responsible engineers, such that they should discuss risks still in the process of innovation within society. To realize this we have developed the holistic and transdisciplinary approach and institutionalized it at a leading technical university in Germany.

We have argued that integrating concepts and theories of diversity and science and technology studies in the regular curricula of engineering education is a promising path to educate socially responsible engineers for tomorrow. In doing so we can create social responsible and sustainable technology, and innovation processes we need to be able to cope with the global challenges in a way, so that our children and grandchildren can live good lives in the future.

REFERENCES

Acar, S., & Lopik, D. (2009). Computational pregnant occupant model, Expecting, for crash simulations. *Proceedings of the Institution of Mechanical Engineers. Part D, Journal of Automobile Engineering*, *223*(7), 891–902. doi:10.1243/09544070JAUTO1072

Andersen, R. D. (2004). Germany and the Humboldtian model. In R. D. Andersen (Ed.), *European Universities from the Enlightenment to 1914*. Oxford, UK: Oxford University Press. doi:10.1093/acprof:oso/9780198206606.003.0004

Andresen, S., Koreuber, M., & Lüdke, D. (Eds.). (2009). *Gender und Diversity: Albtraum oder Traumpaar? Interdisziplinärer Dialog zur "Modernisierung" von Geschlechter- und Gleichstellungspolitik*. Wiesbaden, Germany: VS Verlag für Sozialwissenschaften. doi:10.1007/978-3-531-91387-2

Bakshi-Hamm, P. (2007). Gleichbehandlung, Migration und Statistik. *Migrationund Soziale Arbeit, 3/4*.

Bakshi-Hamm, P., & Lind, I. (2008). Migrationshintergrund und Chancen an Hochschulen: Gesetzliche Grundlagen und aktuelle Statistiken. In *Wissenschaftlerinnen mit Migrationshintergrund* (pp. 11-24). Bonn, Germany: Center for Excellence Women and Science (CEWS). Retrieved July 20, 2016, from http://www.gesis.org/fileadmin/cews/www/download/cews-publik12.pdf

Barnard, S., Hassan, T., Dainty, A., & Bagilhole, B. (2013). Interdisciplinary content, contestations of knowledge and informational transparency in engineering curriculum. *Teaching in Higher Education*, *18*(7), 748–760. doi:10.1080/13562517.2013.836089

Belbin, R. M. (1981). *Management Teams: Why they succeed or fail*. New York: Butterworth-Heinemann.

Belbin, R. M. (2010). *Team Roles at Work* (2nd ed.). New York: Butterworth-Heinemann.

Bildungsberichterstattung, A. (Ed.). (2016). *Bildung in Deutschland 2016. Ein indikatorengestützter Bericht mit einer Analyse zu Bildung und Migration*. Bielefeld, Germany: W. Bertelsmann Verlag GmbH & Co. KG. Retrieved July 25, 2016, from http://www.bildungsbericht.de/de/bildungsberichte-seit-2006/bildungsbericht-2016/pdf-bildungsbericht-2016/bildungsbericht-2016

Bouffier, A., Kehr, P., Lämmerhirt, M., & Leicht-Scholten, C. (2014). Spätes Erwachen an deutschen Hochschulen: Die "Entdeckung" der Lehre und die Berücksichtigung von Gender und Diversity. In C. Tomberger (Ed.), Gender- und Diversity-Kompetenzen in Hochschullehre und Beratung: Institutionelle, konzeptionelle und praktische Perspektiven (pp. 53-66). Hildesheim, Germany: Universitätsverlag Hildesheim.

Boulton, G., & Lucas, C. (2008). *What are universities for?* League of European Research Universities. Retrieved July 25, 2016, from http://www.leru.org/files/general/%E2%80%A2What%20are%20universities%20for%20 (September%20 2008).pdf

Bourdieu, P. (1977). *Outline of a Theory of Practice* (R. Nice, Trans.). Cambridge, UK: Cambridge University Press. doi:10.1017/CBO9780511812507

Brown, T. (2008, June). Design Thinking. *Harvard Business Review*, 84–92.

CESAER. (2016). *Statement on the Modernisation of Higher Education.* Conference of European schools for advanced engineering education and research. Retrieved July 25, 2016, http://www.cesaer.org/content/assets/docs/Docs2016/20160308_CESAER_Statement_on_the_Modernisation_of_Higher_Education.pdf

Chun, E., & Evans, A. (2016). Rethinking Cultural Competence in Higher Education: An Ecological Framework for Student Development. *ASHE Higher Education Report*, *42*(4), 7–162. doi:10.1002/aehe.20102

Faulkner, W. (2007). Nuts and Bolts and People: Gender-troubled engineering identities. *Social Studies of Science*, *3*(3), 331–356. doi:10.1177/0306312706072175

Felder, R. M., & Silverman, L. K. (1988). Learning and teaching styles in engineering education. *Engr. Education*, *78*, 674–681.

Franzke, B. (2010). *Berufswahl heute: Geht es auch weniger stereotyp? Genderaspekte in der Berufswahl und Berufsorientierung.* Retrieved July 20, 2016, from https://lehrerfortbildung-bw.de/allgschulen/alle/gender/m4/01_BF_Gender_in_Berufswahl_und_Berufsorientierung.pdf

Gaethgens, C., & Zervakis, P. A. (2007). *Implementing Bologna. Experience from German Higher Education Institutions.* HRK German Rectors' Conference Bologna Centre. RAABE Fachverlag für Wissenschaftsinformation. Retrieved July 25, 2016, from http://www.ehea.info/uploads/submitedfiles/11_2010/145651.pdf

Grasso, D., & Burkins, M. (2010). *Holistic Engineering Education: Beyond Technology.* New York: Springer Science & Business Media. doi:10.1007/978-1-4419-1393-7

Grots, A., & Pratschke, M. (2009). Design Thinking - Kreativität als Methode. *Marketing Review St. Gallen*, *26*(2), 18–22. doi:10.1007/s11621-009-0027-4

Haraway, D. J. (1991). *Simians, Cyborgs, and Women: The Reinvention of Nature*. New York, NY: Routledge Taylor & Francis Group.

Harding, S. G. (1986). *The Science Question in Feminism*. Ithaca, NY: Cornell University Press.

Hasso Plattner Institute of Design at Stanford. (2010). *An introduction to design thinking. Process Guide*. Retrieved July 18, 2016, from https://dschool.stanford.edu/sandbox/groups/designresources/wiki/36873/attachments/74b3d/ModeGuide-BOOTCAMP2010L.pdf?sessionID=68deabe9f22d5b79bde83798d28a09327886ea4b

Heublein, U., Hutzsch, C., Schreiber, J., Sommer, D. & Besuch, G. (2010). *Ursachen des Studienabbruchs in Bachelor- und in herkömmlichen Studiengängen Ergebnisse einer bundesweiten Befragung von Exmatrikulierten des Studienjahres 2007/08*. Academic Press.

Hewlett, S. A., Marshall, M., & Sherbin, L. (2013). How Diversity Can Drive Innovation. *Harvard Business Review*. December. Retrieved July 28, 2016, from http://www.dzhw.eu/pdf/pub_fh/fh-201002.pdf

Himmel, S., & Ziefle, M. (2016). Smart Home Medical Technologies: Users' Requirements for Conditional Acceptance. *i-com. Journal of Interactive Media*, *15*, 39–50.

Hippler, H. (Ed.). (2011). Ingenieurpromotion – Stärken und Qualitätssicherung Beiträge eines gemeinsamen. Symposiums von acatech, TU9, ARGE TU/TH und 4ING.

House, R. J., Hanges, P. J., Javidan, M., Dorfman, P. W., & Gupta, V. (2004). *Culture, Leadership, and Organizations: The GLOBE Study of 62 Societies*. London: SAGE Publications.

Jamison, A. (2013). *The Making of Green Engineers: Sustainable Development and the Hybrid Imagination*. Morgan & Claypool Publishers.

Jansen-Schulz, B., & van Riesen, K. (Eds.). (2011). *Vielfalt und Geschlecht - relevante Kategorien in der Wissenschaft*. Opladen, Germany: Budrich UniPress Ltd.

Kamp, A. (2014). *Engineering Education in the Rapidly Changing World*. Rethinking the Mission and Vision on Engineering Education at TU Delft. Retrieved July 18, 2016, http://www.lr.tudelft.nl/fileadmin/Faculteit/LR/Nieuws_en_Agenda/Nieuws/2014/docs/visiedocument_TU_def.pdf

Kehr, P., & Leicht-Scholten, C. (2013). Diversity in der Hochschule – in der Mitte angekommen oder zwischen allen Stühlen? *Qualität in der Wissenschaft. Zeitschrift für Qualitätsentwicklung in Forschung. Studium und Administration*, *2*, 30–36.

Knight, D. W., Carlson, L. E., & Sullivan, J. F. (2003). Staying in Engineering: Impact of a Hands-On, Team-Based, First-Year Projects Course on Student Retention.*Proceedings of the 2003 American Society for Engineering Education Annual Conference & Exposition*.

Knoll, B., & Ratzer, B. (2010). *Gender Studies in den Ingenieurwissenschaften*. Wien, Germany: Facultas Verlags- und Buchhandels AG.

Kolb, D. A. (1984). *Experiential Learning*. Englewood Cliffs, NJ: Prentice-Hall.

Krell, G., Riedmüller, B., Sieben, B., & Vinz, D. (Eds.). (2007). *Diversity Studies: Grundlagen und disziplinäre Ansätze.* Campus Verlag GmbH.

Leicht-Scholten, C. (2008). Exzellenz braucht Vielfalt - oder: wie Gender and Diversity in den Mainstream der Hochschulentwicklung kommt - Human Resources, Gender and Diversity Management an der RWTH Aachen. *Journal Netzwerk Frauenforschung NRW Nr. 23*, 33-39.

Leicht-Scholten, C. (2011a). Meeting Global Challenges - Gender and Diversity as Drivers for a Change of Scientific Culture. In C. Leicht-Scholten, E. Breuer, N. Tulodetzki, & A. Wolffram (Eds.), *Going Diverse: Innovative Answers to Future Challenges. Gender and Diversity Perspectives in Science, Technology and Business* (pp. 53–64). Opladen: Budrich UniPress Ltd.

Leicht-Scholten, C. (2011b). Managing Gender und Diversity Mainstreaming in Zeiten der Exzellenz: alte Schwierigkeiten, neue Herausforderungen, geeignete Maßnahmen. In B. Blättel-Mink, A. Franzke, & A. Wolde (Eds.), Gleichstellung im Reformprozess der Hochschulen - Neue Karrierewege von Frauen an Hochschulen? (pp. 185–207). Academic Press.

Leicht-Scholten, C., Breuer, E., Tulodetzki, N., & Wolffram, A. (Eds.). (2011). *Going Diverse: Innovative Answers to Future Challenges. Gender and Diversity Perspectives in Science, Technology and Business*. Opladen: Budrich UniPress.

Leicht-Scholten, C., Steuer, L., & Bouffier, A. (2016). Facing future challenges: Building engineers for tomorrow. *Proceedings of the International Conference "New Perspectives in Science Education"*, (*vol. 5*, pp. 32-37). Padova: libreriauniversitaria.

Leicht-Scholten, C., & Wolffram, A. (2010). Managing Gender and Diversity Mainstreaming an Hochschulen im Spannungsfeld zwischen Theorie und Praxis. *Gender*, *2*, 87–95.

LERU. (2015). *Gendered research and innovation: Integrating Sex and Gender Analysis into the research process*. Retrieved July 26, 2016, from http://www.leru.org/files/publications/LERU_AP18_Gendered_research_and_innovation_final.pdf

Litchfield, K., & Javernick-Will, A. (2015). I Am an Engineer AND: A Mixed Methods Study of Socially Engaged Engineers. *The Journal of Engineering Education*, *104*(4), 363–496. doi:10.1002/jee.20102

Litchfield, K., & Javernick-Will, A. (2016). Socially Engaged Engineers Career Interests and Experiences: A Miners Canary. *Journal of Professional Issues in Engineering Education and Practice*, 04016018. doi:10.1061/(ASCE)EI.1943-5541.0000303

Marra, R. M., Rodgers, K. A., Shen, D., & Bogue, B. (2012). Leaving Engineering: A Multi-Year Single Institution Study. *The Journal of Engineering Education*, *101*(1), 6–27. doi:10.1002/j.2168-9830.2012.tb00039.x

Matthews, M. (n.d.). *Keeping students in engineering: A research-to-practice brief.* Retrieved August 3, 2016, from https://www.asee.org/retention-project/keeping-students-in-engineering-a-research-guide-to-improving-retention

Organisation for Economic Co-operation and Development (OECD) (2014). *OECD Science, Technology and Industry Outlook 2014*. OECD Publishing. 10.1787/sti_outlook-2014-en

Paton, R. A., Wagner, R., & MacIntosh, R. (2012). Engineering education and performance: The German machinery and equipment sector. *International Journal of Operations & Production Management*, *32*(7), 796–828. doi:10.1108/01443571211250086

Plattner, H., Meinel, C., Leifer, L. J., & Leifer, L. (2015). *Design thinking research. Building innovators*. New York: Springer.

Rohrer, D., & Pashler, H. (2012). Learning styles: Wheres the evidence? *Medical Education*, *46*(7), 34–35. doi:10.1111/j.1365-2923.2012.04273.x PMID:22691144

RWTH Aachen University. (2016a). *Zahlenspiegel 2015*. Aachen, Germany: RWTH Aachen University. Retrieved July 6, 2016, from https://www.rwth-aachen.de/cms/root/Die-RWTH/Profil/~enw/Daten-Fakten/

RWTH Aachen University. (2016b). *Rankingreport 2015. Nationale und internationale Hochschulrankings der Jahre 2006 bis 2015*. Aachen, Germany: RWTH Aachen University. Retrieved July 15, 2016, from https://www.rwth-aachen.de/global/show_document.asp?id=aaaaaaaaaaaofvl

Statistisches Bundesamt. (2015a). *Bildung und Kultur. Prüfungen an Hochschulen. 2014* (Fachserie 11, Reihe 4.2). Wiesbaden, Germany: Statistisches Bundesamt. Retrieved June 23, 2016, from https://www.destatis.de/DE/Publikationen/Thematisch/BildungForschungKultur/Hochschulen/PruefungenHochschulen2110420147004.pdf?__blob=publicationFile

Statistisches Bundesamt. (2015b). *Bildung und Kultur. Studierende an Hochschulen. Wintersemester 2014/2015* (Fachserie 11, Reihe 4.1). Wiesbaden, Germany: Statistisches Bundesamt. Retrieved June 23, 2016, from https://www.destatis.de/DE/Publikationen/Thematisch/BildungForschungKultur/Hochschulen/StudierendeHochschulenEndg2110410157004.pdf;jsessionid=3B3B4340D356BD7227C75360B4289C2B.cae1?__blob=publicationFile

Steuer, L. (2014). *Gender und Diversity in MINT-Fächern – Ursachen des Diversity-Mangels*. Berlin: Springer BestMasters Psychologie.

Steuer, L., Berg, T., & Leicht-Scholten, C. (2015). Breaking the habit – New approaches in engineering education. *Proceedings of the International Conference SEFI, Orléans*. Retrieved July 21, 2016, from http://www.sefi.be/conference- 2015/CHAP%2016.%20Education%20concepts%20specific%20for%20Engineering%20Education/60823-%20L.%20STEUER.pdf

Thursby, M. C., Fuller, A. W., & Thursby, J. (2009). An Integrated Approach to Educating Professionals for Careers in Innovation. *Academy of Management Learning & Education*, *8*(3), 389–405. doi:10.5465/AMLE.2009.44287938

VDI. (2016). *15 Jahre Bologna Reform - Quo vadis Ingenieurausbildung?* Retrieved July 21, 2016, from https://www.vdi.de/fileadmin/user_upload/2016_VDI-VDMA-Mercator-Studie-15_Jahre_Bologna-Reform.pdf

Vester, F. (1998). *Denken, Lernen, Vergessen: Was geht in unserem Kopf vor, wie lernt das Gehirn, was lässt es uns im Stich?* München: Dtv Wissen Verlag.

Walther, J., Miller, S. E., & Kellam, N. N. (2012). Exploring the role of empathy in engineering communication through a trans-disciplinary dialogue.*Proceedings of the ASEE Annual Conference and Exposition.*

Weber, M. (1968). *Gesammelte Aufsätze zur Wissenschaftslehre 3*. Tübingen: Mohr-Siebeck.

Willingham, D. T., Hughes, E. M., & Dobolyi, D. G. (2015). The scientific status of learning styles theories. *Teaching of Psychology*, *42*(3), 266–271. doi:10.1177/0098628315589505

ENDNOTES

1 See the broad variety of programs of the German “Komm-mach-MINT”-initiative: http://www.komm-mach-mint.de/ (Retrieved July 24, 2016).

2 In this context it is not just about the content but rather the methodical way of communicating the contents in terms of the methodical expertise: CONTENT - METHOD - SKILLS

3 A further execution of the discussion will not be part of this chapter.

4 For more details see: http://www.dfg.de/download/pdf/dfg_im_profil/geschaeftsstelle/publikationen/exin_broschuere_en.pdf (Retrieved July 25, 2016).

5 For more details see http://www.gdi.rwth-aachen.de/engl/teachings/ (Retrieved July 22, 2016).

6 The organization “Engineers Without Borders” is a worldwide operating institution, with a focus on technical development cooperation. The association is organized in regional groups. The course was conducted with the regional group “Aachen”. Further information see: Home - Engineers Without Borders; http://www.ewb-usa.org/ (Retrieved July 01, 2016).

7 See https://www.ingenieure-ohne-grenzen.org/de/projekte/sle-iog03 (Retrieved August 09, 2016).

8 A non-profit international association of leading European universities of science and technology and engineering schools/faculties. For more information see: http://www.cesaer.org/content/assets/docs/Docs2016/20160308_CESAER_Statement_on_the_Modernisation_of_Higher_Education.pdf (Retrieved July 01, 2016).

Chapter 11
Engineering Pathways in a U.S. Public Institution of Higher Education:
A Strategy for Fostering Student Diversity

Fabiola Ehlers-Zavala
Colorado State University, USA

Anthony Maciejewski
Colorado State University, USA

ABSTRACT

This chapter relates a strategy that emerged from a larger effort of a land-grant institution in the U.S. to more rapidly increase the number of international students on campus and diversify its student body through the development and implementation of pathway programs. Pathway students are international students that do not meet the criteria for direct entry into a university due to lower levels of English language proficiency and/or GPA. The authors discuss strategies for ensuring success in these endeavors.

DOI: 10.4018/978-1-5225-2212-6.ch011

INTRODUCTION

The goal of this chapter is to contribute to the topic of how to increase diversity in engineering programs. Our discussion is in the context of a public private partnership in higher education developed to increase campus internationalization efforts. We begin by briefly situating our chapter in the larger context of the internationalization of engineering programs. Then, we discuss how an internationalization strategy adopted by Colorado State University (CSU), through the formation of a public private partnership, has given an opportunity to its Electrical and Computer Engineering (ECE) program to continue to enhance its diversity. Thus, following the introduction of the public private partnership at our institution, we detail a major development that came with it: the introduction of a *pathway* program in ECE for international students. We explain what a pathway program is in this context, and we focus our discussion on the development of a pathway program at the graduate level. In describing this initiative, we address the topic of turning challenges into successes, and we elaborate on the value of pathway programs in expanding diversity efforts. Finally, we offer our perspective on lessons learned for other institutions and colleagues to consider if found in a similar situation.

BACKGROUND ON INTERNATIONALIZATION IN U.S. ENGINEERING PROGRAMS

In order to make sense of the challenges and potential opportunities for U.S. engineering programs discussed later in this chapter in relation to the topic of diversity, it is important to understand some of the historical context. The exact details regarding the degree of internationalization vary significantly across institutions, engineering disciplines, and type of program (i.e., undergraduate or graduate degree programs). However, it is possible to make some generalizations, especially across large land-grant institutions that traditionally have had large engineering programs and share similar histories.

In particular, the undergraduate engineering programs of land-grant institutions, as is the case at CSU, tend to be well established because they were a major motivation for the creation of such institutions. (For more information on land-grant institutions, see http://www.aplu.org/about-us/history-of-aplu/what-is-a-land-grant-university/). They also have a history of serving the local state population based on the premise of providing a degree that allowed the state's population to improve their standard of living. Thus, many of the early students of such programs were first generation, and this tradition continues. Likewise, until recently, very few of these programs had significant numbers of international students among their ranks. So efforts to internationalize these programs constitute institutional priorities

The landscape for graduate programs in engineering can be quite different from the undergraduate programs. But here too, it varies from discipline to discipline. Using Electrical and Computer Engineering (ECE) as an example, we can say that, for many years, ECE graduate programs have been dominated by the presence of international students. For these students, a U.S. graduate degree in ECE is a very valuable commodity because it frequently opens the doors to employment in the U.S.-based high-tech industry, and definitely gives international students a point of considerable advantage when returning home or going to other countries in pursuit of employment. This results in a situation where the size of an ECE program's graduate population is not limited by the number of qualified students, but by the capacity constraints of delivering a quality educational experience for students in the program. That is, resources must be allocated for departments to properly serve all students. In other words, ideally, as enrollment grows, resources should grow proportionately. Yet we know that this is not always the case. Consequently, when a university, such as CSU, embarks on adopting new initiatives to increase international enrolments, one can foresee not only the benefits of increasing diversity, but also the challenges that can also emerge.

In the next section, we introduce the concept of a public private partnership to increase international enrollments, and thus impact diversity at a university. This discussion is intended to situate the more specific example of CSU that will follow.

STRATEGIC PARTNERSHIPS IN HIGHER EDUCATION

Public Private Partnerships for Campus Internationalization and Diversity

Much has been written on the topic of strategic partnerships in higher education (e.g., Kuder & Banks, 2016; Yang, Ahmad, Lavenda, Robinson, & Seo, 2016; Learch, Paulson, Sander, & Schuch, 2016; Caspar-Terizakis et al., 2016; Sutton & Obst, 2011; Bhandari, Belyavina, & Gutierrez, 2011). However, most of the work has primarily described traditional efforts. In other words, these writings typically relate the experience of universities that form partnerships with other universities around the globe to increase their international efforts, as well as those partnerships that are formed with government agencies to sponsor international students in the completion of their education abroad. To a much lesser extent, we find accounts that relate the experience of a public private partnership formed between a public institution of higher education and a private company. This effort can, at present, be considered a non-traditional way of expanding international student diversity.

In the U.S., the formation of public private partnerships in connection to international education constitutes a new phenomenon that is quickly proliferating. In fact, the landscape of U.S. higher education in this realm has been changing dramatically in the last 10 years with the advent of more private companies forging partnerships with institutions of higher education—a trend that is likely going to continue. For quite some time, Kaplan (a private educational provider in the U.S. best known for test preparation sites on university campuses in the U.S.) had maintained exclusivity in the area of international recruitment in the U.S. But since 2009, Kaplan's market share began to dissipate with the influx of new providers into the U.S. educational system of higher education such as INTO University Partnerships (IUP), Navitas, StudyGroup, Cambridge Education Group, Bridge, Shorelight Education, and Kings Education to offer some examples. (See Appendix A for a sample listing of partnerships formed since 2009 in the U.S.) Though new to the U.S., this type of enterprise connected to international recruitment was not new to others outside the U.S. Australia and the U.K. have been active participants in these types of arrangements for decades. The common outcome of many of these partnerships is "massive growth in the number of fee-paying international students and an increasing private sector role" (see Rodan, 2016, p. 5). At a time when there is less support for public higher education from governments (Altbach, 2016; Johnstone, 2016), this type of strategic venture is becoming increasingly more appealing for a variety of reasons, not just the educational one. As Altbach (2016) pointed out, "International students have become big business, producing billions of dollars for the host countries" (p. 9). Because financial considerations go into the process of decision-making to develop a public private partnership, these partnerships, as discussed later in this chapter when describing the ECE case, are not always well received by those who subscribe to more traditional viewpoints or who resist the idea of what has become the norm in many other parts of the world: the privatization of higher education (see Adamson, Astrand, & Darling-Hamond, 2016; Altbach, 2016; McGee, 2015).

However, for professionals and institutions that are forward thinking, there are many clear reasons for considering a public private partnership. Public private partnerships constitute a type of strategic international partnership that can significantly contribute to enhance diversity efforts of a university to offer a global education to its students. Goodman and Rüland (2016) offer a number of reasons for considering an international partnership and engaging in all the work that this effort entails. They have pointed out that:

- While strategic international partnerships can strengthen ties between the two or more connected entities, the internal process required to develop and sustain these partnerships can also contribute to more synergies within the individual institution—between professors and the administrative leadership of the institution and between different faculties.

- International partnerships lead to the internal globalization of the partners involved as well as of the systems they are rooted in.
- Strategic partnerships are long-term projects. They take effort and time—and resources—to set up, but the benefits are long lasting: They will bring cohort after cohort of exchange students, create fruitful research and collaboration, have political impact on a national and international level, and achieve other positive outcomes.
- Institutions with established international partnerships have an advantage when applying for international research funding. The united assets and strengths of strategic partners make any proposal more compelling. And, in fact, many calls for proposals specifically target consortia.
- When industry and business partners are included, the partnerships can tap into those resources to create more applied curricula or research projects and profit from the different economic systems and environments in the countries represented. (p. ii)

Nevertheless, despite the many benefits that can be derived from the formation of public private partnerships, with new developments, new challenges emerge that need to be both addressed and turned into positive outcomes. In the next section, we describe the challenges that CSU experienced upon entering a public private partnership. We also discuss how these challenges are being turned into opportunities by the ECE department at CSU.

The INTO CSU Partnership

In 2012, CSU took the bold step of embarking on an initiative that would result in the formation of a public private partnership by signing a 30-year contract with INTO University Partnerships (IUP), a British private company. The result of this agreement was the formation of a limited liability corporation, INTO CSU. The goal of INTO CSU was to contribute to the rapid increase of international students on campus and further diversify the university's student body (primarily in the areas of linguistic and cultural diversity, but not exclusively in recognition of the many aspects of diversity—gender, exceptionality, socio-economic status, etc.). However, this effort became quite controversial as it would not only target international students who would be directly admitted to the various degree programs at CSU, but also students who would not typically qualify for direct admission due to lower GPAs or scores in English language proficiency tests required for international admission.

This new development for a research university, which takes much pride in admitting the best and the brightest, was not equally welcomed by all key university players. In some cases, it met a high level of resistance on the part of key academic units, such as ECE.

With the formation of INTO CSU, a new programmatic educational alternative for students who did not qualify for direct admission was introduced. This alternative became known as the *pathway* program, which corresponds to a programmatic alternative for international students during their first year. Most pathway students choose this program because it offers greater support through much individualized attention on the part of INTO CSU personnel (directors, advisors, tutors), and it welcomes students with lower levels of English proficiency and lower GPAs, but whose academic background suggests potential for academic success. It also welcomes those who meet direct-entry requirements, but are pursuing a more supported introduction to university life during that first year. Thus, between February and August 2012, very quickly and with varying degrees of buy-in, pathway programs, both at the undergraduate and graduate level, were developed across key colleges and corresponding departments (e.g., business, engineering, computer science, science, tourism management). These initial programs were deemed to be the most suitable to quickly attract international students and thus get the partnership underway.

Initially at CSU, pathway programs were developed around two options: a two-semester option (standard) and a one-semester option (accelerated). These options were intended to meet the needs of international students who did not speak English as a dominant or primary language and therefore had different levels of linguistic proficiency. These students, for the most part, came with lower GPAs than those required for direct entry (i.e., 2.75 GPA in technical course for the case of ECE at the graduate level), but their academic background was deemed to show potential for success. For students who required more language support and came with lower GPAs, the two-semester option was the more suitable option. For students who required less language and academic support, the one-semester option was presented. After all, the premium cost that students pay for their pathway program allows INTO CSU to provide participants with additional academic support and offer them with expanded student experiences leading to a successful integration at the university. In other words, with the ultimate goal of ensuring academic success, the pathway program is intended to be a highly supported first-year (or semester) experience for international students who for one reason or another may need, or wish to have, greater support in adapting to a new and foreign educational environment.

KEY CHALLENGE FOR THE INTO CSU PARTNERSHIP

Faculty Buy-In

As previously indicated, in the case of CSU, when INTO CSU was formed, pathway programs in engineering were quickly developed both at the undergraduate and graduate levels. This very rapid development and implementation of pathways in engineering naturally met some immediate challenges. The most significant one for development and expansion of the initial proposition has been the *faculty* challenge.

At CSU, where the concept of shared-governance is embraced, a significant and ongoing challenge has been the *faculty* challenge. Sometimes, faculty members who oppose the pathway concept are faculty of international background. Some of them resist the pathway concept as they believe that all international students be subject to meeting the same kinds of expectations they had to meet when pursuing a U.S. education. Therefore, for them it is difficult to observe that other international students would not be subject to the same type of expectation. Other times, it is the U.S. born and educated faculty member with a traditional/conservative mindset who has difficulty in appreciating why the institution should admit international students with lower academic or English proficiency scores into such competitive programs when to do so may mean denying students with better credentials. This situation seems to go against their sense of "fairness," or of a pure meritocracy, as they (based on the feedback shared) firmly believe that only students with the best scores should be admitted to an institution of the statue of CSU—a belief that seen from another angle can also serve to perpetuate other long-standing and still existing inequalities. Some of us, however, know quite well that inequalities exist all over, and that international students also constitute a very diverse group whose experiences may significantly differ, based on opportunities and personal histories and background. Some are more privileged than others. Yet we know that talent can still be identified among those who have not been as fortunate. Among them, we know that there are students who have both the potential and strong motivation to succeed, and those are the students a pathway program hopes to identify and recruit for a university that is committed to global education and to diversify its campus. Oftentimes, this is the case among first-generation international students who come from low-middle socio-economic backgrounds in their home countries.

The negative feelings that faculty may experience in relation to the pathway admissions criteria can be further exacerbated if they believe or perceive that the university is pursuing such pathway programs from a purely financial motive. Faculty members are concerned when they see that the cost of pathway programs is substantially higher than the cost of regular university programs without a full understanding of the additional services pathway students receive. What represents an even greater challenge is when faculty members reject any explanations regarding additional support provided to pathway students, choosing to remain highly skeptical or non-believers. In other words, our experience suggests that oftentimes individuals who resist and oppose the pathway concept do so with little knowledge of what it involves, and they fail to understand, recognize, and therefore acknowledge that the higher cost is a direct result of the much greater effort this type of enterprise entails. They also do not readily accept the reality that our system is set up in such a way that helps those who may have fewer resources to consider the pathway program. At INTO CSU, the success of pathway students who matriculate into degree programs contributes to building scholarship funds that are then put to use in recruiting other international students that may come from less affluent backgrounds. These funds contribute to the diversification of the student body and give students from different regions of the world the opportunity to pursue an education at CSU. In our experience, we have observed that these opponents of the pathway concept may even exhibit a sense of righteous indignation toward the partnership because of the impression that it is taking advantage of these students (those who initially come with lower GPAs and/or language proficiency scores), putting them into a position where they cannot be successful. These opponents have basically pre-determined that pathway students are doomed to fail, and they spend more time arguing against the pathway concept than setting time aside to join the conversation of how to ensure student success. These opponents also do not wish to consider making changes to their pedagogies, and they resist any insinuation that suggests that teaching diverse learners may require learning new ways of supporting the learning that needs to happen in the classroom.

But what these opponents of pathways fail to recognize is, as McGee (2015) reminds us, that the landscape of higher education in the U.S. is no longer what it used to be since the passage of the Higher Education Act in 1965, which represented a shift from what had in years prior to 1945 actually been a luxury good for a privileged few in the U.S. Many have forgotten (or have never known) that, initially, U.S. higher education had been "an earned privilege" or a "macroeconomic luxury good" (McGee, 2015, p. 9). With the developments of 1965 in U.S. higher education, a college education became "a mass-market good supported by changing public policy and undergirded by a changing economy increasingly less reliant on physical capital and manufacturing and more dependent on intellectual capital and services" (McGee, p. 9). This thinking also seemed to have been embraced by many engineering faculty who, despite the fact they may be at a land-grant institution, expect to continue to ideally exclusively serve and educate only the best and the brightest. Opponents of pathways seem to give little consideration to the fact that in other countries some of the best and the brightest may already be at an intellectual disadvantage, as it can also happen in the U.S., when they may have not been privileged enough to have had access to a quality public (or even private) education. Yet we know that international students, regardless of whether they are pathway or not, (as well as their families and/or sponsoring governments, other institutions, etc.), too believe in the value of a U.S. education for securing a better life and being better prepared to contribute to the development of their own societies. In the end, many of us (who are first-generation, speakers of a language other than English, and international) have experienced what Bowen and McPherson (2016) have stated:

Individuals who acquire more education generally benefit economically from doing so and wind up, on the whole, better off than their fellow citizens. .... Not all the benefits of college accrue to individual graduates, however. There is substantial, and growing evidence that there are strong network effects and that a community benefits economically from having a more educated population; there is also evidence that persons with more education are more civic-minded (pp. 75-76).

Thus, the negative attitudes we described above must be addressed with a great deal of care to clearly articulate the goals of pathway programs. There is a need for a clear dialogue between all the parties involved so that everyone understands that such programs are meant to provide more equal access to the benefits of a graduate engineering education, and that they too are in line with the diversity goals of land-grant institutions articulated by the Association of Public and Land-grant Universities (APLU) (see Appendix B). That is, there must be a concerted effort, embraced by all stakeholders, to ensure that a university community is properly educated and prepared to understand that pathway programs are a mechanism for allowing international student populations to gain access and demonstrate that, with sound institutional support, they too can achieve their academic potential, despite coming from disadvantaged backgrounds. In other words, pathway programs represent an opportunity to provide international students (many of whom may have already experienced some level of inequality in their educational systems at home) universal access to quality education. This shift is consistent with what Altbach (2016) has pointed out about the evolution of the American Higher Education system that has gone from elite to mass to universal access.

The ECE Journey in the Context of INTO CSU

Five years after the formation of INTO CSU, in reflecting upon the ECE journey, we are seeing how a pathway program can help an institution achieve internationalization and diversity goals. In Ehlers-Zavala and Maciejewski (2016, "The ECE journey relative to INTO CSU," para. 1), we detailed the stages we identified for the ECE journey in the development of pathway programs:

1. Resistance
2. Awakening
3. Collaboration
4. Innovation

Below, we revisit some of our earlier remarks (see Ehlers-Zavala & Maciejewski, 2016), and offer additional comments when appropriate.

1. **Stage of Resistance:** For the first couple of years, ECE was a reluctant partner with practically "no desire to collaborate with INTO CSU" (Ehlers-Zavala & Maciejewski, 2016, para. 2). The reason for this was a profound lack of understanding of the ultimate goals of the partnership in relation to CSU's strategic plan for internationalizing its campus. ECE was unclear as to its pivotal role in this larger plan that could lead to the success of INTO CSU and CSU, not to mention its own success in further diversifying its student body. The specific concerns that the ECE leadership expressed about the pathway program are listed below:

(1) A "back door" admission for less qualified international students; (2) the "bumping" of students that were directly admissible in favor of the admission of students who could potentially bring down the quality of the education being delivered by ECE faculty; (3) the immediate capacity concerns that emerged in light of fewer resources being re-directed to the department by the central administration; and (4) a possible redirection of university resources to attend to the needs of, what was perceived to be, a "privileged" group of students who could afford this type of very costly education, as is the case of most pathway programs. (Ehlers-Zavala & Maciejewski, 2016, para. 2).

These concerns and sentiments were voiced consistently for the better part of the first two years of the partnership and they became obstacles to the expected and targeted growth. In fact, it took a change in leadership at INTO CSU for an opportunity to re-engineer communication efforts to turn the situation around.

2. **Stage of Awakening:** With the leadership transition at INTO CSU, there was an opportunity for implementing a new approach towards engagement with ECE. For the first time, through clear communication regarding the goals of INTO CSU and the benefits of the pathway program (i.e., to provide greater access to international students of diverse backgrounds), the ECE leadership began to understand the larger goal and what was at stake. Additionally, a greater

involvement on the part of key players from IUP through campus visits helped to bring clarity, transparency and accountability to the still unsettled university community. Specifically, a visit from the head of the Global Recruitment Unit (GRU) at IUP to CSU allowed for the free exchange of information through candid conversations with the faculty and the leadership across academic units. Gradually, the initial myths and misconceptions began to dissipate, which led to the next stage, collaboration.

3. **The Stage of Collaboration:** During the stage of collaboration, new developments started to occur. The initial graduate pathway curriculum in ECE that lacked specificity was revised to offer students with concrete suggested plans of study they could pursue based on their interests. For instance, initially there was one grid that was formulated for students with an interest in ECE. Today, there are eight in the following areas of study: semiconductor devices and processing, robotics/vision, robotics/robot control, high performance computing, control systems, computer engineering, and computer architecture/embedded systems. All of those options have both standard and accelerated alternatives, and each of those options has a well-articulated plan of study. The suggested plans of study for each of the eight concentrations is clearly articulated and specified, and as of AY16-17 each study plan offers a one-semester pathway option as well. The latest development in ECE is to allow students to also consider the M.S. degree as an option after meeting progression requirements with a slightly higher GPA. Additionally, as a gesture of meaningful collaboration, INTO CSU contributes to the funding of an academic advisor in ECE that can devote exclusive attention to the needs of pathway students and can guide them to the point of successful matriculation in ECE.
4. **The Stage of Innovation:** As the collaboration strengthens in time between INTO CSU and ECE, we have begun to look at ways in which we can innovate. In Ehlers-Zavala and Maciejewski (2016, para. 2-3), we stated that, as a result of this collaboration, there have been "tangible ways in which the partnership has begun to influence the ECE department" leading to true educational innovation, which have already included:
    a. Cross-disciplinary collaboration in supporting pathway students;
    b. Faculty development as each partner learns more about the work of the other;
    c. Curricular re-envisioning, as we improve the curriculum of pathway students and inform possible revisions to the existing curriculum that prepares future engineers in general at CSU;
    d. Resource sharing through collaborative practices and through the open exchange of expertise between these two partners; and

e. Research and scholarship through collaborative research and the collaborative dissemination of our findings at our respective professional organizations to both build and expand our knowledge base of pathway students and programs. INTO CSU and ECE, beyond the top leaders in each unit, are working together to research and improve the educational experience of students in engineering. We expect that by developing a better understanding of how diverse international learners perform in graduate courses, we will be able to inform curricular revisions that are truly designed with a group of diverse prospective engineering students in mind.

Assessment Leading to ECE Progress Relative to INTO CSU

In parallel to a more favorable disposition on the part of the ECE leadership and faculty, the number of international students in the graduate pathway program started to significantly increase. This development has been consistent with the growth of direct admits in ECE and national data trends (see Arbeit, Einaudi, & Kang, 2016). CSU's institutional data show that the number of international students at the graduate level has exponentially increased from 56 in Fall 2007, 103 in Fall 2012, and 192 in Fall 2015. As of Fall 2016, the ECE graduate pathway enrollments have been the highest among all programs in the College of Engineering (n= 17), and is only second to those in the College of Business by 1 or 2 students.

Further, in relation to the initial fear on the part of ECE to allow students to come into a graduate pathway program with GPAs lower than 3.0, the data shows that the performance of graduate pathway students is not significantly different from those who have come with higher GPAs and who have been directly admitted to the ECE Masters program. Furthermore, we have found that there is not a significant correlation between a student's undergraduate GPA when admitted into an ECE graduate pathway and the GPA of a student in the graduate ECE courses taken while in the graduate pathway. The majority of the pathway students did better than what would have been predicted by their admission GPA which was at 2.75 or above, but below 3.0. It is important to point out had the graduate pathway program not existed, we would have never welcomed these 10 pathway students who matriculated to the ECE Masters program in AY 15-16. This fact supports the premise that international students, who may appear as not highly qualified to attend a university such as CSU, when given the opportunity and the right support (provided they are up for the challenge) can be successful. As previously stated, international students too constitute a highly diverse group, and not all of them start from a point of advantage (social, educational, economic and/or otherwise). We cannot ignore this fact, as many of them may be like many of us: first-generation (including these two authors).

The analysis of actual pathway student performance has resulted in a greater understanding of how INTO CSU is living up to the initial promise of identifying international students with the potential to succeed by offering enhanced academic and student experience support. This has led to even more opportunities for pathway students. Currently, ECE not only welcomes pathway students into their masters program that leads to the Master of Engineering (as exclusively done in the first four years of the partnership), but also to the Master of Science degree for those students with a matriculation GPA of 3.2 or higher upon completion of the pathway. These students now, officially as of AY 17-18 in our published materials, have the opportunity to take advantage of engaging in research activities that are one of the hallmarks of large land grant universities.

From Challenges to Opportunities

As suggested earlier, there is a real possibility that institutions of higher education getting ready to embrace the pathway model by entering a public private partnership may encounter challenges of the kind introduced and discussed earlier in this chapter. While there may be little that can be done to fully prevent others from experiencing some of this, we believe that we can offer a number of suggestions that can inform their preparation to more quickly turn challenges into opportunities.

One potential challenge occurs when there is inadequate understanding of the overarching goals of the partnership and key players choose to focus on what they may perceive as a threat to the comfortable status quo rather than on the opportunities and long-term rewards that can be experienced from seeing pathway students succeed. For instance, it is very easy to look at pathway students from a *deficit* perspective. That is, it is easier to think about these individuals as students that will take more time on the part of the faculty member than the time faculty would like to spend on these students. Faculty members who view these students under this deficit perspective have opted for thinking about these students as an imposition rather than as an opportunity to guide someone to the attainment of educational and professional success. Thus, rather than viewing these students as individuals who, with the appropriate education and development, can be agents of change in the long run, pathway students are being looked at as individuals with a deficit. They are seen as individuals whose English is not at the level professors would expect and who may not even have content knowledge or the academic skills to perform well in classes.

An alternative to the deficit perspective is the *enrichment* perspective. Under the enrichment perspective, pathway students can be viewed as individuals with more assets because they are more linguistically advanced, and who, with a supportive ethos and environment, can rise to the challenge. Not surprisingly, success is often seen when pathway students come into contact with faculty and staff with high expectations of them and are committed to their success.

We would like to point out that when thinking about international learners (or even those who are born in the U.S. to immigrant families) in the U.S., the *deficit* versus *enrichment* perspective is not new. It has a long-standing tradition often discussed in the field of bilingual and TESOL education, which remain as topics of ongoing educational and political debate.

We fully agree with Cole (2009) who noted that, for universities to stick to the path of greatness, they must ensure that the following is met: Faculty research productivity, quality and impact of research, grant and contract support, honorific awards, access to highly qualified students, excellence in teaching, physical facilities and advanced information technologies, large endowments and plentiful resources, large academic department, free inquiry and academic freedom, location, contributions to the public good, and excellent leadership (pp. 110-113).

One can see from the list above that access to *highly qualified students* is only one of those important ingredients. But, as one can also see, there are at least a dozen others that also matter. Thus, when it comes to pathways, strong leadership at institutions of higher education is needed to help both skeptics and opponents to stop living under the illusion (or delusion), or belief, that pathway programs constitute a threat to academic greatness, and that having pathways students will contribute to bringing down the academic standing of a doctoral institution with high research productivity.

Quite the contrary, as one can easily argue, pathway students actually give an institution that prides itself for its excellence in teaching a real opportunity to demonstrate that in fact the institution merits such distinction and recognition. There is less challenge in teaching the best and the brightest. Pathway students give us an opportunity to demonstrate we can actually successfully teach in diverse environments; they give us an opportunity to engage in research that can help us understand how to effectively educate the graduates of tomorrow for a global world and truly have a global impact; they give us an opportunity to expand our global networks and build greater and stronger academic alliances throughout the various regions of the world; they give us an opportunity to bring the world to those who cannot venture into it; and they remind us of what it means to live in the real world, which is significantly more diverse than the academy.

In sum, pathway programs provide both a challenge and an opportunity. They are challenging because to U.S. students, as well as faculty members and staff, pathway students represent a new type of culturally diverse students. However, this very fact is an opportunity for such students (as well as faculty and staff) to become more culturally and globally aware, which is a prerequisite for operating in today's engineering workforce. The real world is not populated only by the best and the brightest, and graduates need to have the ability to work and excel in diverse backgrounds as well.

LESSONS LEARNED

Strategies for Implementing Pathways within a Public Private Partnership

While there is no single model for success, there are three key ingredients that, if present, can lead to higher levels of success in the development and implementation of pathways resulting from a public private partnership. Below, we discuss some of the key strategies, rooted in our own experience, that we believe are prerequisites for success.

Ensure Knowledge of the Partner and Their Key Responsibilities.

As in the case of any relationship, understanding the partner is of vital importance. This type of understanding needs to be in relation to the self as well as in relation to the other. Having an optimistic outlook and thinking that differences can be dealt with in time, or will get resolved naturally, is probably not advisable, as bringing the private and the public environments together means reconciling two very distinct and different cultures. So understanding the players and their expectations, together with what they can or cannot deliver, is of key importance. In other words, each partner needs to have a very clear understanding of what each needs to contribute to the positive and sustainable development of the partnership. To make this work, there has to be an explicit call for collaboration. Such collaboration needs to unfold internally (within each partner) and externally (across partners) in order to succeed in this pathway education endeavor to ensure student success. Each partner needs to ensure that their respective constituencies are clear on the fact that success depends on a strong willingness, and a deep sense of buy-in, from all stakeholders to effectively collaborate and engage in negotiations throughout the life of the partnership. Each partner will also have to develop a strategic understanding of the self and the other for managing the partnership as well as managing stakeholders' expectations. Failure to do so effectively will likely result in unnecessary levels of resistance that can be avoided (see Ehlers-Zavala & Maciejewski, 2016). In the case of ECE, much of the improvement has resulted from getting to know the concerns of the department head and the faculty. Having regular conversations as to the challenges, making notes, and designing action plans to address and overcome challenges has made a difference.

Through this close working relationship, we have been able to consistently establish and nurture trust—a vital ingredient in any effective collaboration.

Maintain Regular Communication and Adaptability.

Continuous communication is also necessary in fostering knowledge of the partner and is certainly a key to nurturing a positive and healthy partnership. The level of communication that is required in a successful partnership can be fairly sophisticated. Communication is necessary in a variety of forms and levels. So coordinating the communication effort can be a massive undertaking that requires thoughtful and careful consideration to ensure all parties are on the same page at all times, or as much as possible.

Likewise, the ability to exercise flexibility and adaptation to newly emerging and, oftentimes, unforeseeable circumstances can be what differentiates successful from less successful partnerships. Just like other living organisms, partnerships are living entities that need to develop the ability to quickly adapt to new circumstances. In other words, successful partnerships find a way to innovate in the face of new challenges. Partnerships that are resourceful and creative have the potential to withstand the unknown.

In the case of ECE, with the addition of a new staff member in ECE (an academic advisor), as well as the concerted efforts of the leadership of ECE and INTO CSU, today the communication flows smoothly and timely. This was not the case when the partnership started. Today, the communication is continuous and very successful.

Ensure that the Business Plan is Revisited, Assessed, and Evaluated Periodically

As with any plan, the successful implementation depends on continuous monitoring of the operation to assess gains and challenges as well as leverage necessary resources. Therefore, in forming a partnership, it is also important to come up with a plan that can be periodically evaluated in moving forward.

In the case of INTO CSU and ECE, the plan to ensure student success is periodically discussed. There is an effort to continuously assess and understand student performance, and each unit is mutually supported of the other. INTO CSU generates progression reports at the end of each academic year, and has consulted with ECE periodically to ensure the INTO CSU reports capture the student performance data ECE is interested in receiving. ECE input was instrumental in the development of reports that would be well received by not just this department, but by all other engineering departments participating in pathway programs (i.e., civil, mechanical, and chemical engineering).

FINAL OBSERVATIONS ON THE PATHWAY ADVANTAGE FOR ENGINEERING PROGRAMS

The field of engineering confronts some important shortcomings that are well documented and reiterated in this volume and elsewhere. In this context, public partnerships and their pathway programs offer a number of opportunities to overcome some of those shortcomings. Below we account for our top four. By pursuing pathways, engineering programs can:

1. **Actively Contribute to the Comprehensive Internationalization Plan of our Universities:** In the case of CSU, one of the main reasons for forming INTO CSU was that it was in clear alignment with the institution's internationalization goals and diversity efforts. CSU is a land-grant institution that is located in a geographic area that can be easily described as non-diverse (despite being just one hour away from a large-metropolitan city that is significantly more diverse), and whose student body has been characterized as predominantly local (from the state of Colorado and neighboring states). Therefore, to truly prepare graduates for a global market economy, it was imperative to accelerate its internationalization and diversity efforts. INTO CSU would contribute to all of this, and it already has. In general, pathway programs that are born of public private partnerships of the kind described in this chapter allow us to more quickly expand international student diversity on our campuses. As we know, "the positive impact of international students is felt in many areas of the education sector and the economy. International students strengthen U.S. higher education and research and foster cross-border collaboration on shared global problems. Further, for U.S. students who might never have the opportunity to study abroad, international students provide critical global exposure and diverse perspectives" (Farrugia & Bhandari, 2015, p. 7).
2. **Pursue Pedagogical Innovation and Renewal of Our Educational Systems:** Pathways are intended to serve a very specific group of international students—those with greater linguistic, academic, social and transitional needs to integrate well into a new academic environment and excel. These students have the potential for success, but may not be fully equipped yet with all the necessary skills and knowledge to enter educational systems that were initially for a different kind of student (linguistically, culturally, economically, socially, and academically speaking). Oftentimes, as indicated earlier, students in emerging economies may not have had the same access to education that other students in their same countries may have experienced. For example, not all of them may have had the best English language instruction that would later on allow them to be well-placed and considered for a U.S. university educa-

tion. An initial lack of English skills should not be a permanent obstacle for someone who wants to be a quality engineer, and there has to be recognition that any mathematical or technical material presented in English is always an assessment of English. That is, when an international student who does not speak English as a dominant or primary language encounters a math test in English, the student has a double-challenge. This student needs to know not only English but also the language of math in English. International students of non-English speaking background have to learn English for both general and specific academic purposes. There are many international students who are full of motivation to study, but lack the support systems that can guide them in preparing for higher education. In the U.S., welcoming those students into our system gives us a chance to truly innovate in our pedagogies to guide them to success, as faculty have an opportunity to re-examine their pedagogies for meeting the needs of diverse learners, which in the end can benefit all students.

3. **Give International First Generation Students Access to Top U.S. Higher Education:** Sometimes, international students also struggle navigating their own systems. An account of this is provided in Long and Mejia (2016) who remind us that we still live in a world where there are many institutional barriers to the access and success of many students of diverse background. Those of us who are very familiar with this reality have an ethical responsibility to ensure that we work towards eliminating these barriers. As Long and Mejia point out, students "from underrepresented populations are adversely affected not only by inadequate counseling, but also by restrictive admission policies and curricula. Many undergraduate engineering programs have rigid and lengthy course requirements" (2016, p. 212). But we also know that the highly motivated student of international background, more often than not, will rise to the challenge, and their institutions will proudly highlight these graduates. These pathway students going in to engineering may contribute to the diversification of engineering in significant ways. These new graduates that have a global-mindset can contribute to challenge the still existing preconception around the world that a good engineer is someone who is "good at math and science" (Goldberg & Somerville, 2014, p. 97). Unfortunately, there is still too much of this prevailing thinking both in the U.S. and abroad that high quality engineering programs need to have "first- or second-semester courses that are explicitly designed to eliminate students who can't cut it; there's the assumption that either you're good enough or you're not" (Goldberg & Somerville, p. 2014, p. 97). But the twenty-first century engineer is a lot more than this. As indicated in the publication of *The Engineer of 2020* (National Academy of Engineering [NAE], 2004), what the world truly needs is an engineer that

is flexible, capable of problem solving in a multicultural context and socially-oriented to truly contribute to the betterment of communities and societies around the world. Pathway students in engineering may contribute to this ongoing challenge in important ways.

4. **Having a Greater Global Impact:** Pathway programs offer an additional opportunity to the international student to access education to later on have an impact on their own societies around the globe. Education still leads to social mobility, which can have a direct impact on the economic well-being of nations, lowering crime rates, and increasing the overall well-being of humankind (so there is a high potential for an economic, social, and cultural benefit to us and the world). In other words, pathway programs offer all of us a chance to develop and/or expand global social networks that can be very positive in an increasingly interconnected world (social economics), which naturally results from increasing the number of international students our faculty and staff can interact with without having to leave campus.

CONCLUSION

Nowadays, strategic partnerships resulting from private public partnerships contribute an important element to the goal of growing the diversity in the student body of a university through the recruitment of international students (via direct admission or via pathways). Engineering programs that grow in international students of diverse background have the potential to contribute to the betterment of societies through their graduates in significant ways. Their graduates can become agents of change upon graduation. In this context, engineering programs, especially Electrical and Computer Engineering, can lead the way due to the interest in this type of engineering on the part of international students. When it comes to global impact, having an opportunity to educate students from around the world in aspects of diversity can have a real influence in truly growing a diverse workforce.

Developing pathways for students who by definition require greater support is not an easy endeavor. It is also not just one more curricular exercise. It requires a clear understanding of the needs of these pathway students as well as of the exigencies that a degree program will inevitably impose on these students. The educational experience that pathway students experience in that first year of their international academic experience is critical in determining future success. Faculty members who teach pathway students need to have a deep buy-in. For faculty to achieve this mindset, it is critically important that those in leadership positions trying to advance the pathway cause clearly articulate the advantages of adopting pathway. Top leadership (bold and strong) plays an important role in helping faculty understand that pathways not only can benefit the individual student and the institution, but also the communities around the world where these graduates will work. Pathway advocates also need to be ready to candidly share the unique challenges that can be expected if these types of programs are implemented without an appropriate introduction. It is important that leaders and/or stakeholders discuss and clearly articulate the strategies needed to maximize the positive impact that can be derived from the establishment of pathway programs on their campus and beyond.

It is our hope that, in relating what we have learned from our experience both at CSU and INTO CSU, other engineering programs in the U.S. and around the world can be better prepared to assume this possible challenge and reap the potential rewards of a public private partnership that has a pathway component. Since 2009, the US has had an exponential growth of pathway programs, and this trend is likely to continue well into the future. So the question of how to prepare and best manage the development and implementation of these new types of programs that conditionally admit students, both at the undergraduate and graduate levels, is extremely relevant. Pathway programs offer a unique opportunity to impact the world around us. Engineering programs in U.S. institutions of higher education cannot ignore this state of affairs, and leading programs must also lead the way on how to best and successfully incorporate pathway programs to help achieve their institutional internationalization goals.

With strong leadership, adequate education of a university community as well as proper planning for effective pathway development, the implementation phase of pathway programs can yield highly positive results sooner rather than later. There is no reason for enduring growing pains if the experience we share in this chapter is fully considered and the lessons learned implemented. We have developed this chapter as a result of our own experience that organically evolved in time. It is our hope that our experience positively contributes to that of others about to engage in similar undertakings.

REFERENCES

Adamson, F., Astrand, B., & Darling-Hammond, L. (Eds.). (2016). *Global education reform: How privatization and public investment influence education outcomes*. New York, NY: Routledge.

Altbach, P. G. (2016). *Global Perspectives on Higher Education*. Baltimore, MD: Johns Hopkins University Press.

Arbeit, C. A., Einaudi, P., & Kang, K. H. (2016, March). Influx of foreign graduate students and inclusion of newly eligible institutions lead to a significant increase in U.S. graduate enrollment in science and engineering. *InfoBrief*. National Center for Science and Engineering Statistics, National Science Foundation (NSF 16-310). Retrieved from www.nsf.gov/statistics/2016/nsf16310/nsf16310.pdf

Bhandari, R., Belyavina, R., & Gutierrez, R. (Eds.). (2011). Student mobility and the internationalization of higher education: National policies and strategies from six world regions. New York: Institute of International Education (IIE).

Bowen, W. G., & McPherson, M. S. (2016). *Lesson Plan. An agenda for change in American higher education*. Princeton, NJ: Princeton University Press. doi:10.1515/9781400881369

Caspar-Terizakis, C., Bieker, S., Cornel, P., Heinelt, H., Klonecki, K., Wagner, M., . . . Winner, H. (2016). From historical roots to modern-day results: Technische Universität Darmstadt and Tongji University's bi-lateral strategic partnership. In C. Banks, B. Siebe-Herbig, & K. Norton (Eds.), Global perspectives on strategic international partnerships: A guide to building sustainable academic linkages (pp. 25-34). New York, NY: Institute of International Education (IIE).

Cole, J. R. (2009). *The great American university*. New York: Public Affairs.

Ehlers-Zavala, F. P., & Maciejewski, A. A. (2016). The case of an Electrical and Computer Engineering (ECE) Department in the internationalization process of a research I public institution.*Proceedings of the International Forum of the American Society for Engineering Education*.

Farrugia, C. A., & Bhandari, R. (2015). *Open Doors Report on International Educational Exchange*. New York: Institute of International Education.

Goldberg, D. E., & Somerville, M. (2014). A whole new engineer. Douglas, MI: threejoy.

Goodman, A., & Rüland, D. (2016). Foreword. In C. Banks, B. Siebe-Herbig, & K. Norton (Eds.), Global perspectives on strategic international partnerships: A guide to building sustainable academic linkages (pp. i-iii). New York: Institute of International Education (IIE).

Johnstone, D. B. (2016). Financing American higher education: Reconciling institutional financial viability and student affordability. In M. N. Bastedo, P. G. Altbach, & P. J. Gumport (Eds.), *American higher education in the 21st century. Social, political and economic changes* (4th ed.; pp. 310–341). Baltimore, MD: Johns Hopkins University Press.

Kuder, M., & Banks, C. (2016). Current trends in strategic international partnerships. In C. Banks, B. Siebe-Herbig, & K. Norton (Eds.), Global perspectives on strategic international partnerships: A guide to building sustainable academic linkages (pp. xi-xxiv). New York, NY: Institute of International Education (IIE).

Learch, K. M., Paulson, S. K., Sander, H., & Schuch, E. (2016). Strategic relationship development: A German-U.S. university case study. In C. Banks, B. Siebe-Herbig, & K. Norton (Eds.), Global perspectives on strategic international partnerships: A guide to building sustainable academic linkages (pp. 13-24). New York, NY: Institute of International Education (IIE).

Long, L. L. III, & Mejia, J. A. (2016). Conversations about diversity: Institutional barriers for underrepresented engineering students. *The Journal of Engineering Education, 105*(2), 211–218. doi:10.1002/jee.20114

McGee, J. (2015). *Breakpoint: The changing marketplace for higher education.* Baltimore, MD: Johns Hopkins University Press.

National Academy of Engineering (NAE). (2004). *The engineer of 2020: Visions of engineering in the new century.* Washington, DC: The National Academies Press; doi:10.17226/10999

Rodan, P. (2016). Public-private partnership in higher education: Central Queensland University meets campus management services. *Australian Universities Review, 58*, 5–12.

Smale, S. (2015). *Notes from an oral presentation of pathway providers in the U.S. to the Colorado State University community.* Fort Collins, CO.

Sutton, S. B., & Obst, D. (Eds.). (2011). Developing strategic international partnerships: Models for initiating and sustaining innovative institutional linkages. New York, NY: Institute of International Education (IIE).

Yang, T., Ahmad, S., Lavenda, J., Robinson, J. H., & Seo, K. (2016). Beyond the initial partnership: A framework for classifying partnerships as a means to strategize comprehensive internationalization. In C. Banks, B. Siebe-Herbig, & K. Norton (Eds.), Global perspectives on strategic international partnerships: A guide to building sustainable academic linkages (pp. 1-12). New York, NY: Institute of International Education (IIE).

APPENDIX A

Table 1. List of providers and partners since 2009. Adapted from a November 2015 presentation to CSU by Steven Smale, head of the Global Recruitment Unit at IUP.

Private Provider	Higher Education Institution/Partner
Kaplan (U.S.) Since 2009	Northeastern University in Boston, MA; U of Utah, Salt Lake City, UT
INTO (U.K.) Since 2009 to 2016	Oregon State University, University of South Florida; Colorado State University, Marshall University, Drew University, George Mason University, Saint Louis University, University of Alabama at Birmingham
Navitas (Australian) Since 2011	U of Mass, Boston; U of Mass, Darmouth; U of Mass, Lowell
StudyGroup (U.K.) Since 2013	James Madison—Harrisonburg, VA; Long Island U—Long Island, NY; Widener—Chester, PA; Roosevelt University, Chicago, IL; U of Vermont—Burlington, VT; U of Maine—Portland, ME; Long Island U—Brooklyn, NY; Merrimack College—North Andover, MA
Cambridge Education Group (U.K.) Since 2013	Wheelock—Boston, MA; Morrisville State College—Morrisville, NY; U of North Texas—Denton, TX; CAL State Monterey Bay, CA
Shorelight (U.S.) Since 2015	U of Central Florida—Orlando, FL; U of Kansas—Lawrence, KS; Auburn U—Auburn, AL; Florida International U—Miami, FL; Louisiana State U—Baton Rouge, LA; U of South Carolina—Columbia, SC; Adelphi U—Garden City, NY; American U—Washington, DC
Kings Education (U.K.)	Pine Manor—Boston, MA; Canisius College—Buffalo, NY; Concordia College—Bronxville, NY; Marymount California U—Rancho Palos Verdes, CA; Rider U—Lawrence, NJ; Kings Los Angeles—Los Angeles, CA

APPENDIX B

Source: http://www.aplu.org/about-us/diversity-statement.html

Commitment To Diversity

Whereas, the Association of Public and Land-grant Universities (APLU), a voluntary association whose membership consists of public research universities, land-grant institutions, state university systems and other affiliated organizations is committed to advancing diversity and inclusion by exercising the principles of equal access and equal opportunity in education and employment.

Whereas, APLU takes seriously our leadership responsibility to provide equal access and equal opportunity through the development of policies and initiatives that foster academic excellence, diversity and inclusion.

Whereas, APLU prohibits discrimination against any individual on the basis of race, religion, color, sex, age, national origin or ancestry, marital status, parental status, sexual orientation, gender identity or expression, disability, or status as a veteran.

Whereas, APLU believes that human and intellectual diversity contribute to academic excellence, and that the Association and its member institutions benefit from the rich diversity of the persons who comprise our staff, faculty and students.

Whereas, APLU is committed to recruiting and retaining, on a nondiscriminatory basis, people who are members of groups that have been historically underrepresented in higher education and to supporting its members in their efforts to enhance the diversity of their faculty, staff and students.

Whereas, APLU champions diversity, educational equity and the preparation of individuals who can live and work effectively in an increasingly multicultural and interdependent world.

Therefore, be it resolved to approve and adopt the APLU Diversity Statement by the directors of the Association of Public and Land-grant Universities (the "Corporation"), at a properly called meeting on March 24, 2010 at which a quorum was present, did approve the APLU Diversity Statement by a majority or greater vote as required by the bylaws.

M. Peter McPherson
President
March 24, 2010

Chapter 12

Getting Off the Engineering Enrollment Rollercoaster:

Interaction Between Academia and Upstream Petroleum Industry

Tatyana Plaksina
University of Calgary, Canada

ABSTRACT

Traditionally, petroleum engineering programs trained young professionals for the oil and gas industry. After years, this strong tie made the universities completely dependent on petroleum companies for their student placement and research funding. Similar to the operators and service companies, the universities became sensitive to changes in demand. In the last decade, in response to a rapid increase in crude oil prices and unconventional production, the industry experienced a hiring surge which caused a subsequent agitation among engineering freshmen to claim petroleum engineering as their major. As a result, this short period of high demand for petroleum engineering graduates created tremendous pressure on the departments making them expand beyond their educational capacities. In some cases, that led to doubling of the number of undergraduates. Simultaneously, this rapid expansion of the student enrollment created a growing demand for faculty who are now experiencing serious demographic and research gaps.

DOI: 10.4018/978-1-5225-2212-6.ch012

INTRODUCTION

After a short setback of 2008-2009, when oil and gas prices experienced a rapid decline and then a hike back to the price of over 100 $/bbl (Figure 1), the US petroleum engineering programs continued the ongoing student admission frenzy. Inspired by impressive starting salaries of already employed petroleum engineers and their relative scarcity on the job market, first and second year undergraduate students rushed to claim petroleum engineering as their major. In 2013, when oil and gas industry hiring was still booming and excitement among students was growing steadily, several professionals who started their careers in the 1970s and 1980s began to see a number of dangerous signs. Drs. Hill and Holditch advised major petroleum engineering programs in the US against unconstrained growth of student enrollment after investigating the enrollment database compiled by Lloyd Heize of Texas Tech University (Hill & Holditch, 2013). Specifically, they recognized similarities in the enrollment trends in the 2010s and 1980s (Figure 2). In the 1980s, the demand for petroleum engineers created tremendous pressure on the academia which responded with higher numbers of graduates. By the middle of the 1980s, however, the oil and gas industry experienced another bust and most of the petroleum engineering graduates faced difficult job market. In that situation most students and graduates decided to switch to other engineering disciplines and left upstream petroleum industry with bitter disappointment. As a result, some of the universities such as University of Wyoming had to close their petroleum engineering program which resulted in rapid decline in satellite disciplines (for example, geosciences).

In almost three decades after these events, a new generation of aspiring engineering students and faculty members took over the academic stage, on which very few

Figure 1. WTI spot prices in $/bbl reported from 1986 to 2014 (IAE, 2016)

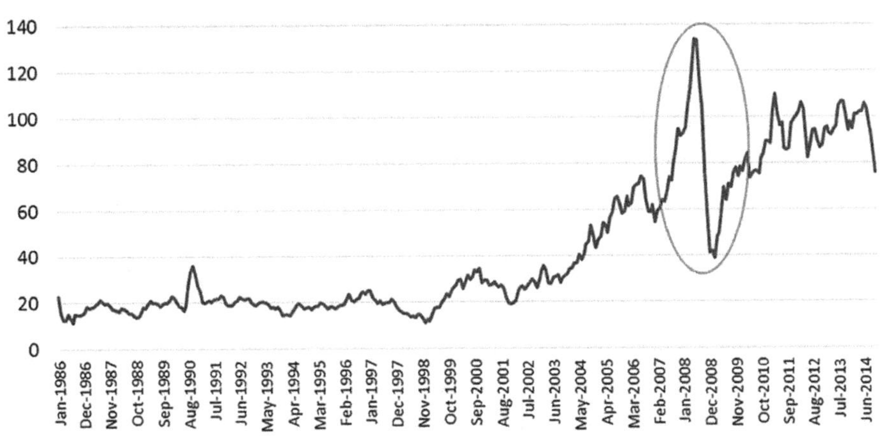

Figure 2. Undergraduate student enrollment from 1972 to 2014 compiled by Dr. Lloyd Heize (Heize, 2014)

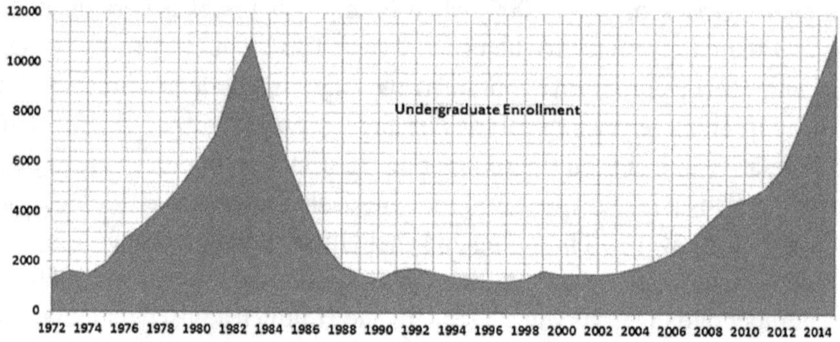

professionals still remembered the consequences of the 1980's bust. As a result, most petroleum engineering programs ramped up their student enrollment to twice or more than twice of the student numbers observed in the mid-2000s. For example, the petroleum engineering department at Louisiana State University – Baton Rouge went from about 400 students in 2009 to approximately 900 students at the end of 2013 (LSU, 2016). Similar rapid expansion is evident from enrollment statistics of Texas Tech University, Oklahoma State University, University of Tulsa, University of Wyoming, University of Louisiana – Lafayette, and Penn State University. Some new programs such as University of North Dakota responded to extra high demand for petroleum engineers with magnitude increases in their enrollment. Per official department statistics, the petroleum engineering department at University of North Dakota went from only 4 students in 2010 to 208 people in 2014 (UND, 2016). In fact, expansion of some petroleum engineering programs (for instance, Louisiana State University) was so rapid that they found themselves on a verge losing ABET accreditation due to extremely small faculty to student ratio (O'Donoghue, 2014).

In 2014, in the middle of this academic expansion, crude oil price started to decline. Operating companies could not generate lavish revenues from oil production and, as a result, started to lay off their employees and curb their internship programs. Within a year the shockwave from shrinking exploration and production (E&P) industry reached the academia. Suddenly, student job placement dropped and the pools of faculty candidates increased. Many PhD holders who were attracted by oil and gas research and development (R&D) departments of operating companies during the booming years found themselves jobless and trying to join the academia. In fact, competition among these highly qualified candidates became so high that some assistant professor positions would have from 60 to 100 applicants. As the chapter discusses later, industrial demand had a dramatic impact on expertise of the junior faculty members. Specifically, pressing interest of the oil and gas companies in unconventional resources caused an explosion of specialists in shale and tight reservoir engineering and simulation, geomechanics for hydraulic fracturing applications, and enhanced oil recovery methods (EOR) for mature, maturing, and unconventional plays. This industrial preference created distortions in overall academic faculty expertise as the pool of potential candidates had very similar research interests.

This chapter investigates the reasons for inefficient interaction between the petroleum E&P industry and the academia, the processes that have led to inability of the industry to accommodate all petroleum graduates, and the impact of all these factors on faculty distribution and expertise. First, a generic example of organizational inefficiency by Peter Senge is used to demonstrate how poor communication and sporadic increases in demand for petroleum graduates have led both industry and academia to a deep crisis. Second, application of qualitative and quantitative analyses to the dataset of Heize helps determine the most influential factors that affect student enrollment including those pursuing BS, MS, and PhD degrees as well as the faculty count. Third, a scrutiny of petroleum engineering faculty distribution by rank, origin, gender, and expertise helps identify demographic, research, and generation gaps among tenure-track professors. The chapter concludes with a summary of the key findings, investigation of possible scenarios for future development, and elaboration on how these gaps and inefficiencies can be avoided in petroleum and other engineering disciplines.

BACKGROUND AND PROBLEM STATEMENT

To understand the root cause of the crisis in petroleum engineering education, it is important to begin with a qualitative analysis of the situation and inspect it as a type of dysfunction or inefficiency in a large organization that includes the academic establishments and upstream petroleum industry. A famous simulation game called the "beer game" and discussed by Peter Senge is one of the best ways to demonstrate why the student enrollment statistics from the 1980s and the recent memory of the price collapse in 2008 did not discourage the US petroleum engineering programs from rapid expansion. In his book about organizations and inefficiencies emerging inside them, Senge (2006) models a situation similar to the one that occurred between the oil and gas industry and academia in the last several years. Figure 3(a) demonstrates the way Senge formulates the "beer game" in which there are several interacting players including the retailer, the wholesaler, and the brewery. None of the participants communicate with each other directly. Instead, they send each other beer request forms with numbers of boxes. Before the beginning of the simulation game, the brewery outputs the product at a steady supply rate (the situation analogous to the number of students graduated by the petroleum engineering programs). Note that the brewery cannot react quickly to any sporadic increases in demand because of physical limitations of its production capacities. Specifically, to expand its beer output, the brewery needs to acquire new buildings, interview and hire more workers, and order additional material. Similarly to the brewery, the universities cannot rapidly increase the number of graduates. Preparation of young professionals usually takes about 4-5 years, requires expansion of the facilities and laboratory spaces for their training as well as hiring more faculty members to advise and teach them. As in the case of the brewery, the process of hiring new university employees is resource and time consuming and might take up to several years (especially, if the university is trying to attract highly qualified international scholars who require extra time and effort to legally enter the country and start working). All these factors result in a very slow response of the petroleum engineering programs to sudden increases in demand.

Figure 3(b) shows the situation of sporadic increase in beer demand during week 2 of the simulation game. More potential customers start coming to the shop and ask for beer. Note that the retailer in the "beer game" only reacts to this sudden increase in demand regardless of whether the wholesaler and the brewery can or cannot increase the production capacity quickly. Because of poor communication and lack of understanding of how the beer production side operates, the retailer quickly gets frustrated with supply shortage and keeps filing more and more requests to the wholesalers to ship larger orders of beer. Analogous situation could be observed in the behavior of the large oil and gas companies. Starting from the boom of 2010-2011, the upstream petroleum firms began to increase pressure on the undergraduate university programs to deliver more young petroleum engineers. Moreover, every year after 2011, the oil and gas companies kept putting increasing numbers of graduates in their "request forms." Similarly to the retailer, they would quickly become frustrated with a very scarce supply of young engineers coming out of universities. The firms disregarded the fact that the universities could not satisfy their demand promptly, especially after the years of 2008-2009 when many petroleum engineering graduates could not find jobs or internships due to previous oil price decline. The industrial cycles of high and low demand for highly trained personnel (and the "roller-coaster" of favorable and unfavorable crude oil prices) became too short for the academia to respond adequately.

Figure 4(a) illustrates the peak of the beer demand after which retailer's request forms with high numbers finally reach the brewery. Due to poor back and forth communication between the participants of the game, the owner of the brewery is not sure about the nature of such unusually high requests and seizes the opportunity to expand. However, brewery expansion and hiring and training new workers takes time, and the wholesaler only sees increased output of beer boxes in his stor-

Figure 3. The setup of the "beer game" by Senge. (a) The game participants include the retailer, the wholesaler, and the brewery. (b) At the beginning of the game sudden spike in beer demand causes disturbance in the beer supply (Senge, 2006).

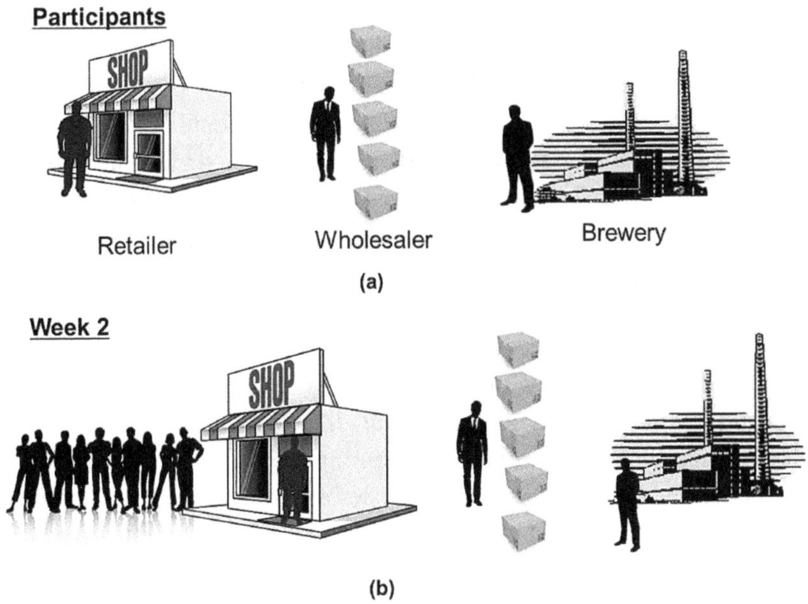

Figure 4. The "beer game" by Senge. (a) Beer demand continues to increase. (b) The brewery finally expands its capacities and sends large orders to the wholesaler (Senge, 2006).

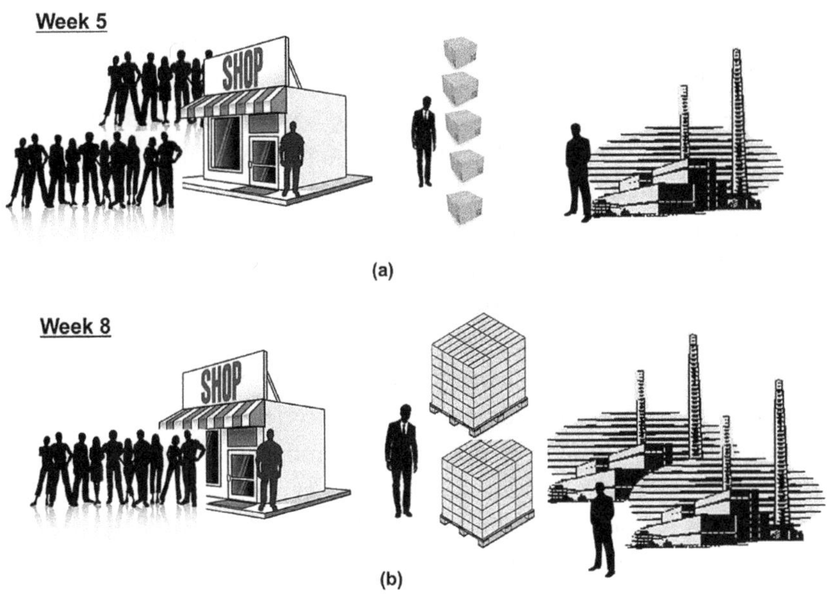

ages in several weeks (Figure 4(b)). Similar situation played out in the petroleum industry and academia interaction. The university programs yielded to increasing industrial demand for petroleum graduates and within 3-5 years ramped up their student enrollment to at least twice the number of students that they used to educate. Such uncontrolled expansion led some programs to be on a verge of falling out of ABET accreditation status due to unacceptably low numbers of faculty members. Around that time, hiring new faculty members was particularly difficult for the petroleum engineering programs because the industry raided graduate programs for PhD students and offered them highly paid industrial jobs. Another factor that contributed to slow faculty hiring response is traditionally poor mobility of PhD holders from industry to academia and visa verse. It is common wisdom in the US academia that once a doctorate student chooses a path of a university professor, he/she will not have any industrial opportunities until maybe 10-20 years later when this scholar might join some consulting business. Naturally, such prospects and lower starting salaries deterred many highly qualified doctorate students from entering the academia. In fact, multiple schools during the boom period (between 2011 and 2014) could not attract petroleum engineering experts to fill a number of faculty positions and had to resort to hiring other engineering doctorate graduates (for instance, Texas Tech University could not find enough candidates for full time faculty positions for several years, and Penn State has been staffing heavily from other engineering disciplines). In attempt to alleviate the faculty shortage, the programs started hiring larger numbers of non-tenure track candidates (like instructors, adjunct professors, and professionals in residence) and faculty candidates who could not be hired by the US based companies due to their country of origin (specifically, most petroleum engineering programs have a significant portion of junior faculty members from Iran based on the country of their bachelor degree, e.g., University of North Dakota, University of West Virginia, LSU, University of Oklahoma, Texas A&M University). This by-product of the dysfunction inside the industry-academia interaction created another distortion in faculty distribution that we will analyze in more detail in the subsequent sections.

Figure 5(a) shows the end of the "beer game" that is unfortunate for all participants. Specifically, the ordering frenzy of the previous weeks resulted in expansion of infrastructure, personnel, and production of the brewery, the wholesaler's storage got flooded with incoming boxes of beer, and the retailer keeps receiving the backlog of his orders while the beer demand declined back to its normal level of week 1. At this point, the retailer puts zero boxes into his order forms and the participants on the supply side are "stuck" with excess of the product that nobody wants to buy. Today, the reader can be a witness of exactly the same finale of the last cycle of boom and bust in petroleum industry and academia. As shown in Figure 5(b), the parallel becomes complete if the retailer is substituted by the oil and gas companies

Figure 5. The "beer game" by Senge. (a) Beer demand drops to week 1 level. (b) Petroleum engineering programs are similar to the brewery and the wholesaler and the industry is similar to the retailer (Senge, 2006).

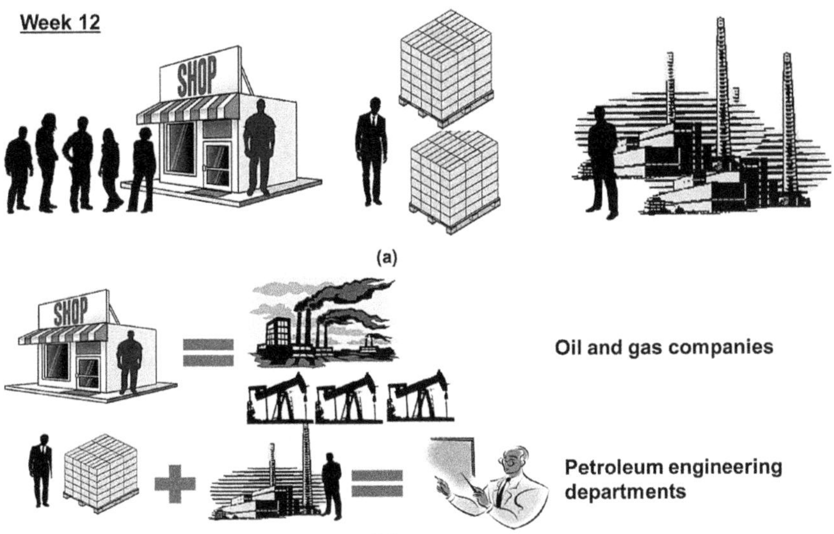

and the wholesaler and the brewery by the academia. Statistics of employees in extractive petroleum industry (Figure 6) vividly demonstrates how fast the industry was increasing pressure on the universities to graduate more petroleum engineers and how rapidly it got rid of tens of thousands of employees in just two years (US Bureau of Labor Statistics, 2016).

Now that the nature of the inefficient interaction between the petroleum industry and academia is clear, it is appropriate to conduct a quantitative analysis. This section provides a statistical analysis of the student enrollment dataset and reveal hidden relations between the major indicators of the upstream petroleum industry (including crude oil spot prices, rig counts, and gasoline prices) and academic indicators (including student enrollment, degrees granted, and faculty counts). Application of statistical tools such as stepwise regression for model selection using Akaike information criterion and ANOVA test is summoned to answer several retrospective and prognostic questions. Among retrospective issues are factors that dominated interaction between the industry and academia in the past. The prognostic section will try to model possible scenarios of the future interaction.

Figure 6. Oil and gas extraction employees (thousands) seasonally adjusted and annually averaged (Us Bureau of Labor Statistics, 2016)

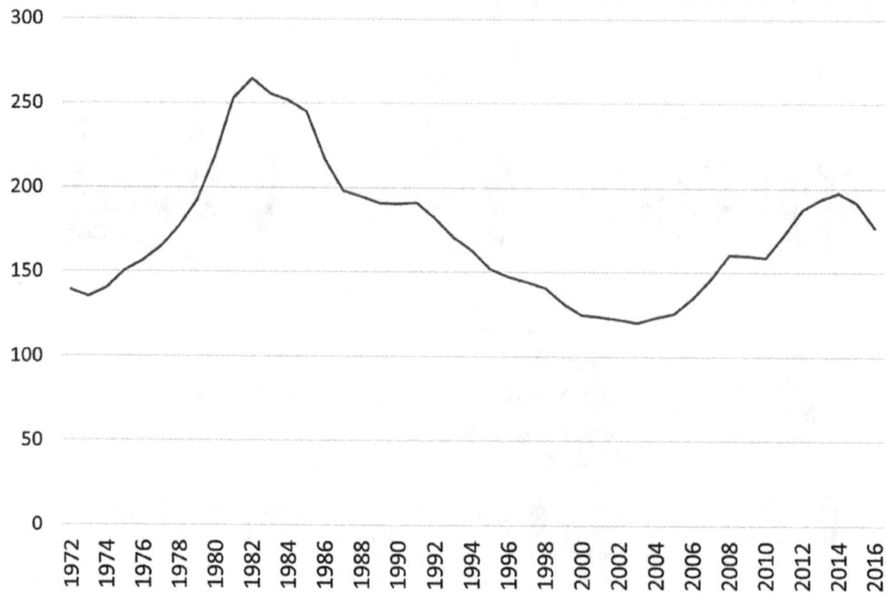

MODERN PETROLEUM ENGINEERING EDUCTION: ISSUES AND POSSIBLE RESOLUTIONS

Retrospective Investigation of Student Enrollment Dataset

Now that the problem in modern petroleum engineering education is formulated and its development during the last decades is tracked, application of more rigorous statistical tools to the student enrollment dataset by Heize can reveal the most significant parameters that influenced both undergraduate and graduate enrollment. A number of plots that illustrate the data collected by Heize during approximately three decades provide the reader with a broader picture. Figure 7 shows the dynamics of undergraduate student enrollment from 1987 plotted against the number of BS degrees awarded and annual average WTI (West Texas Intermediate) crude oil spot price. Note that the WTI data were averaged because they have different resolution than the enrollment data (EIA, 2016). Specifically, WTI spot prices are reported and recorded daily and even hourly, while student enrollment is usually reported once a year. Because of this inconsistency in reporting times between these data series, it is possible to use annual average WTI spot price which might not always be the best representation of its demand volatility. Nonetheless, obvious trends such as the decline of 2008 are obvious even in this low resolution WTI data. Note, that around

Figure 7. Undergraduate student enrollment versus WTI spot price

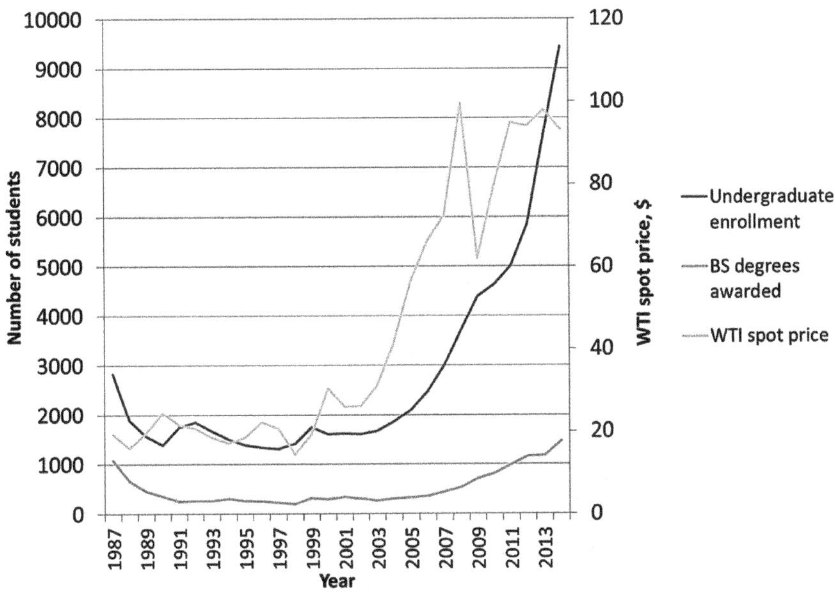

2008 undergraduate student enrollment line changes its slope for about three years indicating decrease in growth of the student number. This can be easily explained by industrial hiring freeze around 2008 when a large portion of fresh graduates could not find a job upon graduation and continuing students had their internships rescinded.

Although WTI spot price is one of the major economic indicators in the upstream petroleum industry that can be used for our statistical investigation, it is not the only one that can illustrate the level of activity in the US crude oil market. Rig count is another popular indicators of "health" of the petroleum industry (Iledare, 2007). High petroleum E&P activity results in proliferation of industrial positions and, thus, has a direct impact on the demand for young professionals. Similarly to WTI spot prices, rig count is reported more frequently than student enrollment and, therefore, we calculate average annual rig count including both on- and offshore rigs (Figure 8).

Student enrollment is not limited only to undergraduates, but also includes graduate students pursuing and receiving MS and PhD degrees. Figure 9 plots PhD enrollment data and the number of PhD degrees awarded juxtaposed with WTI spot price. Although PhD students are the most conservatively growing group of students (because of significant time commitment required by the degree as well as lost financial opportunity of not joining the industry with much higher salaries), even their count shows almost immediate reaction to sudden decline in crude oil price. Around 2008, following the industrial hiring freeze, many doctoratal students de-

Figure 8. Undergraduate student enrollment versus rig count

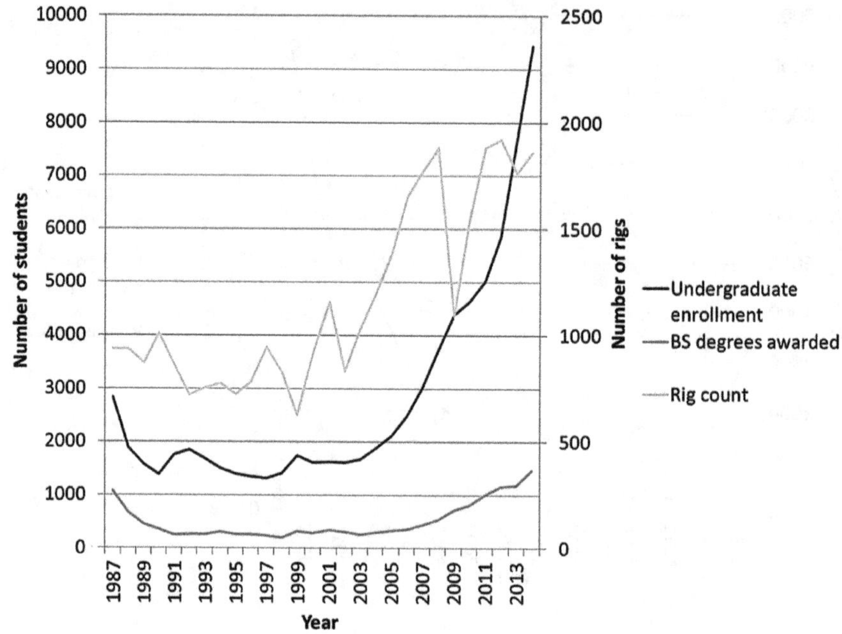

Figure 9. PhD student enrollment versus WTI spot price

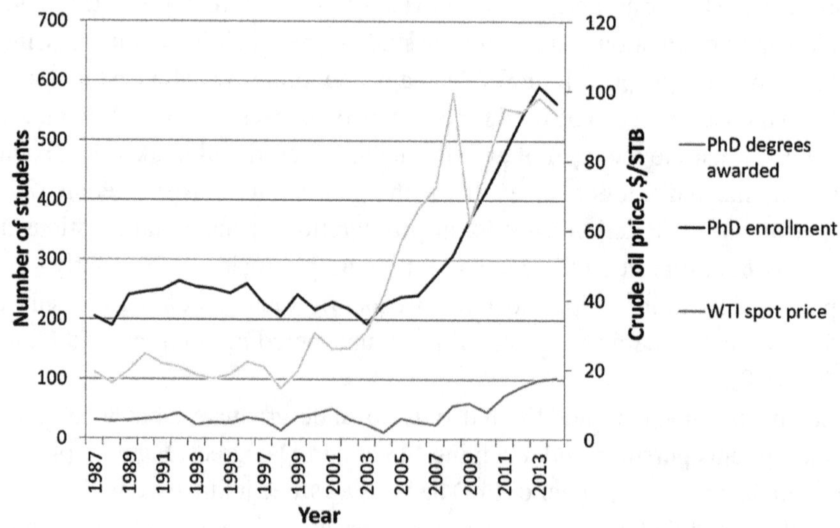

cided to postpone their graduation (illustrated by the change of slope of PhD degrees awarded line after 2008) because most PhD students (more than 80 percent) are of foreign descent. Delaying graduation or accepting post-doctoral positions at the universities for most fresh graduates are the only legal way to survive difficult hiring situation. Moreover, the PhD enrollment continues to maintain high slope around the most unfavorable year because recent master graduates could not join industrial workforce and decided to wait for better industrial opportunities after PhD.

To complete a brief inspection of the dataset, it is important to see the dynamics of the faculty count versus PhD graduates during the last three decades. The preliminary investigation of data in Figure 10 leads to a conclusion that the years with the most favorable crude oil prices lead to the worst shortages of faculty candidates. As mentioned above, PhD graduates prefer to join R&D departments of large petroleum companies than to pursue academic careers. However, the situation changes drastically when the crude oil price declines. In these times, more PhD graduates, who could not postpone their graduation or join post-doctoral programs, decide to apply for junior faculty positions. In fact, this exact scenario is currently playing out. As the reader can observe from the plot (Figure 10), PhD enrollment was growing at a very high rate from 2006 and reached its historic high around 2012. Around that time (specifically, between 2011 and 2014) the market and hiring situations were particularly favorable for highly skilled PhD graduates. After a long wait for the bust end, students started to graduate as shown in PhD degrees awarded curve.

Figure 10. PhD student enrollment versus faculty count

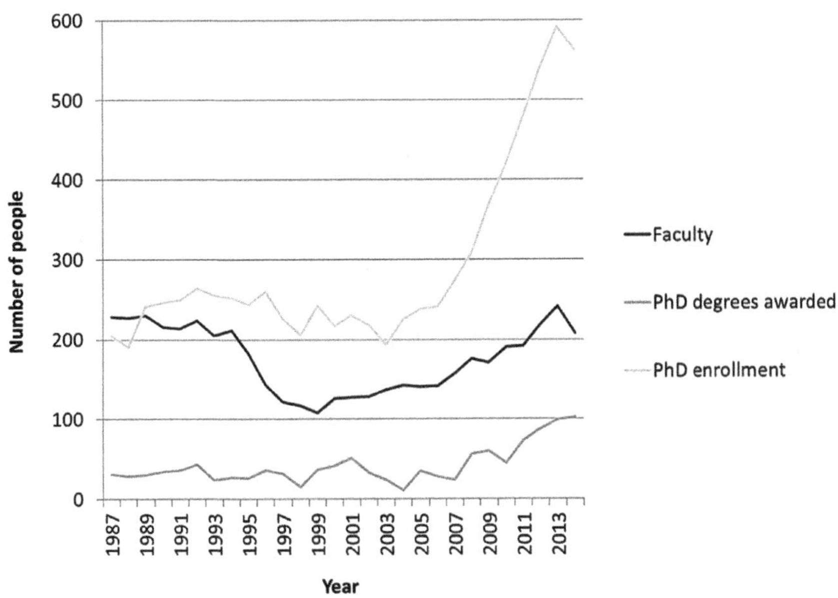

However, not everyone could finish their doctorate degree within those 3 good years to secure industrial employment, and as a result, an unprecedented number of doctorate graduates started to compete for faculty positions in 2014 and on. Some schools such as UT-Austin report record high pool of candidates of about 60 people per position advertised.

Statistical Analysis and Explanatory Model Selection

This section summarizes a statistical analysis of the enrollment dataset to identify the most statistically significant parameters that can explain the enrollment observations and select models that encompass these parameters. Statistical analysis for a dataset that represents a behavior of a certain social group is very different from that for an experimental engineering problem, because for the former we cannot reproduce experiments, easily add observations, and use an experimental design approach. The enrollment data of Heize represents a complex behavior of a socio-economic system (industry-academia) that has multiple participants from both sides with different expectations, policies, and objectives. Therefore, the true model of such interaction is unknown and probably cannot be established with high confidence. Nevertheless, these enrollment data coupled with the surveyed economic parameters give us an opportunity to identify the driving forces within this system.

To analyze the dataset and to select possible explanatory models, one can use stepwise regression method with Akaike Information Criterion (AIC). This approach allows to assess relative quality of the models based on the dataset (Akaike, 1974). In its general form, AIC can be defined as:

$$AIC = 2k - 2\ln\left(L\right)$$

where k is the number of parameters included into the model and L is the optimized likelihood function for the model. Additionally, AIC is an information criterion that penalizes excessive number of parameters used in the estimation process and rewards a better fit. Note that although AIC cannot be used as an absolute quality measurement of the model, it does provide us with the information of how the models from the stepwise regression compare to each other. When using AIC for relative model assessment, it is important to comply with three rules. First, the dataset should contain sufficient number of observations in comparison to the number of parameters for the analysis and the assessment to be valid. In the enrollment dataset and the selected parameters this condition is satisfied. Second, lower value of AIC indicates better model and means that this model minimizes the estimated loss of information. Final, with AIC we can only obtain relative information about "goodness" of the model, the true relationship between the parameters is ultimately unknown (Burnham and Anderson, 2004).

The open source statistical software R was used to perform the analysis using stepwise regression (R, 2013). The dataset contained 28 multidimensional observation points including undergraduate, master, and doctorate enrollment numbers, numbers of awarded BS, MS and PhD degrees, average annual WTI spot price, US Gulf coast conventional gasoline price, rig count, number of faculty members, and number of reporting schools. The data reported observations from 1987 to 2014.

Table 1 provides the summary of stepwise regression analysis performed on the dataset with the initial and final models for undergraduate, master, and doctorate enrollments and the number of faculty members from 1987 to 2014 (for detailed R output refer Appendix 1). For undergraduate enrollment (parameter *und*), the analysis compared several models based on AIC and selected one that shows the relation between enrollment and gasoline price. Table 2 and Table 3 (Appendix 1) provides the details of the final model with corresponding significance codes for explanatory variable. Although the value of R^2 indicates that the fit is far for being considered perfect, the significance codes show that gasoline price is more important than WTI spot price. In connection to Figure 7, we might conclude that undergraduate enrollment slightly has a small lag in response to crude oil prices. Specifically, it appears that undergraduate enrollment does not respond instantaneously to the improved situation in the crude market, but rather waits until the refined products (for example, gasoline) reach the market as a confirmation that the industry is doing well and can hire more young professionals.

The second entry in Table 1 presents similar analysis for MS student enrollment (Table 4). However, this initial model included not only economic parameters and rig count, but also undergraduate enrollment. This parameter was included because many petroleum engineering programs offer so-called fast-track degrees (especially, they start to proliferate during unfavorable economic situation). On one hand, these types of fast-track programs help undergraduate students who could not find employment upon graduation to survive the hiring freeze without substantial loss of engineering competence. On the other hand, this is yet another source of income

Table 1. Stepwise regression results on the enrollment dataset. Initial and final models

Response	Initial Model	Final Model
undergraduate	*wtispot + gulfgasoline + rig*	*gulfgasoline*
master	*wtispot + gulfgasoline + rig + und*	*wtispot + gulfgasoline + und*
doctorate	*wtispot + gulfgasoline + rig + und + ms*	*wtispot + gulfgasoline + rig + und*
faculty	*wtispot + gulfgasoline + rig + und + ms + phd*	*wtispot + und*

for the universities when corporate funding dries out. Keeping undergraduate students in school longer brings many benefits to the universities.

Fast-track programs, however, is not the only explanation for a strong relationship between undergraduate enrollment and master student enrollment. Another significant factor is attraction of graduates from other engineering disciplines. When crude oil and gasoline prices go up and more undergraduates try to claim petroleum engineering as their major, those who already received their BS in another engineering disciple start feeling that there is an opportunity for a big financial gain in the petroleum E&P industry. However, electrical, mechanical, or chemical degrees have certain limitations and disadvantages when applied to highly specialized petroleum discipline. Many engineering students require additional training before they can perform tasks of, for example, petrophysicists, reservoir engineers, production or drilling experts. Thus, many domestic graduates view MS degree as an opportunity to change career and become more specialized and marketable in petroleum field. This academic shift and thinking occurs with the increased demand for petroleum undergraduate students. As the summary of statistical analysis suggests (Table 5 in Appendix 1), the model is better than the final one for undergraduate enrollment as indicated by higher R^2.

The third entry in Table 1 refers to PhD student enrollment and the results of stepwise regression on these observations (Table 6). In connection to Table 7 (Appendix 1), it is possible to conclude that an increasing number of undergraduates stimulates enrollment of PhD students. One explanation for this is that many programs tend to solve the problem of small number of faculty members with extensive teaching responsibilities for PhD students. In many universities doctorate students teach laboratory courses for undergraduates, and help as graders and teaching assistants in both undergraduate and graduate classes. Thus, the reader can observe a very strong relationship with undergraduate enrollment indicated by its high significance. Even when the industry is booming and hiring all petroleum graduates including PhDs, there is an increase in PhD student enrollment. In favorable economic situation, faculty hiring becomes extremely difficult and, therefore, an increase in doctorate student population is inevitable. When the industry freezes hiring and not interested in doctorate students, some of them try to become faculty members or try to postpone graduation.

The final entry of Table 1 is the analysis of faculty numbers. Table 8 (Appendix 1) shows that the final model fit is not particularly good due to its low R^2, but the undergraduate enrollment parameters does appear to be significant for the number of faculty. One reason why it could be hard to observe strong relation between the variables in the presented dataset could be that faculty hiring is the slowest process that is almost insensitive to frenetic changes in crude oil and gasoline prices as well as student enrollment. The data after 2014 (that are yet to be reported) are likely to show an increase in faculty number as most petroleum engineering programs are currently flooded with doctorate applicants seeking academic positions. Rapid increase of the faculty number is the last stage of the "brewery" expansion that is still happening with the sudden decline of demand on the background. However, the "brewery" (reader should understand the petroleum engineering programs) cannot stop at this moment because of the extremely high student enrollment accumulated in previous years. Though the future of such expansion is highly uncertain, the academia has to expand in the next several years to graduate those who are already in the "system."

Petroleum Engineering Faculty: Demographic, Gender, and Expertise Issues

As the previous sections showed, sudden booms and busts in the petroleum E&P industry had very peculiar impact on petroleum engineering faculty, their demographic distribution, and research interests. As no scholar so far ventured into these issues, there is a unique opportunity to investigate the population of petroleum engineering professors both qualitatively and quantitatively. For this data about faculty members of 13 leading US petroleum engineering programs were collected, compiled, and analyzed. This dataset includes only most updated (as of July 2016) information from official department websites and accounted only tenured and tenure-track professors (omitting adjunct professors, instructors, lecturers, etc. who tend to be "less permanent" faculty members).

Figure 11 shows distribution of faculty according to their ranks (assistant, associate, and full professor). Although a superficial inspection might suggest that all three ranks have roughly the same number of professors, this observation is not true. The number of associate professors is by a third less than the number of full professors. This is a direct consequence of industrial raiding for fresh PhD graduates that we discussed earlier and was taking place before 2008. In most instances it takes 5 to 8 years to obtain tenure, thus, this sample of associate professors falls in the correct timeframe. Of course, this ratio is not uniform across all programs. Figure 12 provides a complete picture of faculty rank distribution in selected 13 universities.

As Figure 11 illustrates, full professors comprise the bulk of the faculty body. These are for the most part people who got their degrees around the first crude oil bust in the 1980s in North America. Demographic (specifically, national) distribution of petroleum engineering faculty is another interesting issue. Figure 13 shows the entire national distribution of faculty based on the country where they obtained their bachelor degree (it is usually the country of origin). In this pie chart most professors are from North America and most of them are also full professors. The second and third largest national groups are from Asia and Middle East. Again,

Figure 11. Faculty distribution by rank

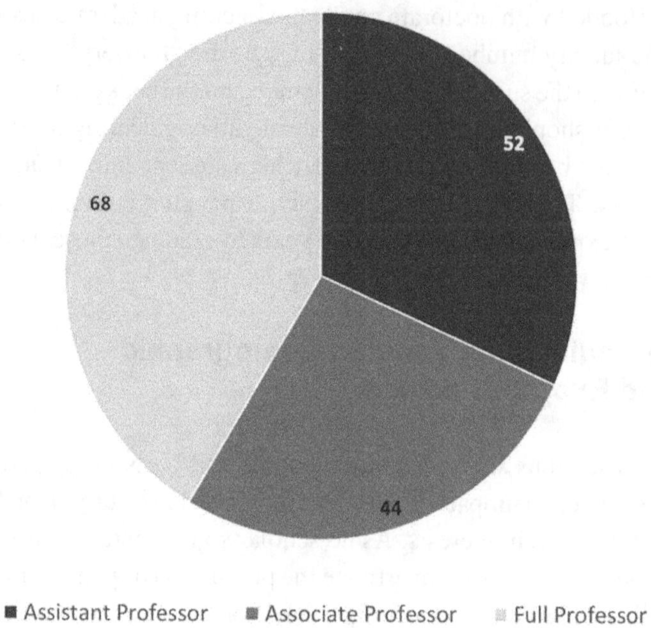

Figure 12. Faculty distribution by rank and university

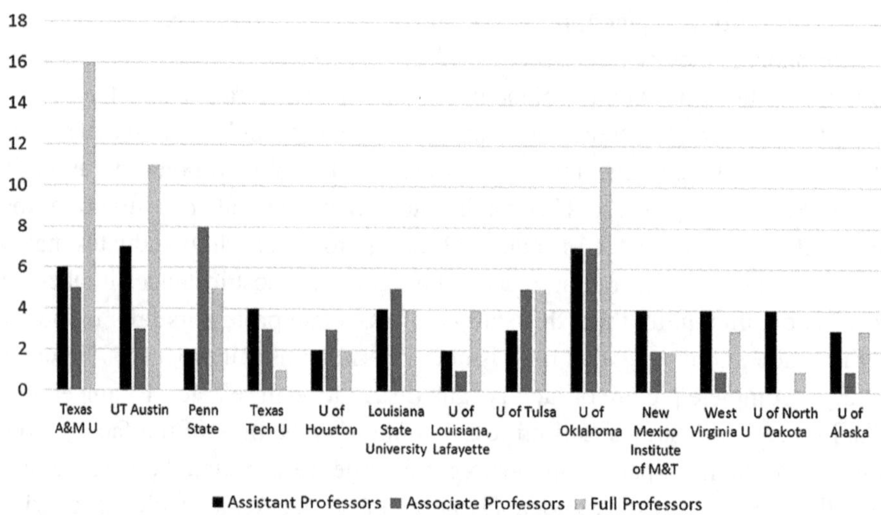

strong presence of Middle Eastern faculty is not surprising as we discussed in the sections above. Many of junior faculty members from Iran as PhD graduates were not eligible for industrial positions due to sanction and decided to join the academia as assistant professors.

Another important demographic factor is gender. Historically, very few women were attracted to petroleum engineering because it was usually associated with drilling and field work. However, in recent years the situation changed and more women started joining graduate programs and procced toward academic careers. Figure 14 shows that today women comprise about 13% of the faculty body which is a great achievement for one of the most male-dominated engineering disciplines. Currently, most petroleum engineering programs try to bring their men-to-women ratios to 4:1.

This section concludes with a discussion about faculty research expertise, preferences, and gaps. In the last decades, the US petroleum industry became particularly interested in development of unconventional oil and gas resources and revitalization of maturing and mature fields. These resources require research in geomechanics (for modeling and performing hydraulic fracturing jobs), EOR methods (to mobilize oil in mature fields and tight formations), and petroleum reservoir engineering and simulation (to help model reservoirs, calculate reserves, and opti-

Figure 13. Faculty distribution by country of origin

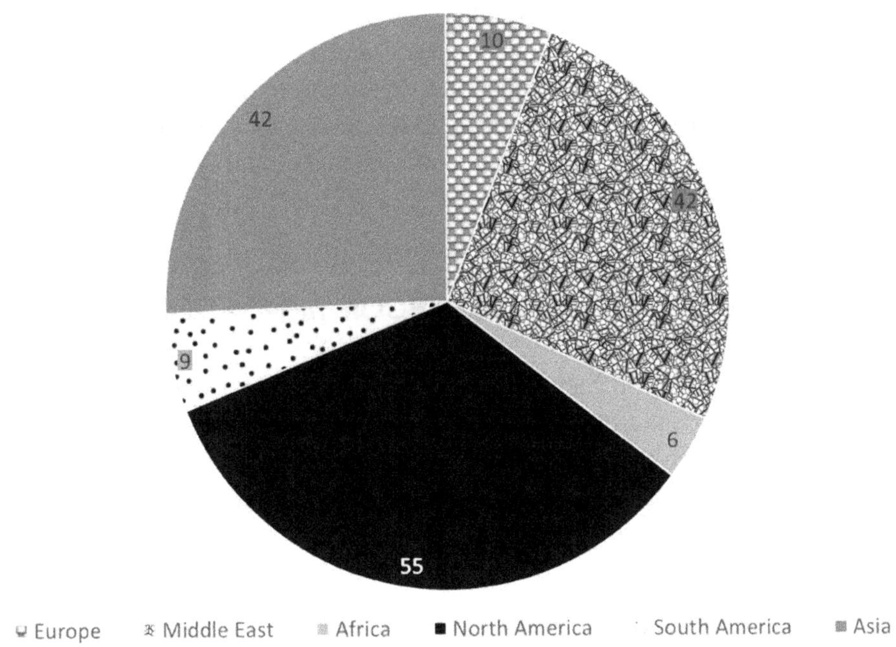

Figure 14. Faculty distribution by gender

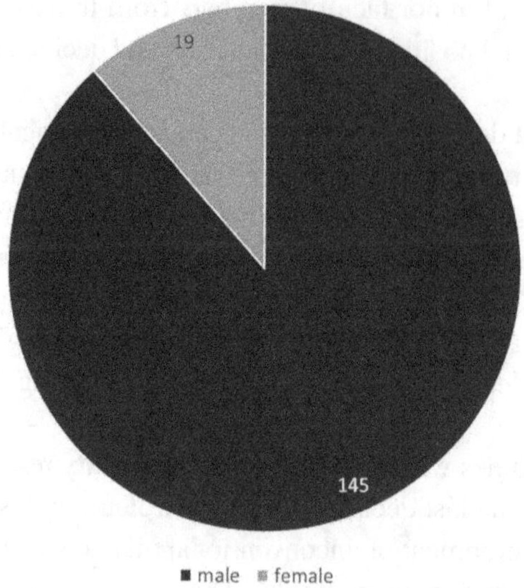

mize production). As as result, large financial streams from major E&P companies started guiding university researchers in certain directions. Figure 15 presents a bar chart with expertise of junior faculty. These "guiding" directions are very obvious and completely coincide with the industrial interests of the last decade. Most assistant professors have virtually identical expertise because their PhD advisors paid them to conduct certain type of research. Once graduated, they simply continued along the known path even though the industry already "pulled back" all financial investments from these research directions. Again, strong industrial demand and interests had a negative impact on the universities as now most petroleum engineering programs are staffed with junior faculty who conduct similar a decade old research without any funding.

POSSIBLE RESOLUTIONS AND RECOMMENDATIONS

Now after retrospective consideration and analysis of the problems that rose inside the industry and academia system from poor communication and blind response to apparently high demand, it is time to forecast and try to model possible scenarios for the future relations between the petroleum E&P industry and the universities. As of today, it is possible to anticipate three scenarios.

Figure 15. Faculty distribution by research expertise

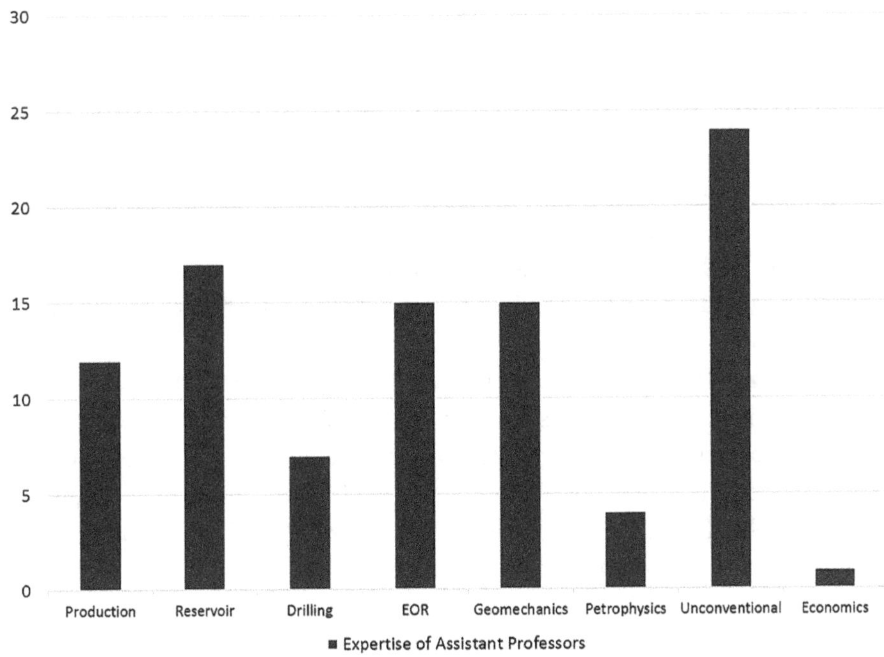

- **Big Bounce:** This scenario might occur if the situation develops along the same route as in 2008. 2008 was the end of a micro-cycle of crude oil price decline and rebounce. In this scenario, rapid expansion of the petroleum engineering programs was justified, because the demand for the graduates is about the increase. Though very encouraging, this scenario is the least likely one to happen, because of overall geo-political and economic situations are not the same as in 2008. In fact, it is still not clear whether domestic operators of unconventional oil and gas resources will recover from the crisis and the large international operating companies will come back to the US to develop these plays. Without these two factors domestic employment is not likely to go back to all times high.
- **Big Crunch:** This scenario is the most pessimistic and yet not the most unlikely. The situation might evolve similarly to that in the 1980s when crude oil prices first caused high demand for petroleum engineers and then suddenly declined leaving an enormous generation gap in the industry. Figure 2 shows this "roller-coaster" and how alike the situations today and 30 years ago. Indeed, the similarities are staggering. If this scenario continues, current expansion of the programs will inevitably lead to yet another generation gap. Very quickly (and it is already happening) students will understand that there

are no opportunities in petroleum engineering and start changing majors or joining other industries. The departments will seize their expansion and some programs might even close down like it happened before (e.g. University of Wyoming). Today the major operating oil and gas companies are only concerned about their own survival on the busting market. Although this attitude is understandable, their withdrawal from the universities creates dissatisfaction and frustration in those who committed to this major and profession. Many university researchers collaborated closely with the industry, hired post-docs and attracted talented PhD students to work on industry projects. Sudden withdrawal of the industry leaves faculty members without funded projects and students without future opportunities.

- **Big Freeze:** The situation might happen as an alternative to the optimistic and pessimistic scenarios discussed above. The industry might recover partially and come back to the universities, but the programs will regulate themselves more strictly and will not allow uncontrolled growth of student enrollment. The supply of young professionals will stabilize around or slightly below the most conservative demand of the industry.

This unfortunate ride on the enrollment "roller-coaster" might be a good lesson for petroleum engineering programs to develop and strictly adhere to their internal policies to avoid over- and under-supplies of graduates in future. It is also valuable for faculty members and their research agendas. Although it might be easier to succumb to large financial investments and engage in research "on demand," the professors need to learn to generate funding internally by generating ideas inside their institutions and then commercializing them. This way will ensure steady flux of funding for other research projects and talented people and make them independent of sudden swings in industrial interests.

FUTURE RESEARCH DIRECTIONS

The problem discussed in this chapter has many more dimensions and ramifications. The dataset of Heize only included enrollment observations from US petroleum engineering programs. Although the US is a locomotive of the global oil and gas industry, the worldwide situation in petroleum engineering education is still pretty much unknown. Larger dataset might reveal even larger and completely hidden internal processes that could be taking place on a global scale.

CONCLUSION

The student enrollment dataset carefully compiled by Dr. Heize over decade provided an opportunity to investigate the historical trends influencing interaction between the petroleum industry and academia. Application of statistical tools allowed to test the hypothesis about sensitivity of student enrollment to some key economic parameters (rig count, WTI spot price, and gasoline price). The analysis suggests that indeed higher demand for petroleum products creates higher demand for young petroleum professionals. However, the academia cannot expand its capacities fast enough to satisfy this volatile demand. Some programs saw instantaneous benefit in this unexpectedly high demand and on the verge of losing ABET accreditation attracted as many as possible students without any guarantee for their future employment. Some programs behaved more conservatively, but still it was difficult not to engage in this frenzy and stay reasonable.

In addition to in-depth investigation of interaction between the industry and academia, the chapter also explored possible scenarios for future development. In any of these scenario, however, similarly to the brewery the petroleum engineering programs should focus on their long-term survival and success. Rapid changes due to sporadic changes in industrial demand have already proved to be detrimental to the health of academia. Loss of interest from the industry already causes termination of several programs in the 1980s. And usually such termination does not come alone because it leads to severe problems in associated disciplines such as environmental engineering and geosciences that proliferate next to petroleum engineering departments.

ACKNOWLEDGMENT

The author of the paper would like to thank Dr. Lloyd Heinze for compiling the student enrollment dataset and providing the spreadsheet for our statistical analysis.

REFERENCES

Akaike, H. (1974). A New Look at the Statistical Model Identification. *IEEE Transactions on Automatic Control, 19*(6), 716–723. doi:10.1109/TAC.1974.1100705

Burnham, K., & Anderson, D. (2004). Multimodel Inference: Understanding AIC and BIC in Model Selection. *Sociological Methods & Research, 33*(2), 261–304. doi:10.1177/0049124104268644

EIA. (2016). *Data.* Retrieved July 30, 2016, from http://www.eia.gov/dnav/pet/pet_crd_drill_s1_m.htm

Hill, A., & Holditch, S. (2013). Déjà vu All Over Again? *Journal of Petroleum Technology, 6*(06), 18–21. doi:10.2118/0613-0018-JPT

Iledare, O. (2007). *An Appraisal of the Global Petroleum Upstream Industry and Its Response to Changes in Crude Oil Prices.* Paper presented at Nigeria Annual International Conference and Exhibition, Abuja, Nigeria. SPE-111901-MS. Available from http://dx.doi.org.lib-ezproxy.tamu.edu:2048/10.2118/111901-MS

LSU. (2016). *Program Statistics.* Retrieved July 30, 2016, from https://www.pete.lsu.edu/about/statistics

R Core Team. (2013). *R: A Language and Environment for Statistical Computing.* Vienna, Austria: R Foundation for Statistical Computing. Available from http://www.R-project.org/

Senge, P. (2006). *The Fifth Discipline: The Art & Practice of the Learning Organization.* New York: Deckle Edge.

UND. (2016). *About the Program.* Retrieved July 30, 2016, from https://engineering.und.edu/petroleum/about.cfm

US Bureau of Labor Statistics. (2016). *Employment, Hours, and Earnings from the Current Employment Statistics survey (National).* Retrieved July 30, 2016, from http://data.bls.gov/timeseries/CES1021100001?data_tool=XGtable

VandenBos, G., Knapp, S., & Doe, J. (2001). *Role of reference elements in the selection of resources by psychology undergraduates.* Retrieved October 13, 2001, from http://jbr.org/articles.html

APPENDIX 1

Stepwise regression analysis and ANOVA test results reported by R (R, 2013).

Table 2. Selection of the final model for undergraduate enrollment

Stepwise Model Path
Analysis of Deviance Table
Initial Model:
und ~ wtispot + gulfgasoline + rig
Final Model:
und ~ gulfgasoline

Table 3. Model testing for undergraduate enrollment response

Call:
lm(formula = bsconf ~ gulfgasoline, data = dt)
Residuals:
Min 1Q Median 3Q Max
-350.36 -103.03 -45.01 55.84 769.52
Coefficients:
Estimate Std. Error t value Pr(>|t|)
(Intercept) 157.90 85.77 1.841 0.0771 .
gulfgasoline 295.62 58.69 5.037 3.05e-05 ***

Signif. codes: 0 '***' 0.001 '**' 0.01 '*' 0.05 '.' 0.1 ' ' 1
Residual standard error: 256 on 26 degrees of freedom
Multiple R-squared: 0.4939, Adjusted R-squared: 0.4744
F-statistic: 25.37 on 1 and 26 DF, p-value: 3.05e-05

Table 4. Selection of the final model for master student enrollment

Stepwise Model Path
Analysis of Deviance Table
Initial Model:
ms ~ wtispot + gulfgasoline + rig + und
Final Model:
ms ~ wtispot + gulfgasoline + und

Table 5. Model testing for master student enrollment response

<table>
<tr><td>Call:</td></tr>
<tr><td>lm(formula = ms ~ wtispot + gulfgasoline + und, data = dt)</td></tr>
<tr><td></td></tr>
<tr><td>Residuals:</td></tr>
<tr><td>Min 1Q Median 3Q Max</td></tr>
<tr><td>-133.63 -51.73 -1.80 67.37 122.06</td></tr>
<tr><td></td></tr>
<tr><td>Coefficients:</td></tr>
<tr><td>Estimate Std. Error t value Pr(>|t|)</td></tr>
<tr><td>(Intercept) 279.55376 26.82680 10.421 2.18e-10 ***</td></tr>
<tr><td>wtispot -10.79646 5.08224 -2.124 0.0441 *</td></tr>
<tr><td>gulfgasoline 389.40333 185.04240 2.104 0.0460 *</td></tr>
<tr><td>und 0.11114 0.01338 8.305 1.62e-08 ***</td></tr>
<tr><td>---</td></tr>
<tr><td>Signif. codes: 0 ‘***’ 0.001 ‘**’ 0.01 ‘*’ 0.05 ‘.’ 0.1 ‘ ’ 1</td></tr>
<tr><td></td></tr>
<tr><td>Residual standard error: 78.74 on 24 degrees of freedom</td></tr>
<tr><td>Multiple R-squared: 0.9028, Adjusted R-squared: 0.8906</td></tr>
<tr><td>F-statistic: 74.27 on 3 and 24 DF, p-value: 2.747e-12</td></tr>
</table>

Table 6. Selection of the final model for PhD student enrollment

Stepwise Model Path				
Analysis of Deviance Table				
Initial Model:				
phd ~ wtispot + gulfgasoline + rig + und + ms				
Final Model:				
phd ~ wtispot + gulfgasoline + rig + und				

Table 7. Model testing for PhD student enrollment response

Call:						
lm(formula = phd ~ wtispot + gulfgasoline + rig + und, data = dt)						
Residuals:						
Min 1Q Median 3Q Max						
-42.098 -25.667 6.019 22.416 45.625						
Coefficients:						
Estimate Std. Error t value Pr(>|t|)						
(Intercept) 241.691468 30.892469 7.824 6.27e-08 ***						
wtispot -5.547900 1.958837 -2.832 0.009443 **						
gulfgasoline 321.374674 76.623945 4.194 0.000347 ***						
rig -0.182174 0.050536 -3.605 0.001492 **						
und 0.041970 0.005316 7.896 5.37e-08 ***						

Signif. codes: 0 '***' 0.001 '**' 0.01 '*' 0.05 '.' 0.1 ' ' 1						
Residual standard error: 30.35 on 23 degrees of freedom						
Multiple R-squared: 0.9414, Adjusted R-squared: 0.9312						
F-statistic: 92.41 on 4 and 23 DF, p-value: 7.977e-14						

Table 8. Model testing for faculty count response

<table>
<tr><td colspan="6">Call:</td></tr>
<tr><td colspan="6">lm(formula = faculty ~ wtispot + und, data = dt)</td></tr>
<tr><td></td><td></td><td></td><td></td><td></td><td></td></tr>
<tr><td colspan="6">Residuals:</td></tr>
<tr><td colspan="6">Min 1Q Median 3Q Max</td></tr>
<tr><td colspan="6">-66.730 -31.329 2.359 31.749 58.035</td></tr>
<tr><td></td><td></td><td></td><td></td><td></td><td></td></tr>
<tr><td colspan="6">Coefficients:</td></tr>
<tr><td colspan="6">Estimate Std. Error t value Pr(>|t|)</td></tr>
<tr><td colspan="6">(Intercept) 158.89996 12.94581 12.274 4.42e-12 ***</td></tr>
<tr><td colspan="6">wtispot -0.66785 0.43611 -1.531 0.1382</td></tr>
<tr><td colspan="6">und 0.01650 0.00651 2.535 0.0179 *</td></tr>
<tr><td colspan="6">---</td></tr>
<tr><td colspan="6">Signif. codes: 0 ‘***’ 0.001 ‘**’ 0.01 ‘*’ 0.05 ‘.’ 0.1 ‘ ’ 1</td></tr>
<tr><td></td><td></td><td></td><td></td><td></td><td></td></tr>
<tr><td colspan="6">Residual standard error: 38.33 on 25 degrees of freedom</td></tr>
<tr><td colspan="6">Multiple R-squared: 0.2321, Adjusted R-squared: 0.1707</td></tr>
<tr><td colspan="6">F-statistic: 3.779 on 2 and 25 DF, p-value: 0.03683</td></tr>
</table>

Compilation of References

Acar, S., & Lopik, D. (2009). Computational pregnant occupant model, Expecting, for crash simulations. *Proceedings of the Institution of Mechanical Engineers. Part D, Journal of Automobile Engineering*, *223*(7), 891–902. doi:10.1243/09544070JAUTO1072

Acker, F. (2005, December). Autism and engineers: Is there a connection? *Ingenia, 25*, 39-42. Retrieved on 25. 8.2016 from http://www.ingenia.org.uk/Ingenia/Articles/343

Acquaye, L. A., & Edmonds Crewe, S. (2012). International Programs: Advancing Human Rights and Social Justice for African American Students. *Journal of Social Work Education*, *48*(4), 763–784. doi:10.5175/JSWE.2012.201100130

Adamson, F., Astrand, B., & Darling-Hammond, L. (Eds.). (2016). *Global education reform: How privatization and public investment influence education outcomes*. New York, NY: Routledge.

Adams, S. G. (2003). Building successful student teams in the engineering classroom. *Journal of STEM Education: Innovations and Research*, *4*(3/4), 1–6.

Adigwe, J. C. (1997). Ethnicity, test anxiety and science achievement in Nigerian students. *International Journal of Science Education*, *19*(7), 773–780. doi:10.1080/0950069970190703

Akaike, H. (1974). A New Look at the Statistical Model Identification. *IEEE Transactions on Automatic Control*, *19*(6), 716–723. doi:10.1109/TAC.1974.1100705

Alliance. (2016). *Washington Accord Signatories*. Retrieved October 15, 2015, form http://www.ieagreements.org/Washington-Accord/signatories.cfm

Altbach, P. G. (2016). *Global Perspectives on Higher Education*. Baltimore, MD: Johns Hopkins University Press.

American Institutes for Research. (2012). *Broadening participation in STEM a call to action*. Washington, DC: American Institutes for Research.

American Society of Engineering Education (ASEE). (2015). *ASEE: Action on Diversity*. Retrieved May 12, 2015 from http://diversity.asee.org/

Andersen, R. D. (2004). Germany and the Humboldtian model. In R. D. Andersen (Ed.), *European Universities from the Enlightenment to 1914*. Oxford, UK: Oxford University Press. doi:10.1093/acprof:oso/9780198206606.003.0004

Andresen, S., Koreuber, M., & Lüdke, D. (Eds.). (2009). *Gender und Diversity: Albtraum oder Traumpaar? Interdisziplinärer Dialog zur "Modernisierung" von Geschlechter- und Gleichstellungspolitik*. Wiesbaden, Germany: VS Verlag für Sozialwissenschaften. doi:10.1007/978-3-531-91387-2

Antonio, A. L., Chang, M. J., Hakuta, K., Kenny, D. A., Levin, S., & Milem, J. F. (2004). Effects of racial diversity on complex thinking in college students. *Psychological Science*, *15*(8), 507–510. doi:10.1111/j.0956-7976.2004.00710.x PMID:15270993

Apple, M. W. (2004). *Ideology and Curriculum*. New York, NY: Routledge Falmer.

Arbeit, C. A., Einaudi, P., & Kang, K. H. (2016, March). Influx of foreign graduate students and inclusion of newly eligible institutions lead to a significant increase in U.S. graduate enrollment in science and engineering. *InfoBrief*. National Center for Science and Engineering Statistics, National Science Foundation (NSF 16-310). Retrieved from www.nsf.gov/statistics/2016/nsf16310/nsf16310.pdf

ASEE. (2012). *Going the distance: best practices and strategies for retaining engineering, engineering technology and computing students*. Retrieved October 15, 2015, from https://www.asee.org/retention-project

Athreya, K., & Eberly, J. C. (2015). *The College Premium, College Noncompletion, and Human Capital Investment*. Federal Reserve Bank of Richmond Working Paper (13-02R).

Auburn University. (n.d.a). *Harold A. Franklin Society: Guidance Through Unity*. Retrieved May 20, 2016, from http://www.auburn.edu/student_info/hafs/mission.html

Auburn University. (n.d.b). *Alabama Power Academic Excellence Program*. Retrieved May 20, 2016, from http://www.eng.auburn.edu/admin/aep/

AUinvolve. (n.d.a). *Black Student Union*. Retrieved May 20, 2016, from https://auburn.collegiatelink.net/organization/bsu/about

AUinvolve. (n.d.b). *International Student Organization*. Retrieved May 20, 2016, from https://auburn.collegiatelink.net/organization/iso

Ayre, M., Mills, J. E., & Gill, J. (2014a). A Gender-Inclusive Engineering Curriculum. In D. Beanland & R. Hadgraft (Eds.), Engineering Education: Transformation and Innovation (pp. 87-89). Melbourne, Australia: RMIT University by arrangement with UNESCO.

Ayre, M. E., Mills, J. E., & Gill, J. (2011). Two steps forward, one step back: Women engineers in Australia. *International Journal of Gender. Science and Technology*, *3*(2), 293–312.

Ayre, M. E., Mills, J. E., & Gill, J. (2013). 'Yes, I do belong': The women who stay in engineering. *Engineering Studies*, *5*(3), 216–232.

Ayre, M. E., Mills, J. E., & Gill, J. (2014b). Family issues for women engineers. In D. Bilimoria & L. Lord (Eds.), *Women in STEM careers: International perspectives on increasing workforce participation, advancement and leadership* (pp. 79–100). Cheltenham, UK: Elgar.

Baker, D. (2014). *The schooled society: The educational transformation of global culture*. Stanford University Press.

Bakshi-Hamm, P. (2007). Gleichbehandlung, Migration und Statistik. *Migrationund Soziale Arbeit, 3/4*.

Bakshi-Hamm, P., & Lind, I. (2008). Migrationshintergrund und Chancen an Hochschulen: Gesetzliche Grundlagen und aktuelle Statistiken. In *Wissenschaftlerinnen mit Migrationshintergrund* (pp. 11-24). Bonn, Germany: Center for Excellence Women and Science (CEWS). Retrieved July 20, 2016, from http://www.gesis.org/fileadmin/cews/www/download/cews-publik12.pdf

Baldiga, K. (2013). Gender differences in willingness to guess. *Management Science, 60*(2), 434–448. doi:10.1287/mnsc.2013.1776

Bandura, A. (1986). *Social foundations of thought and action*. Englewood Cliffs, NJ: Prentice-Hall.

Bandura, A., & Schunk, D. H. (1981). Cultivating competence, self-efficacy, and intrinsic interest through proximal self-motivation. *Journal of Personality and Social Psychology, 41*(3), 586–598. doi:10.1037/0022-3514.41.3.586

Barefoot, B., Fidler, P., & South Carolina Univ, C. (1992, January 1). *National Survey of Freshman Seminar Programming, 1991. Helping First Year College Students Climb the Academic Ladder. The Freshman Year Experience: Monograph Series Number 10*. (ERIC Document Reproduction Service No. ED354842). Retrieved July 26, 2007, from ERIC database.

Barefoot, B. (2004). Higher educations revolving door: Confronting the problem of student dropout in US colleges and universities. *Open Learning, 9*(1), 9–18. doi:10.1080/0268051042000177818

Barnard, S., Hassan, T., Dainty, A., & Bagilhole, B. (2013). Interdisciplinary content, contestations of knowledge and informational transparency in engineering curriculum. *Teaching in Higher Education, 18*(7), 748–760. doi:10.1080/13562517.2013.836089

Baron-Cohen, S. (2007). *Asperger's: The engineer's disease*. Retrieved August 22 2016, from http://wrongplanet.net/forums/viewtopic.php?t=45209

Barry Goldwater Scholarship and Excellence in Education Foundation. (n.d.). *The Barry Goldwater Scholarship and Excellence in Education Foundation*. Retrieved September 2, 2016, from https://goldwater.scholarsapply.org/

Bean, J. P., & Metzner, B. S. (1985). A conceptual model of nontraditional undergraduate student attrition. *Review of Educational Research, 55*(4), 485–540. doi:10.3102/00346543055004485

Beddoes, K. (2015). Professors' Perceptions of How Men and Women Students Experience Engineering Education Differently…Or Not. *Proceedings of European Society for Engineering Education (SEFI)*. Available from http://www.sociologyofengineering.org/wp-content/uploads/2016/01/Beddoes-SEFI-2015.pdf

Belbin, R. M. (1981). *Management Teams: Why they succeed or fail*. New York: Butterworth-Heinemann.

Belbin, R. M. (2010). *Team Roles at Work* (2nd ed.). New York: Butterworth-Heinemann.

Bell, A. P. (1970). Role modelship and interaction in adolescence and young adulthood. *Developmental Psychology*, *2*(1), 123–128. doi:10.1037/h0028613

Benyus, J. (2002). *Biomimicry: Innovation inspired by nature*. New York, NY: Harper Perennial.

Berger, J. B., & Braxton, J. M. (1998). Revising Tintos interactionalist theory of student departure through theory elaboration: Examining the role of organizational attributes in the persistence process. *Research in Higher Education*, *39*(2), 103–119. doi:10.1023/A:1018760513769

Bhandari, R., Belyavina, R., & Gutierrez, R. (Eds.). (2011). Student mobility and the internationalization of higher education: National policies and strategies from six world regions. New York: Institute of International Education (IIE).

Bildungsberichterstattung, A. (Ed.). (2016). *Bildung in Deutschland 2016. Ein indikatorengestützter Bericht mit einer Analyse zu Bildung und Migration*. Bielefeld, Germany: W. Bertelsmann Verlag GmbH & Co. KG. Retrieved July 25, 2016, from http://www.bildungsbericht.de/de/bildungsberichte-seit-2006/bildungsbericht-2016/pdf-bildungsbericht-2016/bildungsbericht-2016

Billington, J., Baron-Cohen, S., & Wheelwright, S. (2007). Cognitive style predicts entry into physical sciences and humanities: Questionnaire and performance tests of empathy and systemizing. *Learning and Individual Differences*, *17*(3), 260–268. doi:10.1016/j.lindif.2007.02.004

Binder, J. F., Baguley, T., Crook, C., & Miller, F. (2015). The academic value of internships: Benefits across disciplines and student backgrounds. *Contemporary Educational Psychology*, *41*, 73–82. doi:10.1016/j.cedpsych.2014.12.001

Blackley, S., & Howell, J. (2015). A STEM Narrative: 15 Years in the Making. *Australian Journal of Teacher Education*, *40*(40), 102–112. doi:10.14221/ajte.2015v40n7.8

Blaschke, J. (2015). *Texas State joins with ASEE in effort to foster diversity in engineering*. San Marcos, TX: Texas State University News Service, Office of Media Relations.

Blickenstaff, J. C. (2005). Women and science careers: Leaky pipeline or gender filter? *Gender and Education*, *17*(4), 369–386. doi:10.1080/09540250500145072

Bloom, B. S., & Krathwohl, D. R. (1956). *Taxonomy of educational objectives: The classification of educational goals. Handbook I: Cognitive domain*. New York: David McKay.

Bonaccio, S., Reeve, C. L., & Winford, E. C. (2012). Text anxiety on cognitive ability test can result in differential predictive validity of academic performance. *Personality and Individual Differences*, *52*(4), 497–502. doi:10.1016/j.paid.2011.11.015

Boone, C., van Olffen, W., & Roijakkers, N. (2004). Selection on the road to a career: Evidence of personality sorting in educational choice. *Journal of Career Development*, *31*(1), 61–78. doi:10.1023/B:JOCD.0000036706.17677.ee

Bouffier, A., Kehr, P., Lämmerhirt, M., & Leicht-Scholten, C. (2014). Spätes Erwachen an deutschen Hochschulen: Die "Entdeckung" der Lehre und die Berücksichtigung von Gender und Diversity. In C. Tomberger (Ed.), Gender- und Diversity-Kompetenzen in Hochschullehre und Beratung: Institutionelle, konzeptionelle und praktische Perspektiven (pp. 53-66). Hildesheim, Germany: Universitätsverlag Hildesheim.

Boulton, G., & Lucas, C. (2008). *What are universities for?* League of European Research Universities. Retrieved July 25, 2016, from http://www.leru.org/files/general/%E2%80%A2What%20 are%20universities%20for%20 (September%202008).pdf

Bourdieu, P. (1977). *Outline of a Theory of Practice* (R. Nice, Trans.). Cambridge, UK: Cambridge University Press. doi:10.1017/CBO9780511812507

Bowen, W. G., & McPherson, M. S. (2016). *Lesson Plan. An agenda for change in American higher education*. Princeton, NJ: Princeton University Press. doi:10.1515/9781400881369

Bowman, M. H., & Stage, F. K. (2002). Personalizing the goals of undergraduate research. *Journal of College Science Teaching*, *32*, 120.

Brainard, S. G., & Carlin, L. (1997, November). *A longitudinal study of undergraduate women in engineering and science*. Paper presented at the Frontiers in Education Conference, Pittsburgh, PA. doi:10.1109/FIE.1997.644826

Brainard, S. G., & Carlin, L. (1998). A Six-Year Longitudinal Study of Undergraduate Women in Engineering and Science. *The Journal of Engineering Education*, *87*(4), 369–375. doi:10.1002/j.2168-9830.1998.tb00367.x

Brookover, W. B., Schweitzer, J. H., Schneider, J. M., Beady, C. H., Flood, P. K., & Wisenbaker, J. M. (1978). Elementary school social climate and school achievement. *American Educational Research Journal*, *15*(2), 301–318. doi:10.3102/00028312015002301

Browne, B., Kaldenberg, D., Brown, W., & Brown, D. (1998). Student as customer: Factors affecting satisfaction and assessment of institutional quality. *Journal of Marketing for Higher Education*, *8*(3), 1–14. doi:10.1300/J050v08n03_01

Brown, T. (2008, June). Design Thinking. *Harvard Business Review*, 84–92.

Brown, T. (2009). *Change by design: How design thinking transforms organizations and inspires innovation*. New York, NY: Harper Business.

Burnham, K., & Anderson, D. (2004). Multimodel Inference: Understanding AIC and BIC in Model Selection. *Sociological Methods & Research*, *33*(2), 261–304. doi:10.1177/0049124104268644

Byars-Winston, A. (2014). Toward a framework for multicultural STEM-focused career interventions. *The Career Development Quarterly*, *62*(4), 340–357. doi:10.1002/j.2161-0045.2014.00087.x PMID:25750480

Byrne, E. M. (1994). *Critical Filters, Hidden Helix: Policies for Advancing Women in Science and Technology*. Brisbane, Australia: University of Queensland.

Cabrera, A. F., Nora, A., & Castaneda, M. B. (1993). College persistence: Structural equations modeling test of an integrated model of student retention. *The Journal of Higher Education*, *64*(2), 123–139. doi:10.2307/2960026

Candy, P. C. (1991). *Self-Direction for Lifelong Learning. A Comprehensive Guide to Theory and Practice*. San Francisco, CA: Jossey-Bass.

Caspar-Terizakis, C., Bieker, S., Cornel, P., Heinelt, H., Klonecki, K., Wagner, M., . . . Winner, H. (2016). From historical roots to modern-day results: Technische Universität Darmstadt and Tongji University's bi-lateral strategic partnership. In C. Banks, B. Siebe-Herbig, & K. Norton (Eds.), Global perspectives on strategic international partnerships: A guide to building sustainable academic linkages (pp. 25-34). New York, NY: Institute of International Education (IIE).

CESAER. (2016). *Statement on the Modernisation of Higher Education*. Conference of European schools for advanced engineering education and research. Retrieved July 25, 2016, http://www.cesaer.org/content/assets/docs/Docs2016/20160308_CESAER_Statement_on_the_Modernisation_of_Higher_Education.pdf

Chandrasekaren, S., Littlefair, G., & Joordens, M. (2012). Learning through projects in engineering education. *Proceedings of the SEFI 40th Annual Conference 2012* (pp. 1-8). Brussels, Belgium: European Society for Engineering Education (SEFI). Retrieved from http://dro.deakin.edu.au/eserv/DU:30048241/chandrasekaran-learningthrough-2012.pdf

Chieffo, L., & Griffiths, L. (2004). Large-Scale Assessment of Student Attitudes after a Short-Term Study Abroad Program. *Frontiers: The Interdisciplinary Journal of Study Abroad*, *10*, 165–177.

Chubin, D. E., May, G. S., & Babco, E. L. (2005). Diversifying the engineering workforce. *The Journal of Engineering Education*, *94*(1), 73–86. doi:10.1002/j.2168-9830.2005.tb00830.x

Chun, E., & Evans, A. (2016). Rethinking Cultural Competence in Higher Education: An Ecological Framework for Student Development. *ASHE Higher Education Report*, *42*(4), 7–162. doi:10.1002/aehe.20102

Coco, M. (2000). Internships: A try before you buy arrangement. *SAM Advanced Management Journal*, *65*, 41.

Cole, J. R. (2009). *The great American university*. New York: Public Affairs.

Concannon, J. P., & Barrow, L. H. (2009). A cross-sectional study of engineering students self-efficacy by gender, ethnicity, year, and transfer status. *Journal of Science Education and Technology, 18*(2), 163–172. doi:10.1007/s10956-008-9141-3

Corbett, C., & Hill, C. (2015). *Solving the equation: The variables for women's success in engineering and computing*. Washington, DC: American Association of University Women.

Cross, J. (2007). *Informal learning rediscovering the natural pathways that inspire innovation and performance* (1st ed.). San Francisco: Pfeiffer/Wiley.

Crow, M. (2016). Why we need more women engineers. *LinkedIn Newsletter*. Available from https://www.linkedin.com/pulse/why-we-need-more-women-engineers-michael-crow

Dahlstrom, E., Walker, J. D., & Dziuban, C. (2013). *ECAR study of undergraduate students and information technology*. Louisville, CO: EDUCASE Center for Analysis and Research.

De Vere, I. (2009). *Developing creative engineers: A design approach to engineering education*. Paper presented at 11th International Conference on Engineering and Product Design Education, Design Society, Brighton, UK. Retrieved September 12, 2016, from https://www.researchgate.net/publication/274718463_Developing_creative_engineers_a_design_approach_to_engineering_education

Dee Fink, L. (2013). *Creating significant learning experiences: An integrated approach to designing college courses*. San Francisco, CA: Jossey Bass.

Demirkan, H., & Spohrer, J. (2015). T-shaped innovators. *Research Technology Management, 58*(5), 12–15. doi:10.5437/08956308X5805007

Denson, C. D., Stallworth, C. A., Hailey, C., & Householder, D. L. (2015). Benefits of informal learning environments: A focused examination of STEM-based program environments. *Journal of STEM Education: Innovations & Research, 16*, 11–15.

Dewart, B., & Rowan, L. (2007). Improving tertiary student outcomes in the first year of study: What your students may not tell you. *Journal of College Student Retention, 9*(3), 259–271. doi:10.2190/CS.9.3.a

DeWitz, S., & Walsh, W. (2002). Self-efficacy and college student satisfaction. *Journal of Career Assessment, 10*(3), 315–326. doi:10.1177/10672702010003003

Diamond, J. B., Randolph, A., & Spillane, J. P. (2004). Teachers Expectations and sense of responsibility for student learning: The importance of race, class, and organizational habitus. *Anthropology & Education Quarterly, 35*(1), 75–89. doi:10.1525/aeq.2004.35.1.75

Doidge, N. (2010). *The brain that changes itself.* Melbourne: Scribe.

Du Plessis, Y. D. (2011). Exploring teamwork paradoxes challenging 21st-century cross-cultural conflict management in a multicultural organizational context. *International Journal of Cross Cultural Management, 12*(1), 49–71. doi:10.1177/1470595811413108

Dwyer, M. M. (2004). More Is Better: The Impact of Study Abroad Program Duration. *Frontiers: The Interdisciplinary Journal of Study Abroad, 10*, 151–163.

Ehlers-Zavala, F. P., & Maciejewski, A. A. (2016). The case of an Electrical and Computer Engineering (ECE) Department in the internationalization process of a research I public institution. *Proceedings of the International Forum of the American Society for Engineering Education.*

EIA. (2016). *Data*. Retrieved July 30, 2016, from http://www.eia.gov/dnav/pet/pet_crd_drill_s1_m.htm

Elliott, K. (2003). Key determinants of student satisfaction. *Journal of College Student Retention, 4*(3), 271–279. doi:10.2190/B2V7-R91M-6WXR-KCCR

Engineer Your life. (2016). Retrieved July 12, 2016, from http://www.engineeryourlife.org/

Erdem, E. (2007). Study of the relationship between test anxiety and the epistemological and problem-solving beliefs of students on a general chemistry course. *World Applied Sciences Journal, 2*, 750–758.

Evans, A., Evans, V., & Evans, A. (2002, Fall). Historically Black colleges and universities (HBCUS). *Education, 123*, 3–17.

Evetts, J. (1996). *Gender and career in science and engineering (Feminist perspectives on the past and present)*. London: CRC Press.

Eyerman, R. (2001). *Cultural trauma: Slavery and the formation of African American identity*. Cambridge Press. doi:10.1017/CBO9780511488788

Eynon, B., Gambino, L. M., & Török, J. (2014). What difference can ePortfolio make? A field report from the connect to learning project. *International Journal of ePortfolio, 4*, 95-114.

Fairweather, H. (1976). Sex differences in cognition. *Cognition, 4*(3), 231–280. doi:10.1016/0010-0277(76)90019-6

Farrugia, C. A., & Bhandari, R. (2015). *Open Doors Report on International Educational Exchange*. New York: Institute of International Education.

Faulkner, W. (2007). Nuts and Bolts and People: Gender-troubled engineering identities. *Social Studies of Science, 3*(3), 331–356. doi:10.1177/0306312706072175

Feisel, L., & Rosa, A. (2005). The role of the laboratory in undergraduate engineering education. *The Journal of Engineering Education, 94*(1), 121–130. doi:10.1002/j.2168-9830.2005.tb00833.x

Felder, R. M., & Brent, R. (2005). Understanding student differences. *The Journal of Engineering Education, 94*(1), 57–72. doi:10.1002/j.2168-9830.2005.tb00829.x

Felder, R. M., & Silverman, L. K. (1988). Learning and teaching styles in engineering education. *Engr. Education, 78*, 674–681.

Felder, R. M., & Silverman, L. K. (1988). Learning and teaching styles in engineering education. *English Education*, *78*, 674–681.

Finson, K. (2002). Drawing a scientist: What we do and do not know after fifty years of drawings. *School Science and Mathematics*, *102*(7), 335–345. doi:10.1111/j.1949-8594.2002.tb18217.x

Focquaert, F., Steven, M., Walford, G., Colden, A., & Gazzaniga, M. (2007). Empathizing and systemizing cognitive traits in the sciences and humanities. *Personality and Individual Differences*, *43*(3), 619–625. doi:10.1016/j.paid.2007.01.004

Fouad, N. A., & Singh, R. (2011). *Stemming the tide: Why women leave engineering*. Milwaukee, WI: University of Wisconsin.

Franzke, B. (2010). *Berufswahl heute: Geht es auch weniger stereotyp?Genderaspekte in der Berufswahl und Berufsorientierung*. Retrieved July 20, 2016, from https://lehrerfortbildung-bw.de/allgschulen/alle/gender/m4/01_BF_Gender_in_Berufswahl_und_Berufsorientierung.pdf

Freeman, K. E., Sharon, T. A., & Duvon, G. W. (2008). Do Learning Communities Enhance the Quality of Students' Learning and Motivation in STEM? *The Journal of Negro Education*, *77*, 227–240.

Gaethgens, C., & Zervakis, P. A. (2007). *Implementing Bologna. Experience from German Higher Education Institutions*. HRK German Rectors' Conference Bologna Centre. RAABE Fachverlag für Wissenschaftsinformation. Retrieved July 25, 2016, from http://www.ehea.info/uploads/submitedfiles/11_2010/145651.pdf

Galloway, P. D. (2008). *The 21st-century engineer: A proposal for engineering education reform*. Reston, VA: ASCE Press.

Gamoran, A. (2001). American Schooling and educational inequality: A forecast for the 21st century. *Sociology of Education*, *74*, 135–153. doi:10.2307/2673258

Gandara, P. & Maxwell-Jolly, J. (1999, December). *Priming the pump: Strategies for increasing the achievement of underrepresented minority undergraduates*. Academic Press.

Georgia State University. (n.d.). *Pressure - Hyperphysics*. Retrieved May 21, 2016, from http://hyperphysics.phy-astr.gsu.edu/hbase/pbuoy.html

Gerbic, P., Lewis, L., & Amin, N. M. (2011). Student perspectives of eportfolios: Change over four semesters. In G. Williams, P. Statham, N. Brown, & B. Cleland (Eds.), *Changing Demands, Changing Directions* (pp. 423–436). Hobart, Australia: Ascilite.

Gibson, D. (2004). Role models in career development: New directions for theory and research. *Journal of Vocational Behavior*, *65*(1), 134–156. doi:10.1016/S0001-8791(03)00051-4

Gill, J. (2004). *Beyond the great divide: Single sex or coeducation?* Sydney, NSW: UNSW Press.

Gill, J., Mills, J. E., Franzway, S., & Sharp, R. (2008a). Oh you must be very clever! High achieving women, professional power and the ongoing negotiation of workplace identity. *Gender and Education*, *20*(3), 223–236. doi:10.1080/09540250801968990

Gill, J., Sharp, R., Mills, J. E., & Franzway, S. (2008b). I still wanna be an engineer! Women, education and the engineering profession. *European Journal of Engineering Education*, *33*(4), 391–402. doi:10.1080/03043790802253459

Girls' Day. (2016). *Griffith University: Women in Engineering*. Retrieved August 4, 2016, from https://www.griffith.edu.au/engineering-information-technology/griffith-school-engineering/women-in-engineering

Goldberg, D. E., & Somerville, M. (2014). A whole new engineer. Douglas, MI: threejoy.

Goodman, A., & Rüland, D. (2016). Foreword. In C. Banks, B. Siebe-Herbig, & K. Norton (Eds.), Global perspectives on strategic international partnerships: A guide to building sustainable academic linkages (pp. i-iii). New York: Institute of International Education (IIE).

Good, T. L. (1981). Teacher expectations and student perceptions: A decade of research. *Educational Leadership*, *38*, 415–422.

Gore, A. (2013). *The future*. London: WH Allen.

Gore, A. (2013). *The future: Six drivers of global change*. London: W. H. Allen.

Grasso, D., & Burkins, M. (2010). *Holistic Engineering Education: Beyond Technology*. New York: Springer Science & Business Media. doi:10.1007/978-1-4419-1393-7

Gray, M., & Lundy, C. (2016). *Implementing and Integrating International Research into the Engineering Curriculum at Lincoln University, Pennsylvania and the University of the West Indies, Trinidad*. Paper presented at the American Society of Engineering Education International Forum, New Orleans, LA.

Grayson, L. P. (1980). A brief history of engineering education in the United States. *IEEE Transactions on Aerospace and Electronic Systems*, *16*(3), 373–392. doi:10.1109/TAES.1980.308907

Grots, A., & Pratschke, M. (2009). Design Thinking - Kreativität als Methode. *Marketing Review St. Gallen*, *26*(2), 18–22. doi:10.1007/s11621-009-0027-4

Hamir, H. B. (2011). *Go Abroad and Graduate On-Time: Study Abroad Participation, Degree Completion, and Time-to-Degree* (Unpublished doctoral dissertation). University of Nebraska, Lincoln, NE.

Hampson, M. E. (1989). *Contributions of gonadal hormones to human cognitive and motor skills: Evidence from the menstrual cycle* (Unpublished doctoral dissertation). University of Western Ontario, Ontario, Canada.

Haraway, D. J. (1991). *Simians, Cyborgs, and Women: The Reinvention of Nature*. New York, NY: Routledge Taylor & Francis Group.

Harding, J. (1996). Science in a Masculine Strait-Jacket. In L. Parker, L. Rennie, & B. Fraser (Eds.), *Gender, Science and Mathematics* (Vol. 2, pp. 3–15). Springer Netherlands. doi:10.1007/978-94-011-0143-1_1

Harding, S. G. (1986). *The Science Question in Feminism*. Ithaca, NY: Cornell University Press.

Harper, S. R. (2009). Niggers no more: A critical race counternarrative on Black male student achievement at predominantly White colleges and universities. *International Journal of Qualitative Studies in Education, 22*(6), 697–712. doi:10.1080/09518390903333889

Hartman, H., & Hartman, M. (2007). Is teamwork a female-friendly pedagogy? In I. Welpe, B. Reschka, & J. Larkin (Eds.), *Gender and engineering: Strategies and possibilities* (pp. 73–91). Frankfurt: Peter Lang.

Hasso Plattner Institute of Design at Stanford. (2010). *An introduction to design thinking. Process Guide*. Retrieved July 18, 2016, from https://dschool.stanford.edu/sandbox/groups/designresources/wiki/36873/attachments/74b3d/ModeGuideBOOTCAMP2010L.pdf?sessionID=68deabe9f22d5b79bde83798d28a09327886ea4b

Haycock, K. (2001). Closing the achievement gap. *Educational Leadership, 58*, 6–11.

Henderson, T. (n.d.). *The Physics Classroom - What Is an Electric Circuit?* Retrieved May 21, 2016, from http://www.physicsclassroom.com/class/circuits/Lesson-2/What-is-an-Electric-Circuit

Heublein, U., Hutzsch, C., Schreiber, J., Sommer, D. & Besuch, G. (2010). *Ursachen des Studienabbruchs in Bachelor- und in herkömmlichen Studiengängen Ergebnisse einer bundesweiten Befragung von Exmatrikulierten des Studienjahres 2007/08*. Academic Press.

Hewlett, S. A., Marshall, M., & Sherbin, L. (2013). How Diversity Can Drive Innovation. *Harvard Business Review*. December. Retrieved July 28, 2016, from http://www.dzhw.eu/pdf/pub_fh/fh-201002.pdf

Hewlett, S. A., Luce, C., Servon, C. B., Sherbin, L., Shiller, P., Sosnovich, E., & Sumberg, K. (2008). *The Athena factor: Reversing the brain drain in science, engineering, and technology*. Boston, MA: Harvard Business Publishing.

Higgins, M. J. & Glanville, M. (1983). *The impact of Act 101 programs in Pennsylvania*. (ERIC Document Reproduction Service No. ED243354)

Hill, A., & Holditch, S. (2013). Déjà vu All Over Again? *Journal of Petroleum Technology, 6*(06), 18–21. doi:10.2118/0613-0018-JPT

Himmel, S., & Ziefle, M. (2016). Smart Home Medical Technologies: Users' Requirements for Conditional Acceptance. *i-com. Journal of Interactive Media, 15*, 39–50.

Hippler, H. (Ed.). (2011). Ingenieurpromotion – Stärken und Qualitätssicherung Beiträge eines gemeinsamen. Symposiums von acatech, TU9, ARGE TU/TH und 4ING.

History Channel. (n.d.). *Dust Bowl*. Retrieved May 21, 2016, from http://www.history.com/topics/dust-bowl

Hoegl, M., & Parboteeah, K. P. (2007). Creativity in innovative projects: How teamwork matters. *Journal of Engineering and Technology Management*, *24*(1-2), 148–166. doi:10.1016/j.jengtecman.2007.01.008

Holmes, M. (2015). Wanted: Diverse cybersecurity workforce. *Magazine of the Society of Women Engineers,61*(1), 24-29. Retrieved August 10, 2016, from http://www.nxtbook.com/nxtbooks/swe/winter15/#/30

Honors Program. (2015). *Fall 2015 course guide*. Retrieved from http://www.honors.ufl.edu/courses/CoursesFall2015.aspx

House, R. J., Hanges, P. J., Javidan, M., Dorfman, P. W., & Gupta, V. (2004). *Culture, Leadership, and Organizations: The GLOBE Study of 62 Societies*. London: SAGE Publications.

Hout, M. (2012). Social and economic returns to college education in the United States. *Annual Review of Sociology*, *38*(1), 379–400. doi:10.1146/annurev.soc.012809.102503

Howard University. (2002). *Annual report*. Retrieved February 4, 2006, from http://www.ed.gov/about/reports/annual/ 2002report/edlite-howard.html

Hyde, J. S., & Linn, M. C. (2006). Gender similarities in mathematics and science. *Science*, *314*(5799), 599–600. doi:10.1126/science.1132154 PMID:17068246

Hyde, J., Lindberg, S., Linn, M., Ellis, A., & Williams, C. (2008). Gender similarities characterize math performance. *Science*, *321*(5888), 494–495. doi:10.1126/science.1160364 PMID:18653867

Ifill, N., Radford, A. W., Wu, J., Cataldi, E. F., Wilson, D., & Hill, J. (2016). *Persistence and Attainment of 2011–12 First Time Postsecondary Students After 3 Years*. Washington, DC: U.S. Department of Education, National Center for Education Statistics.

Iledare, O. (2007). *An Appraisal of the Global Petroleum Upstream Industry and Its Response to Changes in Crude Oil Prices*. Paper presented at Nigeria Annual International Conference and Exhibition, Abuja, Nigeria. SPE-111901-MS. Available from http://dx.doi.org.lib-ezproxy.tamu.edu:2048/10.2118/111901-MS

Jacobi, M. (1991). Mentoring and undergraduate academic success: A literature review. *Review of Educational Research*, *61*(4), 505–532. doi:10.3102/00346543061004505

Jamison, A. (2013). *The Making of Green Engineers: Sustainable Development and the Hybrid Imagination*. Morgan & Claypool Publishers.

Jansen-Schulz, B., & van Riesen, K. (Eds.). (2011). *Vielfalt und Geschlecht - relevante Kategorien in der Wissenschaft*. Opladen, Germany: Budrich UniPress Ltd.

Jenkins, S. (2015). *Electricity*. Retrieved May 21, 2016, from https://prezi.com/3f91mznpuew3/electricity/

Johnson, M. L., Pasquini, L., & Rodems, M. (2013). Connecting first year students to forman and informal learning experiences. In H. Wang & S. Wang (Eds.), *Cases on formal and informal e-learning environments: Opportunities and practices* (pp. 51–67). Hershey, PA: Information Science Reference. doi:10.4018/978-1-4666-1930-2.ch003

Johnstone, D. B. (2016). Financing American higher education: Reconciling institutional financial viability and student affordability. In M. N. Bastedo, P. G. Altbach, & P. J. Gumport (Eds.), *American higher education in the 21st century. Social, political and economic changes* (4th ed.; pp. 310–341). Baltimore, MD: Johns Hopkins University Press.

Jones, B., Paretti, M. C., Hein, S., & Knott, T. (2013). An Analysis of Motivation Constructs with First-Year Engineering Students: Relationships Among Expectancies, Values, Achievement, and Career Plans. *The Journal of Engineering Education*, *99*(4), 319–336. doi:10.1002/j.2168-9830.2010.tb01066.x

Kamp, A. (2014). *Engineering Education in the Rapidly Changing World.* Rethinking the Mission and Vision on Engineering Education at TU Delft. Retrieved July 18, 2016, http://www.lr.tudelft.nl/fileadmin/Faculteit/LR/Nieuws_en_Agenda/Nieuws/2014/docs/visiedocument_TU_def.pdf

Kartoshkina, Y. (2015). Bitter-sweet reentry after studying abroad. *International Journal of Intercultural Relations*, *44*, 35–45. doi:10.1016/j.ijintrel.2014.11.001

Kavanagh, J. (2002). Dame Stephanie Shirley. *The Computer Bulletin*, *44*(3), 20–21. doi:10.1093/combul/44.3.20

Kehr, P., & Leicht-Scholten, C. (2013). Diversity in der Hochschule – in der Mitte angekommen oder zwischen allen Stühlen? *Qualität in der Wissenschaft. Zeitschrift für Qualitätsentwicklung in Forschung. Studium und Administration*, *2*, 30–36.

Kena, G., Hussar, W., McFarland, J., de Brey, C., Musu-Gillette, L., Wang, X., & Dunlop Velez, E. et al. (2016). *The Condition of Education 2016 (NCES 2016-144)*. Washington, DC: U.S. Department of Education, National Center for Education Statistics.

King, R. (2012). *Engineers for the future*. Sydney, NSW: Engineers Australia, Australian Council of Deans of Engineering.

Kirkup, G., Zalevski, A., Maruyama, T., & Batool, I. (2010). *The UKRC statistics guide on women in science, engineering, technology and the built environment (SET).* Leeds, UK: WISE Campaign.

Kirkwood, J. (2015). *STEM Activities for Kids: How strong is a piece of paper?* Retrieved March 2, 2016, from http://creeksidelearning.com/stem-activities-for-kids-how-strong-is-a-piece-of-paper/

Knight, D. W., Carlson, L. E., & Sullivan, J. F. (2003). Staying in Engineering: Impact of a Hands-On, Team-Based, First-Year Projects Course on Student Retention.*Proceedings of the 2003 American Society for Engineering Education Annual Conference & Exposition.*

Knoll, B., & Ratzer, B. (2010). *Gender Studies in den Ingenieurwissenschaften*. Wien, Germany: Facultas Verlags- und Buchhandels AG.

Kolb, D. (2014). *Experiential learning: Experience as the source of learning and development* (2nd ed.). Upper Saddle River, NJ: Pearson FT Press.

Kolb, D. A. (1984). *Experiential Learning*. Englewood Cliffs, NJ: Prentice-Hall.

Koppi, T., Roberts, M., & Naghdy, G. (2012). *Perceptions of a gender-inclusive curriculum amongst Australian Information and Communications Technology Academics*. Paper presented at the Fourteenth Australasian Computing Education Conference, ACE 2012, Melbourne, Australia.

Korpershoek, H., Kuyper, H., & van der Werf, M. P. C. (2012). The role of personality in relation to gender differences in school subject choices in pre-university education. *Sex Roles*, *67*(11), 630–645. doi:10.1007/s11199-012-0222-7

Kraiger, K., Ford, J. K., & Salas, E. (1993). Application of cognitive, skill-based, and affective theories of learning outcomes to new methods of training evaluation. *The Journal of Applied Psychology*, *78*(2), 311–328. doi:10.1037/0021-9010.78.2.311

Krathwohl, D. R., Bloom, B. S., & Masia, B. B. (1964). *Taxonomy of educational objectives, handbook ii: affective domain*. New York: David McKay Company.

Kreinberg, N., & Lewis, S. (1996). The Politics and Practice of Equity: Experiences from Both Sides of the Pacific. In L. Parker, L. Rennie, & B. Fraser (Eds.), *Gender, Science and Mathematics* (Vol. 2, pp. 177–202). Springer Netherlands. doi:10.1007/978-94-011-0143-1_15

Krell, G., Riedmüller, B., Sieben, B., & Vinz, D. (Eds.). (2007). *Diversity Studies: Grundlagen und disziplinäre Ansätze*. Campus Verlag GmbH.

Kuder, M., & Banks, C. (2016). Current trends in strategic international partnerships. In C. Banks, B. Siebe-Herbig, & K. Norton (Eds.), Global perspectives on strategic international partnerships: A guide to building sustainable academic linkages (pp. xi-xxiv). New York, NY: Institute of International Education (IIE).

Kug, G. (2008). *High-impact educational practices: What they are, who has access to them, and why they matter*. Washington, DC: Association of American Colleges and Universities.

Kuniavsky, M. (2010). *Smart things: Ubiquitous computing user experience design*. Burlington, MA: Morgan Kaufmann.

LaFave, J. M., Kang, H. S. & Kaiser, J. D. (2015). Cultivating intercultural competencies for civil engineering students in the era of globalization: Case study. *Journal of Professional Issues in Engineering Education and Practice, 141*(3).

Landivar, L. C. (2013). *Disparities in STEM Employment by Sex, Race, and Hispanic Origin*. Washington, DC: United States Census Bureau.

Lappapadolce. (n.d.). *Science experiment on soil erosion*. Retrieved March 30, 2016, from http://www.lapappadolce.net/en/science-experiment-on-soil-erosion-2/5/

Lashley, C., & Best, W. (2003). *12 Steps to Study Success*. Cengage Learning EMEA.

Lassibille, G., & Navarro Gómez, L. (2008). Why do higher education students drop out? Evidence from Spain. *Education Economics*, *16*(1), 89–105. doi:10.1080/09645290701523267

Learch, K. M., Paulson, S. K., Sander, H., & Schuch, E. (2016). Strategic relationship development: A German-U.S. university case study. In C. Banks, B. Siebe-Herbig, & K. Norton (Eds.), Global perspectives on strategic international partnerships: A guide to building sustainable academic linkages (pp. 13-24). New York, NY: Institute of International Education (IIE).

Legge, K. (2012). *Doing it differently*. Paddington, NSW, Australia: Place Partners.

Leicht-Scholten, C. (2008). Exzellenz braucht Vielfalt - oder: wie Gender and Diversity in den Mainstream der Hochschulentwicklung kommt - Human Resources, Gender and Diversity Management an der RWTH Aachen. *Journal Netzwerk Frauenforschung NRW Nr. 23*, 33-39.

Leicht-Scholten, C. (2011b). Managing Gender und Diversity Mainstreaming in Zeiten der Exzellenz: alte Schwierigkeiten, neue Herausforderungen, geeignete Maßnahmen. In B. Blättel-Mink, A. Franzke, & A. Wolde (Eds.), Gleichstellung im Reformprozess der Hochschulen - Neue Karrierewege von Frauen an Hochschulen? (pp. 185–207). Academic Press.

Leicht-Scholten, C., Steuer, L., & Bouffier, A. (2016). Facing future challenges: Building engineers for tomorrow. *Proceedings of the International Conference "New Perspectives in Science Education", (vol. 5*, pp. 32-37). Padova: libreriauniversitaria.

Leicht-Scholten, C. (2011a). Meeting Global Challenges - Gender and Diversity as Drivers for a Change of Scientific Culture. In C. Leicht-Scholten, E. Breuer, N. Tulodetzki, & A. Wolffram (Eds.), *Going Diverse: Innovative Answers to Future Challenges. Gender and Diversity Perspectives in Science, Technology and Business* (pp. 53–64). Opladen: Budrich UniPress Ltd.

Leicht-Scholten, C., Breuer, E., Tulodetzki, N., & Wolffram, A. (Eds.). (2011). *Going Diverse: Innovative Answers to Future Challenges. Gender and Diversity Perspectives in Science, Technology and Business*. Opladen: Budrich UniPress.

Leicht-Scholten, C., & Wolffram, A. (2010). Managing Gender and Diversity Mainstreaming an Hochschulen im Spannungsfeld zwischen Theorie und Praxis. *Gender*, *2*, 87–95.

Lenning, O. T., Sauer, K., & Beal, P. E. (1980). *Student retention strategies*. American Association for Higher Education.

LERU. (2015). *Gendered research and innovation: Integrating Sex and Gender Analysis into the research process*. Retrieved July 26, 2016, from http://www.leru.org/files/publications/LERU_AP18_Gendered_research_and_innovation_final.pdf

Lincoln University. (2016). *2015 Fact Book*. Retrieved July 25, 2016, from http://www.lincoln.edu/sites/default/files/pdf/institutional-research/2015FactBook.pdf

Linnehan, F., Weer, C., & Uhl, J. (2005). African-American students early trust beliefs in work-based mentors. *Journal of Vocational Behavior*, *66*(3), 501–515. doi:10.1016/j.jvb.2004.02.003

Litchfield, K., & Javernick-Will, A. (2015). I Am an Engineer AND: A Mixed Methods Study of Socially Engaged Engineers. *The Journal of Engineering Education, 104*(4), 363–496. doi:10.1002/jee.20102

Litchfield, K., & Javernick-Will, A. (2016). Socially Engaged Engineers Career Interests and Experiences: A Miners Canary. *Journal of Professional Issues in Engineering Education and Practice*, 04016018. doi:10.1061/(ASCE)EI.1943-5541.0000303

Locke, E. A., Shaw, K. N., Saari, L. M., & Latham, G. P. (1981). Goal setting and task performance: 1969–1980. *Psychological Bulletin, 90*(1), 125–152. doi:10.1037/0033-2909.90.1.125

Long, L. L. III, & Mejia, J. A. (2016). Conversations about diversity: Institutional barriers for underrepresented engineering students. *The Journal of Engineering Education, 105*(2), 211–218. doi:10.1002/jee.20114

Lotkowski, V., Robbins, S., & Noeth, R. (2004). The role of academic and non-academic factors in improving college retention. *ACT Policy Report.* Retrieved January 12, 2006, from http://www.act.org/research/policy/index.html

Loy, J., & Tatham, P. (2016). Redesigning production systems. In S. S. Muthu & M. M. Savalani (Eds.), *Handbook of sustainability in additive manufacturing* (Vol. 1, pp. 145–168). New York, NY: Springer. doi:10.1007/978-981-10-0549-7_7

Loy, J., Tatham, P., Healy, R., & Tapper, C. L. (2016). 3D printing meets humanitarian design research: Creative technologies in remote regions. In A. Connor & S. Marks (Eds.), *Creative technologies for multidisciplinary applications* (pp. 54–75). Hershey, PA: IGI Global. doi:10.4018/978-1-5225-0016-2.ch003

LSU. (2016). *Program Statistics.* Retrieved July 30, 2016, from https://www.pete.lsu.edu/about/statistics

Lucas, S. R., & Berends, M. (2002). Sociodemographic diversity, correlated achievement, and De Facto tracking. *Sociology of Education, 75*(4), 328–348. doi:10.2307/3090282

Lunneborg, P. W. (1975). *Longitudinal Criteria of College Achievement Predicted from Aptitude, Achievement, and Interest Measures.* Seattle, WA: Educational Assessment Center, University of Washington.

Maccoby, E., & Jacklin, C. (1976). *The psychology of sex differences.* Stanford, CA: Stanford University Press.

Marra, R. M., Rodgers, K. A., Shen, D., & Bogue, B. (2012). Leaving Engineering: A Multi-Year Single Institution Study. *The Journal of Engineering Education, 101*(1), 6–27. doi:10.1002/j.2168-9830.2012.tb00039.x

Marshall Aid Commemoration Commission. (n.d.). *Marshall Scholarships.* Retrieved September 2, 2016, from http://www.marshallscholarship.org/

Martin, A. J., & Marsh, H. W. (2008). Academic buoyancy: Towards an understanding of students everyday academic resilience. *Journal of School Psychology*, *46*(1), 53–83. doi:10.1016/j.jsp.2007.01.002 PMID:19083351

Martin, P., & Barnard, A. (2013). The experience of women in male-dominated occupations: A constructivist grounded theory inquiry. *South African Journal of Industrial Psychology*, *2*(39), 1–12. doi:10.4102/sajip.v39i2.1099

Mathias, J., Bruce, M., & Newton, D. P. (2013). Challenging the western stereotype: Do Chinese international foundation students learn by rote? *Research in Post-Compulsory Education*, *18*(3), 221–238. doi:10.1080/13596748.2013.819257

Matthews, M. (2012). *Keeping students in engineering: A research-to-practice brief.* American Society of Engineering Education. Retrieved from http://www.asee.org/retention-project/keeping-students-in-engineering-a-research-guide-to-improving-retention

Matthews, M. (n.d.). *Keeping students in engineering: A research-to-practice brief.* Retrieved August 3, 2016, from https://www.asee.org/retention-project/keeping-students-in-engineering-a-research-guide-to-improving-retention

Mattis, M. (2007). Upstream and downstream in the engineering pipeline: What's blocking US women from pursuing engineering careers? In R. Burke & M. Mattis (Eds.), *Women and minorities in science, technology, engineering and mathematics* (pp. 334–362). Cheltenham, UK: Edward Elgar. doi:10.4337/9781847206879.00025

McDonald, M. (2008). Developing relationships in a learning and teaching environment within Te Wānanga o Aotearoa: A philosophy of teaching/learning, a Pakeha perspective. In R. Hunia (Ed.), Toroa Te Nukuroa (3): Reflections from Within. Te Awamutu, New Zealand: Te Wānanga o Aotearoa University.

McGee, J. (2015). *Breakpoint: The changing marketplace for higher education.* Baltimore, MD: Johns Hopkins University Press.

McIlwee, J., & Robinson, G. (1992). *Women in engineering: Gender, power and workplace culture.* Albany, NY: State University of New York Press.

McLoughlin, L. (2005). Spotlighting: Emergent gender bias in undergraduate engineering education. *The Journal of Engineering Education*, *94*(4), 373–381. doi:10.1002/j.2168-9830.2005.tb00865.x

Meyers, K., Ohland, M., Pawley, A., Silliman, S., & Smith, K. (2012). Factors relating to Engineering Identity *Global. The Journal of Engineering Education*, *14*(1), 119–131.

Miller, M. J., Lent, R. W., Lim, R. H., Hui, K., Martin, H. M., Jezzi, M. M., & Williams, K. et al. (2015). Pursuing and adjusting to engineering majors: A qualitative analysis. *Journal of Career Assessment*, *23*(1), 48–63. doi:10.1177/1069072714523084

Miller, R., & Morgaine, W. (2009). The benefits of e-portfolios for students and faculty in their own words. *Peer Review*, *11*, 8.

Mills, J. E., & Treagust, D. F. (2003). Engineering education: Is problem-based or project-based learning the answer? *Australasian Journal of Engineering Education, 3*(2), 2–16.

Mills, J., Ayre, M., & Gill, J. (2010). *Gender inclusive engineering education.* New York, NY: Routledge.

Mills, J., Ayre, M., & Gill, J. (2012). *Gender Inclusive Curriculum in Engineering and Construction Management.* Canberra, Australia: Australian Learning and Teaching Council.

Mills, J., Franzway, S., Gill, J., & Sharp, R. (2014). *Challenging knowledge, sex and power: Gender, work and engineering.* New York, NY: Routledge.

Milstein, T. (2005). Transformation abroad: Sojourning and the perceived enhancement of self-efficacy. *International Journal of Intercultural Relations, 29*(2), 217–238. doi:10.1016/j.ijintrel.2005.05.005

Mohan, T., Archee, R., Gurney, M., Wood, J. T., Tamparo, C., & Lindh, W. Q. (2013). COMU1030 Communication skills: Spoken language and interpersonal. Albany, NY: Cengage Learning Australia Pty Ltd.

Moore, J. S., III. (2013). *Self-Regulated Learning and Ethnic/Racial Variables: Predicting Minority First-Generation College Students' Persistence* (Ph.D. Thesis). University of Rhode Island, Kingston, RI.

Moore, T. J., & Smith, K. A. (2014). Advancing the state of the art of STEM Integration. *Journal of STEM Education, 15*, 5–10.

Mouros, N. (2003). *Defining, teaching and assessing lifelong learning skills.* Paper presented at the Frontiers in Education, 2003. FIE 2003 33rd Annual.

Museus, S. D., & Ravello, J. N. (2010). Characteristics of academic advising that contribute to racial and ethnic minority student success at predominantly White institutions. *NACADA Journal, 30*(1), 47–58. doi:10.12930/0271-9517-30.1.47

Myers, J. (2010, November 9). Why more women aren't becoming engineers. *The Globe and Mail.*

NAEP. (2014). *High School Coursetaking. The Condition of Education.* Retrieved on October 15, 2015, from http://nces.ed.gov/programs/coe/indicator_cod.asp

Narayanan, V. K., Olk, P., & Fukami, C. (2006). Determinants of internship effectiveness: an exploratory model. In Academy of Management Proceedings (Vol. 2006, pp. E1-E6). Academy of Management. doi:10.5465/AMBPP.2006.22898555

National Academy of Engineering (NAE). (2004). *The engineer of 2020: Visions of engineering in the new century.* Washington, DC: The National Academies Press; doi:10.17226/10999

National Center for Education Statistics. (2016). *National postsecondary student aid study.* Retrieved from https://nces.ed.gov/surveys/npsas/

National Geographic. (n.d.). *Deforestation Facts, Deforestation Information, Effects of Deforestation.* Retrieved March 31, 2016, from http://environment.nationalgeographic.com/environment/global-warming/deforestation-overview/

National Science Foundation Graduate Research Fellowship Program. (n.d.). *About.* Retrieved September 2, 2016, from https://www.nsfgrfp.org/general_resources/about

National Science Foundation. (n.d.). *Historically Black Colleges and Universities Undergraduate Program.* NSF 02-160. Retrieved June 19, 2007, from http://www.nsf.gov/pubs/2002/nsf02160/nsf02160.pdf

National Science Teachers Association (NSTA). (2016). *Principles of Professionalism for Science Educators.* Available from http://www.nsta.org/docs/PositionStatement_Professionalism.pdf

NCES. (2014a). *Children Living in Poverty. The Condition of Education.* Retrieved on October 15, 2015, from http://nces.ed.gov/programs/coe/indicator_cce.asp

NCES. (2014b). *Concentration of Public School Students Eligible for Free or Reduced-Price Lunch. The Condition of Education.* Retrieved on October 15, 2015 from from http://nces.ed.gov/programs/coe/indicator_clb.asp

NCES. (2014c). *Percentage of public high school teachers with a college major and standard certification in their respective main teaching assignment in mathematics, English, and science, by race/ethnicity concentration of schools: 2007–08. Status and trends in the education of racial and ethnic minorities.* Retrieved on October 15, 2015, from http://nces.ed.gov/programs/coe/indicator_cod.asp

NCHEMS. (2016). *National Center for Higher Education Management Systems. Retention Rates: First-Time College Freshman Returning their Second Year.* Retrieved August 9, 2016, from http://www.higheredinfo.org/dbrowser/index.php?submeasure=223&year=2010&level=nation&mode=data&state=0

Nettles, M., Wagener, U., Millett, C., & Killenbeck, A. (1999, Winter). Student retention and progression: A special challenge for private historically black colleges and universities. *New Directions for Higher Education*, (108), 51-67. Retrieved June 2, 2007, from Academic Search Premier database.

Nichols, J. (2004). Unique characteristics, leadership styles, and management of historically Black colleges and universities. *Innovative Higher Education*, *28*(3), 219–229. doi:10.1023/B:IHIE.0000015109.49156.fb

Nicholson, E. (1973). *Predictors of Graduation from College.* Iowa City, IA: American College Testing Program, Research and Development Division.

Nicol, D. (2009). Assessment for learner self-regulation: Enhancing achievement in the first year using learning technologies. *Assessment & Evaluation in Higher Education*, *34*(3), 335–352. doi:10.1080/02602930802255139

Noel, L., & Levitz, P. (2006). *National Research Report: How students rate the quality service climate on campus*. Retrieved June 22, 2007, from http://www.noellevitz.com

Noel, L., & Levitz, P. (n.d.). Satisfaction-Priorities Surveys Interpretive Guide. Cedar Rapids, IA. *Ruffalo Noel Levitz, LLC.*

Noel-Levitz Student Satisfaction Inventory. (n.d.). Retrieved November 18, 2008 from www.paulsmiths.edu/offices/president/SSI%20InterpretiveGuide%202008-09.pdf

Nolan, B., & van Lier, B. (2015). *31 Brilliant Ideas for a Better World.* Amsterdam, Netherlands: What Can Design Do.

NSF. (2015). *Women, Minorities, and Persons with Disabilities in Science and Engineering.* Retrieved January 02, 2016, from http://www.nsf.gov/statistics/wmpd/

Nwachukwu, P. (2014). Influence of Gender and Socio-Economic Factors on Students Performance in Mathematics in Shomulu Local Government Area of Lagos State. *Journal of Education and Practice*, *5*, 69–72.

Oakes, J., & Guiton, G. (1995). Matchmaking: The dynamics of high school tracking decisions. *American Educational Research Journal*, *32*(1), 3–33. doi:10.3102/00028312032001003

Oakley, B., Felder, R. M., Brent, R., & Elhajj, I. (2004). Turning student groups into effective teams. *Journal of Student Centered Learning, 2*, 9-34.

Ohland, M. W., Sheppard, S. D., Lichtenstein, G., Eris, O., Chachra, D., & Layton, R. A. (2008). Persistence, engagement, and migration in engineering programs. *The Journal of Engineering Education*, *97*(3), 259–278. doi:10.1002/j.2168-9830.2008.tb00978.x

Olitsky, N. (2014). How do academic achievement and gender affect the earnings of STEM majors? A propensity score matching approach. *Research in Higher Education*, *55*(3), 245–271. doi:10.1007/s11162-013-9310-y

Oreopoulos, P., & Petronijevic, U. (2013). Making college worth it: A review of the returns to higher education. *The Future of Children*, *23*(1), 41–65. doi:10.1353/foc.2013.0001 PMID:25522645

Organisation for Economic Co-operation and Development (OECD) (2014). *OECD Science, Technology and Industry Outlook 2014.* OECD Publishing. 10.1787/sti_outlook-2014-en

Orwell, G. (2011). 1984. London: Penguin.

Osborne, J., Simon, S., & Collins, S. (2003). Attitudes towards science: A review of the literature and its implications. *International Journal of Science Education*, *25*(9), 1049–1079. doi:10.1080/0950069032000032199

Parsons, R. L. (2009). The effects of an internationalized university experience on domestic students in the United States and Australia. *Journal of Studies in International Education*, *14*(4), 313–334. doi:10.1177/1028315309331390

Paton, R. A., Wagner, R., & MacIntosh, R. (2012). Engineering education and performance: The German machinery and equipment sector. *International Journal of Operations & Production Management*, *32*(7), 796–828. doi:10.1108/01443571211250086

Payne, R. S. (1994). The relationship between teachers beliefs and sense of efficacy and their significance to urban LSES minority students. *The Journal of Negro Education*, *63*(2), 181–196. doi:10.2307/2967382

Petzold, K., & Peter, T. (2014). The social norm to study abroad: Determinants and effects. *Higher Education*, *69*(6), 885–900. doi:10.1007/s10734-014-9811-4

Pew Charitable Trusts. (n.d.). Retrieved June 23, 2007, from http://www.pewtrusts.org/about_us

Phinney, J. S., & Alpuria, L. L. (1996). At the interface of cultures: Multiethnic/multiracial high school and college students. *The Journal of Social Psychology*, *136*(2), 139–158. doi:10.1080/00224545.1996.9713988 PMID:8691825

Phipps, A. (2008). *Women in Science, Engineering and Technology: three decades of UK initiatives*. Stoke on Trent, UK: Trentham Books.

PISA. (2015). *A closer look at gender gaps in education and beyond.* Program for International Student Assessment. Retrieved July 15, 2016, from http://oecdinsights.org/2015/03/05/a-closer-look-at-gender-gaps-in-education-and-beyond/

Plant and Soil Sciences eLibrary. (n.d.). *Water Erosion*. Retrieved May 21, 2016, from http://passel.unl.edu/pages/informationmodule.php?idinformationmodule=1086025423&topicorder=18&maxto=7

Plattner, H., Meinel, C., Leifer, L. J., & Leifer, L. (2015). *Design thinking research. Building innovators*. New York: Springer.

Prescott, J., & Bogg, J. (2011). Career attitudes of men and women working in the computer games industry. *Eludamos (Göttingen)*, *5*(1), 7–28.

Pullin, G. (2011). *Design meets disability*. Cambridge, MA: MIT.

R Core Team. (2013). *R: A Language and Environment for Statistical Computing*. Vienna, Austria: R Foundation for Statistical Computing. Available from http://www.R-project.org/

Race, P. (2007). *The lecturer's toolkit: A practical guide to learning, teaching and assessment* (3rd ed.). Abingdon, UK: Routledge.

Ramsden, P. (2003). *Learning to teach in higher education* (2nd ed.). Abingdon, UK: Routledge Falmer.

Reeves, M. (n.d.). *Engineer your life*. Retrieved August 23 2016. http://www.engineeryourlife.org/cms/makingithappen/6274/7039.aspx

Rhodes Trust. (n.d.). *Rhodes House - Home of The Rhodes Scholarships*. Retrieved September 2, 2016, from www.rhodeshouse.ox.ac.uk/

Riegle-Crumb, C., & Moore, C. (2013). The gender gap in high school physics: Considering the context of local communities. *Social Science Quarterly*, *95*(1), 253–268. doi:10.1111/ssqu.12022 PMID:25605978

Riley, D. (2008). *Engineering and social justice*. San Rafael, CA: Morgan and Claypool.

Riley, D., & Claris, L. (2009). From persistence to resistance: Pedagogies of liberation for inclusive science and engineering. *International Journal of Gender Science and Technology*, *1*(1), 35–60.

Rodan, P. (2016). Public-private partnership in higher education: Central Queensland University meets campus management services. *Australian Universities Review*, *58*, 5–12.

Rogers, L. (2001). *Sexing the Brain*. London: Weidenfeld and Nicolson.

Rohrer, D., & Pashler, H. (2012). Learning styles: Wheres the evidence? *Medical Education*, *46*(7), 34–35. doi:10.1111/j.1365-2923.2012.04273.x PMID:22691144

Roscigno, V. J., & Ainsworth-Darnell, J. W. (1999). Race, cultural capital, and educational resources: Persistent inequalities and achievement returns. *Sociology of Education*, *72*(3), 158–178. doi:10.2307/2673227

RWTH Aachen University. (2016a). *Zahlenspiegel 2015*. Aachen, Germany: RWTH Aachen University. Retrieved July 6, 2016, from https://www.rwth-aachen.de/cms/root/Die-RWTH/Profil/~enw/Daten-Fakten/

RWTH Aachen University. (2016b). *Rankingreport 2015. Nationale und internationale Hochschulrankings der Jahre 2006 bis 2015*. Aachen, Germany: RWTH Aachen University. Retrieved July 15, 2016, from https://www.rwth-aachen.de/global/show_document.asp?id=aaaaaaaaaaaofvl

Salisbury, M., Paulsen, M., & Pascarella, E. (2011). Why do All the Study Abroad Students Look Alike? Applying an Integrated Student Choice Model to Explore Differences in the Factors that Influence White and Minority Students Intent to Study Abroad. *Research in Higher Education*, *52*(2), 123–150. doi:10.1007/s11162-010-9191-2

Sappleton, N., Takruri-Rizk, H., Dhar-Bhattacharjee, S., & Bezer, R. (2009). *The organisational culture of NW engineering workplaces: The influence on women engineers.* Paper presented at the 6th Annual IAS-STS Conference: Critical Issues in Science and Technology, Graz, Austria.

Schön, D. A. (1983). *The reflective practitioner: How professionals think in action*. New York, NY: Basic books.

Schunk, D. H. (1990). Goal setting and self-efficacy during self-regulated learning. *Educational Psychologist*, *25*(1), 71–86. doi:10.1207/s15326985ep2501_6

Schunk, D. H. (1991). Self-efficacy and academic motivation. *Educational Psychologist*, *26*(3-4), 207–231. doi:10.1080/00461520.1991.9653133

Senge, P. (2006). *The Fifth Discipline: The Art & Practice of the Learning Organization*. New York: Deckle Edge.

Seron, C., Silbey, S. S., Cech, E., & Rubineau, B. (2016). Persistence Is Cultural Professional Socialization and the Reproduction of Sex Segregation. *Work and Occupations*, *43*(2), 178–214. doi:10.1177/0730888415618728

Shada, A., Kelly, K., Cox, R., & Malik, S. (2011). Growing a new culture of assessment: Planting eportfolios in the metro academies program. *International Journal of ePortfolio, 1*, 71-83.

Sharp, R., Franzway, S., Mills, J. E., & Gill, J. (2012). Flawed policy, failed politics? Challenging the sexual politics of managing diversity in engineering organisations. *Gender, Work and Organization*, *19*(6), 555–572. doi:10.1111/j.1468-0432.2010.00545.x

Shih, K. (2009b). Study abroad participation up, except among minority students. *Diverse Education, November, 16.*

Shih, K. (2009a). Study abroad participation up, except among minority students. *Diverse Issues in Higher Education*, 16.

Sinclair, A. (2007). *Leadership for the disillusioned: Moving beyond myths and heroes to leading that liberates*. Crows Nest, NSW: Allen & Unwin.

Singer-Freeman, K., & Bastone, L. (2016). *Pedagogical Choices Make Large Classes Feel Small* (NILOA Occasional Paper No. 27). Urbana, IL: University of Illinois and Indiana University, National Institute for Learning Outcomes Assessment.

Singer-Freeman, K., Bastone, L., & Skrivanek, J. (2014). ePortfolios Reveal an Emerging Community of Underrepresented Minority Scholars. *International Journal (Toronto, Ont.)*, *4*, 85–94.

Singer, K., Foutz, T., Navarro, M., & Thompson, S. (2015). Investigating the Extent that an integrative learning module broadens the perception of first-year students about the engineering profession. *American Journal of Engineering Education*, *6*, 99–112.

Smale, S. (2015). *Notes from an oral presentation of pathway providers in the U.S. to the Colorado State University community*. Fort Collins, CO.

Smerdon, E. T. (1996). Lifelong learning for engineers: Riding the whirlwind. *American Ceramic Society Bulletin*, *75*, 51–52.

Somerville, M. (2008, August). *Collaborative design and assessment: learning 'with and for' users*. Paper presented at the Building effective, sustainable, practical assessment, Seattle, WA.

Statistisches Bundesamt. (2015a). *Bildung und Kultur. Prüfungen an Hochschulen. 2014* (Fachserie 11, Reihe 4.2). Wiesbaden, Germany: Statistisches Bundesamt. Retrieved June 23, 2016, from https://www.destatis.de/DE/Publikationen/Thematisch/BildungForschungKultur/Hochschulen/PruefungenHochschulen2110420147004.pdf?__blob=publicationFile

Statistisches Bundesamt. (2015b). *Bildung und Kultur. Studierende an Hochschulen. Wintersemester 2014/2015* (Fachserie 11, Reihe 4.1). Wiesbaden, Germany: Statistisches Bundesamt. Retrieved June 23, 2016, from https://www.destatis.de/DE/Publikationen/Thematisch/Bildung-ForschungKultur/Hochschulen/StudierendeHochschulenEndg2110410157004.pdf;jsessionid=3B3B4340D356BD7227C75360B4289C2B.cae1?__blob=publicationFile

Stäuble, B. (2004). *Towards gender inclusiveness in the teaching of science and engineering subjects: A case study.* Paper presented at the 27th HERDSA Annual Conference: Transforming Knowledge into Wisdom, Miri, Sarawak.

Steuer, L., Berg, T., & Leicht-Scholten, C. (2015). Breaking the habit – New approaches in engineering education. *Proceedings of the International Conference SEFI, Orléans*. Retrieved July 21, 2016, from http://www.sefi.be/conference- 2015/CHAP%2016.%20Education%20concepts%20specific%20for%20Engineering%20Education/60823-%20L.%20STEUER.pdf

Steuer, L., Berg, T., & Leicht-Scholten, C. (2015). Breaking the habit–new approaches in engineering education. *Proceedings of Annual Conference of the European Society for Engineering Education*. Available from: http://www.sefi.be/conference-2015/CHAP%2016.%20Education%20concepts%20specific%20for%20Engineering%20Education/60823-%20L.%20STEUER.pdf

Steuer, L. (2014). *Gender und Diversity in MINT-Fächern – Ursachen des Diversity-Mangels*. Berlin: Springer BestMasters Psychologie.

Strauss, L., & Volkwein, J. F. (2004). Predictors of student commitment at two-year and four-year institutions. *The Journal of Higher Education*, *75*(2), 203–226. doi:10.1353/jhe.2004.0007

Strayhorn, T. L., Long, L. L. III, Williams, M. S., Dorimé-Williams, M. L., & Tillman-Kelly, D. L. (2014). Measuring the educational benefits of diversity in engineering education: A multi-institutional survey analysis of women and underrepresented minorities. In *Proceedings of the 121st ASEE Annual Conference and Exposition*. Indianapolis, IN: American Society of Engineering Education.

Sutton, S. B., & Obst, D. (Eds.). (2011). Developing strategic international partnerships: Models for initiating and sustaining innovative institutional linkages. New York, NY: Institute of International Education (IIE).

Te ManaKōrero: Relationships for learning. (2007). Retrieved August 7, 2016, from http://temangoroa.tki.org.nz/tags/content?tags=te_mana_korero

The Harry S. Truman Scholarship Foundation. (n.d.). *A Living Memorial & the Future of Public Service Leadership*. Retrieved September 2, 2016, from https://www.truman.gov/our-mission/who-we-are

Thomas, E., & Galambos, N. (2004). What satisfies students?. *Research in Higher Education*, *45*(3), 251–269. RetrievedJune22007 doi:10.1023/B:RIHE.0000019589.79439.6e

Thompson, J. (2016). The sometimes neglected study of the role of women in World War II. *Ethos (Berkeley, Calif.)*, *24*(1), 24–25.

Thursby, M. C., Fuller, A. W., & Thursby, J. (2009). An Integrated Approach to Educating Professionals for Careers in Innovation. *Academy of Management Learning & Education*, *8*(3), 389–405. doi:10.5465/AMLE.2009.44287938

Tinto, V. (1975). Dropout from Higher Education: A Theoretical Synthesis of Recent Research. *Review of Educational Research*, *45*(1), 89–125. doi:10.3102/00346543045001089

Tinto, V. (2012). *Completing college: Rethinking institutional action*. University of Chicago Press. doi:10.7208/chicago/9780226804545.001.0001

Tobias, S. (1985). Test anxiety: Interference, defective skills, and cognitive capacity. *Educational Psychologist*, *20*(3), 135–142. doi:10.1207/s15326985ep2003_3

Tobolowsky, B. F. (2005). *The 2003 national survey on first-year seminars: Continuing innovations in the collegiate curriculum* (Monograph No. 41). Columbia, SC: University of South Carolina, National Resource Center for The First-Year Experience and Students in Transition.

Tocci, C. M., & Engelhard, G. Jr. (1991). Achievement, Parental Support, and Gender Differences in Attitudes Toward Mathematics. *The Journal of Educational Research*, *84*(5), 280–287. doi:10.1080/00220671.1991.10886028

Tryggvason, G., & Apelian, D. (2011). *Shaping our world: Engineering education for the 21st century*. Hoboken, NJ: John Wiley & Sons.

Tryon, G. S. (1980). The measurement and treatment of test anxiety. *Review of Educational Research*, *50*(2), 343–372. doi:10.3102/00346543050002343

Turnbo, D. (2002). *A study of the impact on a new student intrusive advising model on undergraduate student attrition at a private higher education institution in Delaware* (Doctoral dissertation). Wilmington College.

UND. (2016). *About the Program*. Retrieved July 30, 2016, from https://engineering.und.edu/petroleum/about.cfm

US Bureau of Labor Statistics. (2016). *Employment, Hours, and Earnings from the Current Employment Statistics survey (National)*. Retrieved July 30, 2016, from http://data.bls.gov/timeseries/CES1021100001?data_tool=XGtable

US-Ireland Alliance. (n.d.). *The George J. Mitchell Scholarship*. Retrieved September 2, 2016, from http://www.us-irelandalliance.org/content/3/en/George%20Mitchell%20Scholarship%20Program%20%7C%20US-Ireland%20Alliance.html

van 't Klooster, E., van Wijk, J., Go, F., & van Rekom, J. (2008). Educational travel: The overseas internship. *Annals of Tourism Research*, *35*, 690–711.

VandenBos, G., Knapp, S., & Doe, J. (2001). *Role of reference elements in the selection of resources by psychology undergraduates*. Retrieved October 13, 2001, from http://jbr.org/articles.html

Vaz, R., & Quinn, P. (2014) Long term impacts of off-campus project work on student learning and development. In *Proceedings of the 44th Annual IEEE Frontiers in Education Conference*. Madrid, Spain: Institute of Electronics and Electrical Engineers. doi:10.1109/FIE.2014.7044128

VDI. (2016). *15 Jahre Bologna Reform - Quo vadis Ingenieurausbildung?* Retrieved July 21, 2016, from https://www.vdi.de/fileadmin/user_upload/2016_VDI-VDMA-Mercator-Studie-15_Jahre_Bologna-Reform.pdf

Veloo, A., Ali, R. M., & Krishnasamy, H. N. (2014). Affective Determinants of Additional Mathematics Achievement in Malaysian Technical Secondary Schools. *Procedia: Social and Behavioral Sciences*, *112*, 613–620. doi:10.1016/j.sbspro.2014.01.1208

Vester, F. (1998). *Denken, Lernen, Vergessen: Was geht in unserem Kopf vor, wie lernt das Gehirn, was lässt es uns im Stich?* München: Dtv Wissen Verlag.

Victoria University of Wellington Strategic Plan 2009–2014. (n.d.). Retrieved July 26, 2016, from http://www.victoria.ac.nz/about/publications/strategic-plan2/strategic-plan.pdf

Wächter, C. (2012). Interdisciplinary Teaching and Learning for Diverse and Sustainable Engineering Education. In A. Béraud, A.-S. Godfroy, & J. Michel (Eds.), *GIEE 2011: Gender and Interdisciplinary Education for Engineers: Formation Interdisciplinaire des Ingénieurs et Problème du Genre* (pp. 47–63). Rotterdam: SensePublishers. doi:10.1007/978-94-6091-982-4_5

Walther, J., Miller, S. E., & Kellam, N. N. (2012). Exploring the role of empathy in engineering communication through a trans-disciplinary dialogue.*Proceedings of the ASEE Annual Conference and Exposition.*

Wang, J., Korczykowski, M., Rao, H., Yong, F., Pluta, J., Ruben, C., & Detre, J. et al. (2007). Gender difference in neural response to psychological stress. *Social Cognitive and Affective Neuroscience*, 2(3), 227–239. doi:10.1093/scan/nsm018 PMID:17873968

Weber, M. (1968). *Gesammelte Aufsätze zur Wissenschaftslehre 3*. Tübingen: Mohr-Siebeck.

Weimer, M. (2002). *Learner-centered teaching: Five key changes to practice*. San Francisco, CA: Jossey Bass.

Wheelwright, S., Baron-Cohen, S., Goldenfeld, N., Delaney, J., Fine, D., Smith, R., & Wakabayashi, A. et al. (2006). Predicting autism spectrum quotient (AQ) from the systemizing quotient-revised (SQ-R) and empathy quotient (EQ). *Brain Research*, *1079*(1), 47–56. doi:10.1016/j.brainres.2006.01.012 PMID:16473340

Wilcox, P., Winn, S., & Fyvie-Gauld, M. (2005). The role of social support in the first year experience. Studies in Higher Education, 30, 707-722.

Wilde, D. (2011). Personalities into teams. *IEEE Engineering Management Review*, *1*(1), 20–24. doi:10.1109/EMR.2011.5729969

Willingham, D. T., Hughes, E. M., & Dobolyi, D. G. (2015). The scientific status of learning styles theories. *Teaching of Psychology*, *42*(3), 266–271. doi:10.1177/0098628315589505

Wolfe, J. (2010). *Team writing: A guide to working in groups.* Bedford: St. Martin's.

Wolfe, J., & Powell, E. (2009). Biases in interpersonal communication: How engineering students perceive gender typical speech acts in teamwork. *The Journal of Engineering Education*, *98*(1), 5–16. doi:10.1002/j.2168-9830.2009.tb01001.x

Woodford, C. (n.d.). *Electric motors*. Retrieved March 31, 2016, from http://www.explainthatstuff.com/electricmotors.html

Woodson, C. G. (1933). *The mis-education of the negro*. Trenton, NJ: African World Press, Inc.

Xu, Y. (2013). Career outcomes of STEM and non-STEM college graduates: Persistence in majored-field and influential factors in career choices. *Research in Higher Education*, *54*(3), 349–382. doi:10.1007/s11162-012-9275-2

Yang, T., Ahmad, S., Lavenda, J., Robinson, J. H., & Seo, K. (2016). Beyond the initial partnership: A framework for classifying partnerships as a means to strategize comprehensive internationalization. In C. Banks, B. Siebe-Herbig, & K. Norton (Eds.), Global perspectives on strategic international partnerships: A guide to building sustainable academic linkages (pp. 1-12). New York, NY: Institute of International Education (IIE).

Yang, M., Webster, B., & Prosser, M. (2011). Travelling a thousand miles: Hong Kong Chinese students study abroad experience. *International Journal of Intercultural Relations*, *35*(1), 69–78. doi:10.1016/j.ijintrel.2010.09.010

Zaharna, R. S. (1989). Self-shock: The double-binding challenge of identity. *International Journal of Intercultural Relations*, *13*(4), 501–525. doi:10.1016/0147-1767(89)90026-6

About the Contributors

Monica Gray, Ph.D., M.P.H., C.P.H., is an Assistant Professor of Civil and Environmental Engineering at The Lincoln University. She simultaneously received her Ph.D. in Civil and Environmental Engineering (Water Resources concentration) and Masters of Public Health (Environmental & Occupational Health concentration) from the University of South Florida, Tampa. She also received a Masters in Biological Engineering from the University of Georgia, Athens and B.S. in Agricultural Engineering from the University of the West Indies, Trinidad. Prior to arriving at The Lincoln University, Dr. Gray managed the global curriculum portfolio for over a hundred and twenty programs in sixteen countries at Arcadia University and work as an Environmental Engineer and Environmental Health Program Director at the Philadelphia Housing Authority and Organica Biotechnology. Her research interest is in the area of systems approaches for sustainable environmental management and community health with particular focus on water resources and sanitation systems. This includes water and wastewater treatment, reuse and management. Her current research project is the identification and modelling of Pharmaceutical and Personal Care Products (PPCPs) in community water.

Ken D. Thomas is the Associate Director for Student Affairs in the Honors College and Honors Lecturer in Biosystems Engineering of both the Honors College and Department of Biosystems Engineering at Auburn University. He received both his B.Sc. Chemical & Process Engineering and MSc Environmental Engineering from the University of the West Indies, St. Augustine and Ph.D. Civil Engineering (Environmental & Water Resources Engineering [EWRE]) from the University of South Florida (USF). His current research interests are in sustainability science, sustainable development and sustainable development engineering, especially in a West Indian context, and engineering education.

***

Mary Ayre has had a long involvement working at tertiary level in the area of women and math-science education across several universities in the UK. In the late 1990s she led a project on gender inclusive curriculum at the University of South Australia, a project which inspired her to pursue her interest in equity in science and engineering leading to her recently completed doctoral study of women and engineering work. She has several publications in her area and was involved as a joint author in the 2010 book Gender Inclusive Engineering Education.

Anna Bouffier acquired her final degree "Magistra Artium" in the subjects sociology, philosophy and political sciences. Since 2012 she is research associate at the research unit "Gender and Diversity in Engineering" (GDI) in the faculty of Civil Engineering at RWTH Aachen University in Aachen, Germany. She works within the field of applied gender and diversity research in science and technology. Currently her work focuses on gender and diversity perspectives in higher education in engineering.

Rae Cooper is a designer, lecturer and researcher at Griffith University with a commercial background in Graphics and Communication Design. Her focus is on first year education and she is the first year co-ordinator across all the design disciplines, including Industrial students from the Engineering school. Cooper is very committed to innovative teaching practice to support diversity and to foster inclusion through the development of strategic extra-curricular activities. She is currently completing her doctorate, focusing on protest and agitation within design and media - an area that combines her professional experience and expertise for uncovering alternatives.

Gretchen Dietz is an undergraduate student at the University of Mount Union, Alliance, Ohio, USA, Gretchen is a senior in the civil engineering program.

Fabiola P. Ehlers-Zavala was named INTO Colorado State University (CSU)'s Center Director in November 2014, having previously fulfilled the role of INTO CSU Academic Director (March 2013-November 2014). She earned her B.A. in English Language and Literature together with her teaching certificate from the Pontificia Universidad Católica de Valparaíso in 1992. She then pursued graduate education in the U.S., and she earned both her M.A. (1994) in English and Ph.D. in English Studies (1999) from Illinois State University. Upon her graduation, she worked for Illinois State University (ISU) as Assistant Professor in Bilingual/Bicultural Education until she received her tenure and promotion to Associate Professor. While at ISU, in her last year, she directed the Bilingual/Bicultural Education Program in the College of Education. In 2006, she moved to Colorado to teach in the M.A.

in English at Colorado State University where she received her second tenure and promotion to Associate Professor in 2009. Between 2009 and 2013, she directed the M.A. in English (TESL/TEFL). Her areas of expertise include: second language/ bilingual reading, second language assessment, and ESL/bilingual teacher preparation. She is the coauthor of Reading Strategies for Spanish Speakers. Her publications in books and journals include "Meeting the reading comprehension challenges of diverse English language learners in K-12: Key contributions from reading research" (2016), "Advocacy in Language Teaching" (2013), "History of Bilingual Special Education" (2011), "Bilingualism and Education: Educating At-Risk Learners" (2010), "How Can Teachers Help Adolescent English Language Learners Attain Academic Literacy?" (2009), "Teaching Adolescent English Language Learners" (2008), "Assessing English Language Learners (ELLs) in Mainstream Classrooms" in The Reading Teacher (2006 & reprinted in 2010); "Bilingual Reading from a Dual Coding Perspective" (2005) in Proceedings of the 4th International Symposium on Bilingualism; "Preparing Quality Bilingual/Bicultural Teachers in the 21st Century: A PDS Model for Educational Change and Success" (2004), "Use of Lexical Borrowings in Sonoran Border Spanish" (2003). She serves on the editorial board of The International Multilingual Research Journal (IMRJ) and TESOL Journal. She is Past President of Illinois TESOL/Bilingual Education, Past Chair TESOL International Bilingual Education Interest Section and Past Chair of TESOL International's Nominating Committee. Most recently, Dr. Ehlers-Zavala served as member of the International TESOL Diversity and Inclusion Committee, and currently serves in the Editorial Review Board for The Reading Teacher (RT) Volume 70 review year (2015-2016). She was recently named editor of the AAAL Newsletter.

Sonja Gaedicke works as research associate at the research group Gender and Diversity in Engineering (GDI) at RWTH Aachen University. In addition, she is writing her PhD thesis in the Sociology. Within the scope of her disertation she analyses the connection between urban spaces and deviant, criminal behaviour in Berlin and Cologne. She studied Sociology, Political Science and English Literature at RWTH Aachen University and graduated with a Magister degree in 2007. Focus of Work: Gender and Diversity Studies, Gender and Diversity in Engineering Education, Urban Sociology.

Judith Gill, currently Adjunct Associate Professor at the University of South Australia, is a former senior mathematics teacher who came into education research to investigate girls and math learning. For the past 30 years she has worked in teacher education, focussing on issues of gender in education and work. Her research has concerned school gender contexts, workplace learning, women and engineering and gender and citizenship. To date she has published 7 books and numerous journal articles in these areas.

Jennifer Groh joined the Purdue Women in Engineering Program (WIEP) team in 2009. She received a B.S. in microbiology from Purdue University, and a Ph.D. in microbiology from the University of Oklahoma. Prior to joining WIEP, she was the Graduate Program Coordinator in the Purdue Weldon School of Biomedical Engineering. As Associate Director of WIEP, Jennifer administers the undergraduate Mentee & Mentor Program and the Graduate Mentoring Program, teaches two Women in Engineering seminars (ENGR 19400 and 49400), and oversees WIEP's K-12 outreach programming. Dr. Groh has also invested over 100 hours of training in academic coaching to become a certified Affiliate Coach with LifeBound, Inc. with a specialized focus in serving Science, Technology, Engineering, and Math (STEM) populations nationwide. Additionally, she has facilitated numerous national workshops on academic coaching which have been well received by a variety of audiences, including undergraduate and graduate students, faculty and staff in higher education, and corporate representatives. In addition to leading these engaging sessions, Dr. Groh integrates coaching into WIEP programming, student mentoring, and her personal life.

Juile Hessedence is a senior undergraduate student studying civil engineering at the University of Mount Union in Alliance, Ohio. She currently works at Chagrin Valley Engineering and serves as the Vice-President of the Society of Women in Engineering. She lead the efforts to get a Society of Women in Engineering Chapter recognized on campus in order to help women reach their full potential as students, leaders, and engineers. Over the course of her engineering career, Julie has had the opportunity to study abroad in Guatemala and Panama. Both experiences required Julie to apply her engineering knowledge to solve real-world problems. Her experience in Panama gave her the opportunity to encourage women and minorities to go into STEM related fields. Upon graduation, Julie hopes to obtain a job in the civil engineering field as an engineer in training.

Yvonne Hilton, Ed.D. has worked in higher education for over two decades. She holds a Bachelor's Degree in Therapeutic Recreation from Lincoln University (1988), a Master's Degree in Physical Education from the University of Delaware (1990) and a Doctor of Education degree in Organizational Leadership and Innovation from Wilmington University (2009). Dr. Hilton began her tenure at Lincoln in the fall of 2000. She demonstrates a passion and willingness to address the needs of a diverse population of students in a challenging and caring learning environment. Dr. Hilton has given herself to the service of the university community: she has served on various university faculty standing committees; been a panelist in a women's health program; served as a facilitator in an anti-drug workshop for student athletes; been a faculty advisor to the department's major's club.

Simon Howell is a lecturer in the Griffith School of Engineering with a particular responsibility for teaching professional skills in the engineering program. He is also a member of the teaching team for Creative Engineering. Howell has a focus on experiential learning and cross-cultural communication. He is heavily involved with the development of the first year experience in engineering at Griffith University and has a background in student support and academic skills development.

Melissa L. Johnson currently serves as the Associate Director of the University of Florida Honors Program. She has the unique opportunity to work with students from their admission into the program through alumni development activities. She advises honors students on academic, co-curricular, and career development matters, while also teaching innovative courses in professional development, civic engagement, and leadership. Melissa received her Ph.D. in educational technology from the University of Florida. Professionally she has served as an editorial board member for Honors in Practice, the Journal of Student Affairs Research and Practice, the NACADA Journal, and the Journal of The First-Year Experience and Students in Transition. She has served as the chair of the Advising High Achieving Students Commission through the National Academic Advising Association and recently was elected to the Board of Directors for the National Collegiate Honors Council.

Carmen Leicht-Scholten, PhD, holds a bridging professorship at the faculty of Civil Engineering. Political scientist by training she has been published in the field of gender and diversity in higher education and science and technology studies for many years. Carmen Leicht-Scholten was responsible for the development of the Gender- and Diversity-Strategy of RWTH University that succeeded in the German Excellence Initiative. She has been the first German gender expert to discuss gender and diversity in the process of the excellence initiative. With the development of the concept of gender and diversity for Aachen University she has been a pioneer and best practice for many other German universities. Carmen Leicht-Scholten has been expert in many national and international projects and is evaluator for the European Commission. She is German representative of the cost action GenderSTE and representing Aachen University in CDIO, an innovative educational framework for producing the next generation of engineers. Currently she is working in different national and international projects to integrate gender perspectives in engineering education (see http://www.gdi.rwth-aachen.de/engl/research/). She is coeditor of "Gender- Zeitschrift für Geschlecht, Kultur und Gesellschaft" and author of various publications on gender in science and engineering.

Terry Long is an undergraduate student at the University of Mount Union in Alliance, Ohio, USA. Terry is a senior studying mechanical engineering.

Jennifer Loy has a background in design for manufacturing in the UK and Australia. She teaches Creative Engineering across all engineering disciplines in the first year at Griffith University on the Gold Coast in Queensland. Loy has a particular interest in teaching experiential learning and interdisciplinary education. She is the Program Leader of Industrial Design and Deputy Director of the Griffith Centre for Creative Arts Research.

Constance Lundy, MA, has over three decades of experience in study abroad and higher education administration, and as Director of International Programs and Services at Lincoln University, the nation's oldest degree granting HBCU. In this position, she is responsible for promoting and preparing students for study abroad experiences. Under her direction, Lincoln has seen a steady increase in the number of students opting to participate in an international education experience. She has served as an evaluator for the International Gilman Scholarship, the Boren Fellowship, Fulbright Pre-Academic and Gateway Orientation programs, and the National Council for International Visitors. For three years, she directed the Institute of International Education's South African Education Program Orientation which was held at Lincoln University during the post-apartheid era, and from 2006-2008, she directed a Fulbright program that brought 30 foreign scholars to Lincoln University for a summer pre-academic orientation experience, and Lincoln was again selected to host a 2010 Fulbright Gateway orientation that will bring 60 scholars to the campus for academic and cultural experiences. She serves on the board of directors of the College Consortium for International Studies (CCIS), and under her management, Lincoln University sponsors the KCP International Language program, based in Tokyo, Japan, and a Liberal Arts, Business and Economics, and Legal Studies program at the Anglo American University in Prague, Czech Republic.

Anthony A. Maciejewski received the B.S. (summa cum laude), M.S., and Ph.D. degrees in Electrical Engineering in 1982, 1984, and 1987, respectively, all from The Ohio State University under the support of a National Science Foundation (NSF) graduate fellowship. From October of 1985 to September of 1986 he was an American Electronics Association Japan Research Fellow at the Hitachi Central Research Laboratory in Tokyo, Japan where he performed work on the development of parallel processing algorithms for computer graphic imaging and simulation. In 1988, Prof. Maciejewski joined the faculty of Purdue University as an Assistant Professor of Electrical Engineering. He was promoted to Associate Professor in 1993 and Full Professor in 1998. In August of 2001 he joined Colorado

State University (CSU) where he is currently a Professor and Head of the Department of Electrical and Computer Engineering. His research and teaching interests center on the analysis, simulation, and control of kinematically redundant robotic systems. His current work focuses on how kinematic redundancy can be utilized to design failure tolerant robotic systems for remote operations. He has over 200 publications, directs a research laboratory and has developed graduate courses in these areas. His commitment to education resulted in his receiving four teaching awards. His research has been supported by NSF, Sandia Nat'l Lab, Oak Ridge Nat'l Lab, DARPA, NASA, Nat'l Imagery and Mapping Agency, Missile Defense Agency, Non-lethal Technology Innovation Center, the NEC Corporation, Caterpillar, AT&T, H-P, Intel, Wolf Robotics, and the TRW Foundation. Prof. Maciejewski has served on eight editorial boards including the IEEE Transactions on Robotics and Automation and the IEEE Transactions on Systems, Man, and Cybernetics. He is a Fellow of the Institute of Electrical and Electronics Engineers "for contributions to the design and control of kinematically redundant robots," and has served as the VP for Financial Activities, AdCom member and Secretary for its Robotics and Automation Society. He was Technical Program Co-chair for the 1997 International Conference on Robotics and Automation (ICRA), Technical Program Chair for ICRA 2002 and has served on over 100 other conference program committees. He is also a member of the Association for Computing Machinery.

Julie Mills, after graduating with an Honours degree in Civil Engineering, spent over a decade working as a professional engineer across two Australian states. She returned to University where she completed her PhD in engineering education and has taught engineering for the past 20 years. She is currently Head of School in engineering and has achieved national and international recognition for her commitment to gender equity in engineering teaching and learning. She was also involved in the jointly authored publications "Gender Inclusive Engineering Education" (2010) and "Challenging Knowledge, Sex and Power: Gender, Work and Engineering" (2013).

Helen Muga, Ph.D., is an Assistant Professor in civil engineering at the University of Mount Union, Alliance, Ohio. She worked as a postdoctoral researcher in the Civil & Environmental Engineering department at the University of South Florida, Tampa, Florida. She received a Ph.D. in Environmental Engineering from Michigan Technological University, Michigan, USA, a Masters degree in Environmental Engineering from Michigan Technological University, Michigan, USA, a Masters degree in Chemical Engineering from Curtin University of Technology, Australia, and a Bachelors degree in Chemistry from the University of Papua New Guinea. Her research interests include water and wastewater engineering, sustainability, life cycle engineering, green engineering, diffusion and adoption of green technology, and international development work.

Amy Peterson is an Assistant Professor in the Department of Chemical Engineering at Worcester Polytechnic Institute (WPI). She earned her PhD in Chemical Engineering at Drexel University and completed her postdoctoral research in the Interfaces Department at the Max Planck Institute of Colloids and Interfaces, where she was the recipient of an NSF Graduate Research Fellowship and Alexander von Humboldt Postdoctoral Research Fellowship. Her research focuses on functional polymer surfaces and interfaces and she has received a WPI Teaching Innovation Grant.

Tatyana Plaksina is an assistant professor of petroleum engineering at the department of Chemical and Petroleum Engineering at University of Calgary. She received her PhD in Petroleum Engineering at Texas A&M University, TX, USA. She also holds an MS in Petroleum Engineering from Louisiana State University, Baton Rouge, and a BA in Math/Computer Science from Lawrence University, WI. In addition to reservoir engineering and optimization, her research interests include geothermal energy, CO_2 sequestration, advanced petroleum economics and risk analysis, and petroleum engineering education. Dr. Plaksina is an author of multiple peer-reviewed publications in reputable journals and a frequent speaker at numerous SPE and AAPG conferences.

Aaron Sakulich is an Assistant Professor in the Department of Civil and Environmental Engineering at Worcester Polytechnic Institute (WPI). His professional background is in the investigation of concrete systems that incorporate various industrial wastes in order to improve durability and reduce environmental impact. As a Fulbright Scholar to Morocco, Aaron wrote the only widely available English-language grammar textbook for colloquial Moroccan Arabic in addition to performing research focused on Moroccan mining wastes. Recent experiences as the principal investigator on grants funded through the National Science Foundation's REU and IRES programs have demonstrated to Aaron the value of diversity in the research and industrial settings.

Kristy Spear is currently an Advisor for the University of Florida Honors Program. She serves as an instructor for the program as well, teaching Honors Professional Development, in addition to other courses. Prior to joining the Honors Program, she spent several years in human resources with responsibilities ranging from training and development to global faculty recruitment. Capitalizing on the experience she gained in HR, Kristy transitioned to a role in career services. As the Assistant Director for Career Development she designed and implemented programs to aid students in career exploration and preparation. Many of these programs were presented at national conferences and used as model initiatives at other institutions. Kristy earned an interdisciplinary B.S. from the University of Central Florida with a focus on Health and Communication and graduated with a minor in Hospitality Management. She received her M.Ed. from the University of Florida in Educational Leadership and is currently finishing her Ph.D. in Curriculum and Instruction with a specialization in Educational Technology and a certificate in Digital Humanities.

Chandra Austin Stallworth graduated from North Carolina A & T State University in Graphic Communications Systems with an emphasis in CAD and received a M.Ed and Ph.D. from the University of Minnesota Twin Cities. Chandra has been inducted into Phi Beta Kappa, Epsilon Phi Tau, and Gold Key Honor Society and has also been involved with VICA. She has several years of experience working with different engineering firms as a CAD technician. The title of her dissertation is Factors influencing African-American high school students in career decision self-efficacy and engineering-related goal intentions. Chandra was a post-doctoral research associate at Utah State University during academic year 2009-2010 and then she was post-doctoral teaching fellow in the Honors College at Auburn University. Dr. Stallworth left this position to further her education and is currently in Law School.

Linda Steuer is research associate at the research group "Gender and Diversity in Engineering" (GDI), RWTH Aachen University. The gender and diversity chair investigates, places and publishes new themes in gender and diversity research. As "bridging professorship", the chair is designed to strengthen the interdisciplinary research between technological, engineering, natural, human, social and economical sciences. In the course of numerous projects in the field of interdisciplinary and diverse teaching, such as a transatlantic course with the Stanford University, Linda Steuer has gained scientific experiences.

Index

A

ABET 43, 45, 50-51, 59, 262, 266, 281
academic advising 6, 9-10, 12, 15, 21-22, 25, 97, 183, 195, 198
Academic Buoyancy or Resilience 59
Academic Experiential Learning 185, 187, 200
Affective Domain 46, 56, 59
affective organizational 43, 48, 59
African American 4-5, 28, 31, 40, 44, 50-51, 55
assessment 20, 22-23, 44, 58, 61, 105, 109, 116, 118, 125, 128, 131, 133-136, 153, 164-165, 173, 178, 181, 197, 199, 247, 253, 272
At-Risk 1, 4, 25
attrition 1-4, 21-22, 25, 44, 50, 54, 97, 106, 117, 123, 138, 150, 163, 165

B

beer game 263-267
Bouffier 201, 211, 217, 220, 230, 232

C

Career Development 40, 104-105, 151, 190
Career Trajectory 164, 175
cognitive analysis 43, 46, 48, 59
Cognitive Domain 46, 55, 59
communication 65, 101, 109, 111, 116, 118, 120-121, 124, 128-129, 137-138, 149, 153, 168, 184-185, 190, 193-194, 200, 213, 224, 234, 245, 251, 263-264, 278
conflict 106, 110, 124, 128
Content Integration 61, 65
context integration 65
Creative Thinking Skills 176, 184, 186, 200
Cultural Competence 42, 46, 50, 59, 230

D

digital revolution 107, 109, 111, 122, 127, 129
distributed manufacturing 109, 111, 129
Domain Relevant Skills 176, 200
dominant majority 115, 118, 131, 138, 141, 144, 149, 154

E

educational strategies 196
empathy 108, 113-114, 118-119, 128-129, 136, 145, 151, 153-154, 217-218, 234
empowering 61, 122, 214-215
engineering 27-28, 30-31, 40, 42-52, 54-57, 59, 61-70, 75, 88-92, 94-99, 101-140, 142-153, 156-175, 181, 186, 198-199, 201-216, 219-220, 222-238, 241-244, 246-249, 251-257, 260-264, 266-268, 272-275, 277-282
engineering faculty 113, 116, 130, 132, 135, 165, 243, 263, 275
Ethnocentric 129
ethnocentricity 109, 144, 154
experiences 1, 3, 5-6, 8, 19-21, 50, 68, 91, 93, 95, 97, 99, 104, 107-110, 115, 117-118, 120, 123, 126-127, 129-132,

135-136, 138, 140-141, 145, 148, 151, 159, 166, 172, 177, 183, 185-187, 190-195, 200, 207, 214-216, 224-226, 233, 241-242
experiential 105, 107, 110, 130, 132, 135, 137, 140, 146, 152, 154, 176, 178, 183-187, 190-191, 193-195, 200, 216, 228, 232
experiential learning 107, 110, 132, 135, 137, 140, 146, 152, 154, 176, 178, 183-185, 187, 190-191, 193-195, 200, 216, 228, 232

F

First-Year Experience 25, 94, 105

G

Gaedicke 201
gender and diversity 165, 201-202, 206, 208, 210-213, 215, 217-219, 222, 228, 232-233
gender bias 130, 132, 134, 145, 154, 173
gender equity 158, 169-170, 175
Gender Inclusive Engineering Curriculum 165, 175
Germany 65, 201-202, 204, 206, 209-211, 213, 219, 224, 228-234
global competencies 177-178, 184-185, 187-188, 195-197, 200
global education 239, 242, 255
global engineering 61
graduate attributes 106-108, 144

H

HBCU 1-5, 17, 39
high-impact practices 92-94, 105
Historically Black Colleges and Universities (HBCUs) 23, 28
Honors College 26, 29-35, 38-39

I

inclusivity 135, 156-157, 161, 164-166, 169, 175
Individual Development Plans 176, 178, 181
Industrial Experiential Learning 200
Informal Learning Community 105
institutional environment 1-3
institutional factors 1, 6-7, 9-12, 20-21, 25
internationalization 51, 53, 107, 237-238, 244, 252, 255-257
internships 43, 88, 96, 99-101, 176-178, 190-191, 197, 200, 264, 269

K

K-5 61
K-12 61, 66

L

leadership 24, 27, 29, 35, 38, 40-41, 50, 53, 56, 96, 99, 117-118, 139, 144, 150, 169-170, 174, 202, 206, 211, 231, 240, 245-247, 249, 251, 254-255, 259
Leicht-Scholten 165, 174, 201, 203, 207, 209, 217, 219-220, 226, 230, 232-234
LGBTI 115, 121, 130, 138, 154
lifelong learning 43, 45-47, 49, 55, 57-59
Lloyd Heize 261-262

M

management 24, 95, 101, 108, 128, 143, 151, 156, 161, 164-165, 168-170, 173, 184, 197-200, 206, 225, 229, 232-234, 241, 257
Masculine Workplace Culture 167, 175
mathematics 27, 43-45, 57-58, 61, 64-68, 90, 99, 112, 120, 130, 136, 148, 158, 161-162, 171-173, 210
mentoring 3, 23, 88, 90, 95, 181-182, 190
minorities 3, 26, 30-31, 36, 38-39, 42, 45, 49-50, 57, 90, 122, 124, 131, 141, 144, 147, 158, 169-170, 173, 199, 204-205
module 61, 69-70, 73, 75, 77, 84, 87, 105, 212, 220

P

paradigm 54, 108, 131, 135, 137, 146, 154, 200, 208
pathways 104, 109, 115, 236, 242-243, 249-250, 252, 254-255
Peter Senge 263
petroleum engineering 260-264, 266-268, 273-275, 277-281
Philip Candy 46
Predominantly White Institutions (PWIs) 26
Private-Public Partnership 236
professional skills 129, 177-178, 184-185, 187-188, 190-191, 197, 200
Project-based Experiential Learning 187, 200
projects 43, 93, 100, 109, 111-112, 115-117, 123, 130, 132, 134-140, 142-143, 145-146, 149-151, 166, 176, 178, 181, 185, 187, 189, 193-195, 198, 200, 222-223, 226, 228, 232, 240, 280

R

retention 2-6, 19-21, 23-26, 29, 31, 35, 42-44, 51, 89, 91-92, 96, 102, 117, 123, 128, 137, 139, 151, 156, 163, 168, 183, 196, 232
RWTH Aachen University 201-202, 210, 212, 218, 221, 223, 227, 233-234

S

Self-Direction 42-43, 47, 54-55
self-efficacy 23, 42-45, 47-49, 54-55, 57-59, 66, 89, 91-94, 98, 103, 105, 171
seminar 22, 88-89, 94, 98-99, 101, 103, 217, 219
Social Cognitive Theory 92, 105
Social Dynamics 164, 175
socially responsible engineers 201-202, 208, 217, 228
STEM 30-31, 39, 50, 53-54, 56, 61-71, 89-90, 96, 98-100, 102-105, 110, 127, 147, 161, 170, 176-180, 183-184, 186, 188-189, 191-192, 195-198, 203, 207
stepwise regression 267, 272-274, 283
Steuer 165, 174, 201, 203, 217, 219-220, 226, 232, 234
strategies 2, 23, 42-43, 47, 51, 54-55, 67-69, 91, 101, 106-107, 112, 114-119, 122-124, 126-127, 130-132, 134-135, 137, 139, 142, 156, 169, 172, 193-194, 196-197, 202, 236, 250, 255-256
Student Organization 29, 35, 40
study abroad 42-44, 46-54, 56, 58-59, 96, 100-101, 177, 187, 194, 197, 252
systemizing 113-115, 122, 125, 128-129, 136, 151, 153

T

teaching approach 114, 201, 213, 220, 225
teamwork 110, 117, 128-129, 143, 153, 156, 164, 172, 175, 186, 198, 214, 220, 225-226
technical university 201, 228
technology 16, 55, 59, 61-65, 67, 69, 90, 94, 98-99, 101, 107, 111, 126, 128, 133, 140, 144-145, 161, 170-174, 181, 189, 197-198, 200, 202, 206, 208-211, 214, 217-220, 228, 230, 232-233, 235, 282
transferable skills 106, 110, 117-118, 123, 126, 129, 137, 143

U

undergraduate research 38, 55, 88, 93, 96, 99-100, 181
Undergraduate student enrollment 260, 262, 268-270

W

Workplace Culture 156, 167, 169, 173, 175
workplace skills 184, 186, 200

www.ingramcontent.com/pod-product-compliance
Lightning Source LLC
LaVergne TN
LVHW080557300926
849162LV00005B/980